Probability and Its Applications

Published in association with the Applied Probability Trust

Editors: J. Gani, C.C. Heyde, P. Jagers, T.G. Kurtz

T0190461

Probability and Its Applications

O.L.V. Costa, M.D. Fragoso and
R.P. Marques

Discrete-Time Markov Jump Linear Systems

 Springer

Oswaldo Luiz do Valle Costa, PhD
Ricardo Paulino Marques, PhD

Department of Telecommunications and Control Engineering,
University of São Paulo, 05508-900 São Paulo, SP, Brazil

Marcelo Dutra Fragoso, PhD

Department of Systems and Control, National Laboratory for Scientific Computing,
LNCC/MCT, Av. Getúlio Vargas, 333, 25651-075, Petrópolis, RJ, Brazil

Series Editors

J. Gani
Stochastic Analysis Group, CMA
Australian National University
Canberra ACT 0200
Australia

C.C. Heyde
Stochastic Analysis Group, CMA
Australian National University
Canberra ACT 0200
Australia

P. Jagers
Mathematical Statistics
Chalmers University of Technology
SE-412 96 Göteborg
Sweden

T.G. Kurtz
Department of Mathematics
University of Wisconsin
480 Lincoln Drive
Madison, WI 53706
USA

Mathematics Subject Classification (2000): 93E11, 93E15, 93E20, 93B36, 93C05, 93C55, 60J10, 60J75

British Library Cataloguing in Publication Data
Costa, O. L. V.
    Discrete-time Markov jump linear systems. – (Probability and its applications)
    1. Jump processes   2. Markov processes   3. Discrete-time systems   4. Linear systems
    I. Title   II. Fragoso, M. D.   III. Marques, Ricardo Paulino
    519.2′33

Library of Congress Cataloging-in-Publication Data
Costa, Oswaldo Luiz do Valle.
    Discrete-time Markov jumb linear systems / O.L.V. Costa, M.D. Fragoso, and R. P. Marques.
        p. cm. — (Probability and its applications)
    Includes bibliographical references and index.

    1. Stochastic control theory.   2. Stochastic systems.   3. Linear systems.   4. Control theory.
    5. Markov processes.   I. Fragoso, M.D. (Marcelo Dutra)   II. Marques, Ricardo Paulino.
    III. Title.   IV. Probability and its applications (Springer-Verlag)
    QA402.37.C67 2005
    003′.76—dc22   2004059023

ISBN 978-1-84996-908-6   e-ISBN 978-1-84628-082-5
Springer is a part of Springer Science+Business Media
springeronline.com

© Springer-Verlag London Limited 2010
Printed in the United States of America

12/3830-543210 Printed on acid-free paper

Preface

This book is intended as an introduction to discrete-time Markov Jump Linear Systems (MJLS), geared toward graduates and researchers in control theory and stochastic analysis. This is not a textbook, in the sense that the subject is not developed from scratch (i.e., we do not go into detail in the classical theory of topics such as probability, stochastic optimal control, filtering, H_∞-control, etc.). Also, this is not an exhaustive or even comprehensive treatise on MJLS. In order to write a book of moderate size, we have been obliged to pass over many interesting topics like, for instance, adaptive control of MJLS, hidden Markov chain filtering, etc.

Besides the fact that the subject of MJLS is by now huge and is growing rapidly, of course it would be impossible to cover all aspects of MJLS here, for various reasons, including the fact that the subject is, to some extent, too young. It will take yet a long time to put MJLS theory on the same footing as linear systems theory. The book reflects the taste of the authors and is essentially devoted to putting together their body of work on the subject. Certainly, there will be somebody who will find that some of his favorite topics are missing, but this seems unavoidable. We have tried to put together in the Historical Remarks a representative bibliographic sample, but again, the rapidly burgeoning list of publications renders futile any attempt to be exhaustive.

In this book we emphasize an operator theoretical approach for MJLS, which aims to devise a theory for MJLS which parallels that for the linear case, and differs from that known as *multiple model* and *hidden Markov model*. We confine our attention to the discrete-time case. Although the operator theoretical methodology has served as an inspiration to extensions to the continuous-time case, the many nuances in this scenario pose a great deal of difficulties to treat it satisfactorily in parallel with the discrete-time case.

We undertook this project for two main reasons. Firstly, in recent years, the scope of MJLS has been greatly expanded and the results are scattered over journal articles and conference proceedings papers. We felt there was a lack of an introductory text putting together systematically recent results.

Secondly, it was our intention to write a book that would contribute to open the way to further research on MJLS.

Our treatment is theoretically oriented, although some illustrative examples are included in the book. The reader is assumed to have some background in stochastic processes and modern analysis. Although the book is primarily intended for students and practitioners of control theory, it may be also a valuable reference for those in fields such as communication engineering and economics. Moreover, we believe that the book should be suitable for certain advanced courses or seminars.

The first chapter presents the class of MJLS via some application-oriented examples, with motivating remarks and an outline of the problems. Chapter 2 provides the bare essential of background. Stability for MJLS is treated in Chapter 3. Chapter 4 deals with optimal control (quadratic and H_2). Chapter 5 considers the filtering problem, while Chapter 6 treats the quadratic optimal control with partial information. Chapter 7 deals with the H_∞-control of MJSL. Design techniques, some simulations and examples are considered in Chapter 8. Finally, the associated coupled algebraic Riccati equations, and some auxiliary results are considered in Appendices A, B, and C.

This book could not have been written without direct and indirect assistance from many sources. We are very grateful to our colleagues from the Laboratory of Automation and Control – LAC/USP at the University of São Paulo, and from the National Laboratory for Scientific Computing – LNCC/MCT. We had the good fortune of interacting with a number of special people. We seize this opportunity to express our gratitude to our colleagues and research partners, specially to Profs. C.E. de Souza, J.B.R. do Val, J.C. Geromel, E.M. Hemerly, E.K. Boukas, M.H. Terra, and F. Dufour. Many thanks go also to our former PhD students. We acknowledge with great pleasure the efficiency and support of Stephanie Harding, our contact at Springer. We wish to express our appreciation for the continued support of the Brazilian National Research Council – CNPq, under grants 472920/03-0, 520169/97-2, and 304866/03-0, and the Research Council of the State of São Paulo – FAPESP, under grant 03/06736-7. We also acknowledge the support of PRONEX, grant 015/98, and IM-AGIMB.

We (OLVC and MDF) have been fortunate to meet Prof. C.S. Kubrusly, at the very beginning of our scientific career. His enthusiasm, intellectual integrity, and friendship were important ingredients to make us continue. To him we owe a special debt of gratitude.

Last, but not least, we are very grateful to our families for their continuing and unwavering support. To them we dedicate this book.

São Paulo, Brazil   O.L.V. Costa
Petrópolis, Brazil   M.D. Fragoso
São Paulo, Brazil   R.P. Marques

Contents

1

Markov Jump Linear Systems

One of the main issues in control systems is their capability of maintaining an acceptable behavior and meeting some performance requirements even in the presence of abrupt changes in the system dynamics. These changes can be due, for instance, to abrupt environmental disturbances, component failures or repairs, changes in subsystems interconnections, abrupt changes in the operation point for a non-linear plant, etc. Examples of these situations can be found, for instance, in economic systems, aircraft control systems, control of solar thermal central receivers, robotic manipulator systems, large flexible structures for space stations, etc. In some cases these systems can be modeled by a set of discrete-time linear systems with modal transition given by a Markov chain. This family is known in the specialized literature as Markov jump linear systems (from now on MJLS), and will be the main topic of the present book. In this first chapter, prior to giving a rigorous mathematical treatment and present specific definitions, we will, in a rather rough and non-technical way, state and motivate this class of dynamical systems.

1.1 Introduction

Most control systems are based on a mathematical model of the process to be controlled. This model should be able to describe with relative accuracy the process behavior, in order that a controller whose design is based on the information provided by it performs accordingly when implemented in the real process. As pointed out by M. Kac in [148], *"Models are, for the most part, caricatures of reality, but if they are good, then, like good caricatures, they portray, though perhaps in a distorted manner, some of the features of the real world."* This translates, in part, the fact that to have more representative models for real systems, we have to characterize adequately the *uncertainties*.

Many processes may be well described, for example, by time-invariant linear models, but there are also a large number of them that are subject to uncertain changes in their dynamics, and demand a more complex approach.

If this change is an abrupt one, having only a small influence in the system behavior, classical sensitivity analysis may provide an adequate assessment of the effects. On the other hand, when the variations caused by the changes significantly alter the dynamic behavior of the system, a stochastic model that gives a quantitative indication of the relative likelihood of various possible scenarios would be preferable. Over the last decades, several different classes of models that take into account possible different scenarios have been proposed and studied, with more or less success.

To illustrate this situation, consider a dynamical system that is, in a certain moment, well described by a model $\mathcal{G}_1$. Suppose that this system is subject to abrupt changes that cause it to be described, after a certain amount of time, by a different model, say $\mathcal{G}_2$. More generally we can imagine that the system is subject to a series of possible qualitative changes that make it switch, over time, among a countable set of models, for example, $\{\mathcal{G}_1, \mathcal{G}_2, \ldots, \mathcal{G}_N\}$. We can associate each of these models to an *operation mode* of the system or just *mode* and will say that the system *jumps* from one mode to the other or that there are *transitions* between them.

The next question that arises is about the jumps. What hypotheses, if any at all, have to be made on them? It would be desirable to make none, but it would also strongly restrict any results that might be inferred. We will assume in this book that the jumps evolve stochastically according to a Markov chain, that is, given that at a certain instant k the system lies in mode i, we know the jump probability for each of the other modes, and also the probability of remaining in mode i (these probabilities depend only on the current operation mode). Notice that we assume only that the jump probability is known: in general, we do not know a priori when, if ever, jumps will occur.

We will restrict ourselves in this book to the case in which all operation modes are discrete-time, time-invariant linear models. With these assumptions we will be able to construct a coherent body of theory, develop the basic concepts of control and filtering and present controller and filter design procedures for this class of systems, known in the international literature as *discrete-time Markov jump linear systems* (MJLS for short). The Markov state (or operation mode) will be denoted throughout the book by $\theta(k)$.

Another main question that arises is whether or not the current operation mode $\theta(k)$ is known at each time k. Although in engineering problems the operation modes are not often available, there are enough cases where the knowledge of random changes in system structure is directly available to make these applications of great interest. This is the case, for instance, of a non-linear plant for which there are a countable number of operating points, each of them characterized by a corresponding linearized model, and the abrupt changes would represent the dynamics of the system moving from one operation point to another. In many situations it is possible to monitor these changes in the operating conditions of the process through appropriate sensors. In a deterministic formulation, an adaptive controller that changes its parameters in response to the monitored operating conditions of the process is termed a gain scheduling controller (see [9], Chapter 9). That is, it is a linear feedback controller whose parameters are changed as a function of operating conditions in a preprogrammed way. Several examples are presented in [9] of this kind of controller, and they could also be seen as examples for the optimal control problem of systems subject to abrupt dynamic changes, with the operation mode representing the monitored operation condition, and transition between the models following a Markov chain. For instance, in ship steering ([9]), the ship dynamics change with the ship's speed, which is not known a priori, but can be measured by appropriate speed sensors. The autopilot for this system can be improved by taking these changes into account. Examples related to control of pH in a chemical reactor, combustion control of a boiler, fuel air control in a car engine and flight control systems are also presented in [9] as examples of gain scheduling controllers, and they could be rewritten in a stochastic framework by introducing the probabilities of transition between the models, and serve as examples of the optimal control of MJLS with the operation mode available to the controller.

Another example of control of a system modeled by a MJLS, with $\theta(k)$ representing abrupt environmental changes measured by sensors located on the plant, would be the control of a solar-powered boiler [208]. The boiler flow rate is strongly dependent upon the receiving insolation and, as a result of this abrupt variability, several linearized models are required to characterize the evolution of the boiler when clouds interfere with the sun's rays. The control law described in [208] makes use of the state feedback and a measurement of $\theta(k)$ through the use of flux sensors on the receiver panels. A numerical example of control dependent on $\theta(k)$ for MJLS using Samuelson's multiplier-accelerator macroeconomic model can be found in [27], [28] and [89]. In this example $\theta(k)$ denotes the situation of a country's economy during period k, represented by "normal," "boom" and "slump." A control law for the governmental expenditure was derived.

The above examples, which will be discussed in greater detail in Chapter 8, illustrate the situation in which the abrupt changes occur due to variation of the operation point of a non-linear plant and/or environmental disturbances or due to economic periods of a country. Another possibility would be changes due to random failures/repairs of the process, with $\theta(k)$ in this case indicating the nature of any failure. For $\theta(k)$ to be available, an appropriate failure detector (see [190], [218]) would be used in conjunction with a control reconfiguration given by the optimal control of a MJLS.

Unless otherwise stated, we will assume throughout the book that the current operation mode $\theta(k)$ is known at each time k. As seen in the above discussion, the hypothesis that $\theta(k)$ is available is based on the fact that the operating condition of the plant can be monitored either by direct inspection or through some kind of sensor or failure detector. Some of the examples mentioned above will be examined under the theory and design techniques developed here.

Only in Chapters 3 and 5 (more specifically, Subsection 3.5.2 of Chapter 3, and Sections 5.4 and 5.5 of Chapter 5) shall we be dealing with the situation in which $\theta(k)$ is not known. In Chapter 3 we will consider the case in which there is an estimate $\hat{\theta}(k)$ of the real value $\theta(k)$, and we will study an important question that arises. Under what condition on the probability of a correct reading of $\theta(k)$ (that is, on $\mathcal{P}(\hat{\theta}(k) = i \mid \theta(k) = i)$) does one have that stability of the closed loop system of a MJLS will be maintained when we replace the real value $\theta(k)$ by its estimate $\hat{\theta}(k)$? That is, is stability maintained when we use a feedback control based on $\hat{\theta}(k)$ instead of $\theta(k)$? Conditions for this will be presented in Subsection 3.5.2. In Section 5.4 the problem of linear filtering for the case in which $\theta(k)$ is not known will be examined. Tracing a parallel with the standard Kalman filtering theory, a linear filter for the state variable is developed. The case in which there are uncertainties on the operation mode parameters is considered in Section 5.5.

1.2 Some Examples

In this section we present examples to motivate and illustrate some characteristics of MJLS. Example 1.1 below exhibits a sample, so to speak, of how a system like this behaves, and Example 1.2 illustrates when MJLS should be used.

Example 1.1. Consider a system with two operation modes, described by models $\mathcal{G}_1$ and $\mathcal{G}_2$ given by

$$
\begin{aligned}
(\mathcal{G}_1) &: x(k+1) = 0.8x(k) \\
(\mathcal{G}_2) &: x(k+1) = 1.2x(k).
\end{aligned}
\tag{1.1}
$$

When System (1.1) is in operation mode 1, its state evolves according to $\mathcal{G}_1$, and when in mode 2, according to $\mathcal{G}_2$, with jumps from one mode to the other. When a jump will occur (or if we had more operation modes, to which of the modes the system would jump to) is not known precisely. All we know is the probability of occurrence of jumps, given that the system is in a given mode. For this example, we will assume that when the system is in mode 1, it has a probability of 30% of jumping to mode 2 in the next time step. This of course means that there is always a probability of 70% of the system remaining in mode 1. We will also assume that when the system is in mode 2, it will have a probability of 40% of jumping to mode 1 and of 60% of remaining in mode 2. This is expressed by the following transition probability matrix

$$
\mathbf{P} = [p_{ij}] = \begin{bmatrix} 0.7 & 0.3 \\ 0.4 & 0.6 \end{bmatrix}
\tag{1.2}
$$

where each element p_{ij} in (1.2) gives the probability of transition from mode i to mode j. The Markov chain defined by matrix P and the dynamics given by $\mathcal{G}_1$ and $\mathcal{G}_2$ give us only a hint of the specific behavior of the system. Since any sequence of operation modes is essentially stochastic, we cannot know a priori the system trajectories, but as will be seen throughout this book, much information can be obtained from this kind of structure.

For the sake of illustration, consider one of the possible sequences of operation modes for this system when $k = 0, 1, \ldots, 20$ and the initial state of the Markov chain is $\theta(0) = 1$. For these timespan and initial conditions, the Markov chain may present $2^{20} = 1\,048\,576$ different realizations. The most probable is the following:

$$\overline{\Theta}_{0\ldots20} = \{1,1,1,1,1,1,1,1,1,1,1,1,1,1,1,1,1,1,1,1,1\},$$

while the least probable is

$$\underline{\Theta}_{0\ldots20} = \{1,2,1,2,1,2,1,2,1,2,1,2,1,2,1,2,1,2,1,2,1\}.$$

Notice that the probability of occurrence of $\overline{\Theta}_{0\ldots20}$ is 1 in $1\,253$, while the probability of occurrence of $\underline{\Theta}_{0\ldots20}$ is merely 1 in 1.62×10^9. Figure 1.1 presents the trajectory of the system for a randomly chosen realization of the Markov chain, given by

$$\Theta_{0\ldots20} = \{1,2,1,2,1,2,2,2,2,1,1,1,1,1,1,2,2,1,1,2\}$$

when the initial state is set as $x(0) = 1$.

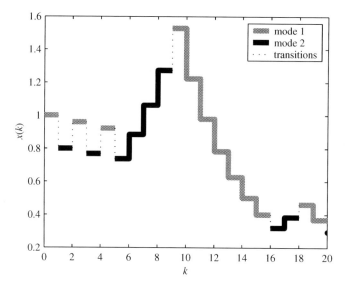

Fig. 1.1. A randomly picked trajectory for the system of Example 1.1.

Different realizations of the Markov chain lead to a multitude of very dissimilar trajectories, as sketched in Figure 1.2. The thick lines surrounding the gray area on the graphic are the extreme trajectories. All 1 048 574 remaining trajectories lie within them. The thin line is the trajectory of Figure 1.1. One could notice that some trajectories are unstable while others tend to zero as k increases. Stability concepts for this kind of system will be discussed in detail in Chapter 3.

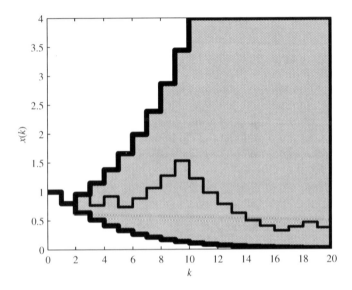

Fig. 1.2. Trajectories for Example 1.1. The gray area contains all possible trajectories. The thin line in the middle is the trajectory of Figure 1.1.

Example 1.2. To better illustrate when MJLS should be used, we will consider the solar energy plant described in [208], and already mentioned in Section 1.1. This same example will be treated in greater detail in Chapter 8. It consists of a set of adjustable mirrors, capable of focusing sunlight on a tower that contains a boiler, through which flows water, as sketched in Figure 1.3.

The power transferred to the boiler depends on the atmospheric conditions, more specifically on whether it is a sunny or cloudy day. With clear skies, the boiler receives more solar energy and so we should operate with a greater flow than on cloudy conditions. Clearly, the process dynamics is different for each of these conditions. It is very easy to assess the current weather conditions, but their prediction is certainly a more complex problem, and one that can only be solved on probabilistic terms.

Given adequate historical data, the atmospheric conditions can be modeled as a Markov chain with two states: 1) sunny; and 2) cloudy. This is a reasonable model: we can assume that the current state is known, that we do not know how the weather will behave in the future, but that the probabilities of it being sunny or cloudy in the immediate future are known. Also, these probabilities will depend on whether it is sunny or cloudy now. Given that the process dynamics behaves differently according to the current state of this Markov chain, considering a model like the one in Section 1.1 seems natural.

Since the current state of the Markov chain is known, we could consider three approaches to control the water flow through the boiler.

1. One single control law, with the differences in the dynamic behavior due to the changes in the operation mode treated as disturbances or as model uncertainties. This would be a kind of *Standard Robust Control* approach.

2. Two control laws, one for each operation mode. When the operation mode changes the control law used also changes. Each of the control laws is independently designed in order to stabilize the system and produce the best performance while it is in the corresponding operation mode. Any information regarding the transition probabilities is not taken into account. This could be called a *Multimodel* approach.

3. Two control laws. The same as above, except that both control laws were designed considering the Markov nature of the jumps between operation modes. This will be the approach adopted throughout this book, and for now let's call it the *Markov jump* approach.

When using the first approach, one would expect poorer performance, especially when system dynamics for each operation model differs significantly. Also, due to the random switching between operation modes, the boiler is markedly a time-variant system, and this fact alone can compromise the stability guarantees of most of the standard design techniques.

The second approach has the advantage of, at least on a first analysis, present better performance, but the stability issues still remain. With relation to the performance, there is a very important question: with the plant+controller dynamics changing from time to time, would this approach give, on the long run, the best results? Or rephrasing the question, would the best controllers for each operation mode, when associated, result in the best controller for the overall system? The answer to this question is: not necessarily.

Assuming that the sequence of weather conditions is compatible with the Markov chain, a Markov jump approach would generate a controller that could guarantee stability in a special sense and could also optimize the expected performance, as will be seen later. We should stress the term *expected*, for the sequence of future weather conditions is intrinsically random.

Whenever the problem can be reasonably stated as a Markov one, and theoretical guarantees regarding stability, performance, etc, are a must, the Markov jump approach should be considered. In many situations, even if the switching between operation modes does not follow rigidly a Markov chain, it is still possible to estimate a likely chain based on historical data with good results, as is done in the original reference [208] of the boiler example.

Notice that the Markov jump approach is usually associated with an *expected* (in the probabilistic sense) performance. This implies that for a specific unfavorable sequence of operation modes, the performance may be very poor, but in the long run, or for a great number of sequences, it is likely to present the best possible results.

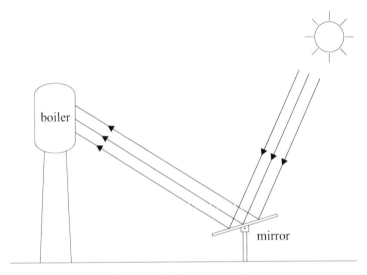

Fig. 1.3. A solar energy plant

Finally, we refer to [17] and [207], where the readers can find other applications of MJLS in problems such as tracking a maneuvering aircraft, automatic target recognition, decoding of signals transmitted across a wireless communication link, inter alia.

1.3 Problems Considered in this Book

Essentially, we shall be dealing with variants of the following class of dynamic systems throughout this book:

$$\mathcal{G} = \begin{cases} x(k+1) = A_{\theta(k)}x(k) + B_{\theta(k)}u(k) + G_{\theta(k)}w(k) \\ y(k) = L_{\theta(k)}x(k) + H_{\theta(k)}w(k) \\ z(k) = C_{\theta(k)}x(k) + D_{\theta(k)}u(k) \\ x(0) = x_0, \theta(0) = \theta_0. \end{cases} \tag{1.3}$$

As usual, $x(k)$ represents the state variable of the system, $u(k)$ the control variable, $w(k)$ the noise sequence acting on the system, $y(k)$ the measurable variable available to the controller, and $z(k)$ the output of the system. The specification of the actors of $\mathcal{G}$ will depend upon the kind of problem we shall be dealing with (optimal control, H_2-control, H_∞-control, etc.) and will be more rigorously characterized in the next chapters. For the moment it suffices to know that all matrices have appropriate dimensions and $\theta(k)$ stands for the state of a Markov chain taking values in a finite set $\mathbb{N} \triangleq \{1, \ldots, N\}$. The initial distribution for θ_0 is denoted by $\upsilon = \{\upsilon_1, \ldots, \upsilon_N\}$, and the transition probability matrix of the Markov chain $\theta(k)$ by $\mathrm{P} = [p_{ij}]$. Besides the additive stochastic perturbation $w(k)$, and the uncertainties associated to $\theta(k)$, we will also consider in some chapters parametric uncertainties acting on the matrices of the system.

Generically speaking, we shall be dealing with variants of $\mathcal{G}$, defined according to the specific types of problems which will be presented throughout the book. Unless otherwise stated, $\theta(k)$ will be assumed to be directly accessible. The problems and respective chapters are the following:

1. *Stability* (Chapter 3)

    In this case we will consider just the first equation of $\mathcal{G}$ in (1.3), with $w(k)$ either a sequence of independent identically distributed second order random variables or an ℓ_2-sequence of random variables. Initially we will consider a homogeneous version of this system, with $u(k) = 0$ and $w(k) = 0$, and present the key concept of mean square stability for MJLS. In particular we will introduce some important operators related to the second moment of the state variable $x(k)$, that will be used throughout the entire book. The non-homogeneous case is studied in the sequence. When discussing stabilizability and detectability we will consider the control law $u(k)$ and output $y(k)$, and the case in which we use a feedback control based on an estimate $\hat{\theta}(k)$ instead of $\theta(k)$. The chapter is concluded analyzing the almost sure stability of MJLS.

2. *The Quadratic and H_2-Optimal Control Problems* (Chapter 4)

    We will initially consider that the state variable $x(k)$ in (1.3) is observable. It is desired to design a feedback control law (dependent on $\theta(k)$) so as to minimize the quadratic norm of the output $z(k)$. Tracing a parallel with the standard theory for the optimal control problem of linear systems (the so-called LQ regulator problem), we will obtain a solution for the finite horizon and H_2-control problems through respectively a set of control coupled difference and algebraic Riccati equations.

3. *Filtering* (Chapter 5)

    In this chapter we will consider in (1.3) that $w(k)$ is a sequence of independent identically distributed second order random variables, and the controller has only access to $y(k)$. Two situations are considered.

First we will assume that the jump variable $\theta(k)$ is available, and it is desired to design a Markov jump linear filter for the state variable $x(k)$. The solution for this problem is obtained again through a set of filtering coupled difference (finite horizon) and algebraic (infinite horizon) Riccati equations. The second situation is when the jump variable $\theta(k)$ is not available to the controller, and again it is desired to design a linear filter for the state variable $x(k)$. Once again the solution is derived through a difference and algebraic Riccati-like equation. The case with uncertainties on the parameters of the system will be also analyzed by using convex optimization.

4. *Quadratic Optimal Control with Partial Information* (Chapter 6)

   We will consider in this chapter the linear quadratic optimal control problem for the case in which the controller has only access to the output variable $y(k)$ (besides the jump variable $\theta(k)$). Tracing a parallel with the standard LQG control problem, we will obtain a result that resembles the separation principle for the optimal control of linear systems. A Markov jump linear controller is designed from two sets of coupled difference (for the finite horizon case) and algebraic (for the H_2-control case) Riccati equations, one associated with the control problem, and the other one associated with the filtering problem.

5. *The H_∞-control Problem* (Chapter 7)

   This chapter deals with an H_∞-like theory for MJLS, following an approach based on a worst-case design problem. We will analyze the special case of state feedback, tracing a parallel with the time domain formulation used for studying the standard H_∞ theory. In particular a recursive algorithm for the H_∞-control CARE is deduced.

6. *Design Techniques and Examples* (Chapter 8)

   This chapter presents and discusses some applications of the theoretical results introduced earlier. It also presents design-oriented techniques based on convex optimization for control problems of MJLS with parametric uncertainties on the matrices of the system. This final chapter is intended to conclude the book, assembling some problems and the tools to solve them.

7. *Coupled Algebraic Riccati Equations* (Appendix A)

   As mentioned before, the control and filtering problems posed in this book are solved through a set of coupled difference and algebraic Riccati equations. In this appendix we will study the asymptotic behavior of this set of coupled Riccati difference equations, and some properties of the corresponding stationary solution, which will satisfy a set of coupled algebraic Riccati equations. Necessary and sufficient conditions for the existence of a stabilizing solution for the coupled algebraic Riccati equations will be established.

Appendices B and C present some auxiliary results for Chapters 5 and 6.

1.4 Some Motivating Remarks

When $\mathbb{N} = \{1\}$ we have that system $\mathcal{G}$ described in (1.3) reduces to the standard state space linear equation. A well known result for this class of system is that a structural criterion for stability of $\mathcal{G}$ is given in terms of the eigenvalues of the matrix $A_1 = A$. Keeping this in mind, the following questions regarding multiple operation mode systems come up naturally:

- Is it possible to adapt to $\mathcal{G}$, in a straightforward manner, the structural concepts from the classical linear theory such as *stabilizability, detectability, controllability,* and *observability?*
- What happens if we assume in a "natural way," that all the matrices $\{A_i\}, i \in \mathbb{N}$ have their eigenvalues less than one? Will $\|x(k)\| \to 0$ as $k \to \infty$?
- If at least one of the $\{A_i\}$ is unstable, will $\|x(k)\| \to \infty$ as $k \to \infty$?
- What happens if all the matrices $\{A_i\}$ are unstable? Will $\|x(k)\| \to \infty$ as $k \to \infty$?
- In short, is it adequate to consider stability for each mode of operation $A_i, i \in \mathbb{N}$ as a stability criterion for $\mathcal{G}$? The same question applies, mutatis mutandis, for stabilizability, detectability, controllability, and observability.
- If the above criterion is not valid, is it still possible to have a criterion based upon the eigenvalues of the matrices $\{A_i\}$?
- Is it possible to get, explicitly, an optimal control policy for the quadratic cost case? Is it, as in the classical linear case, a function of Riccati equations?
- Will a *Separation Principle* hold, as in the LQG case?
- Is it possible to carry on a H_∞ (H_2 or H_2/H_∞) synthesis with explicit display of the control policy?

Although it may appear, prima facie, that $\mathcal{G}$ is a straightforward extension of the linear case, it is perhaps worthwhile to point out from the outset that $\mathcal{G}$ carries a great deal of subtleties which distinguish it from the linear case and provide us with a very rich structure. In a nutshell, we can say that $\mathcal{G}$ differs from the linear case in many fundamental issues. For instance:

- Stability issues are not straightforwardly adapted from the linear case.
- It seems that a separation principle, as in the LQG case, does not hold anymore for the general setting in which (θ, x) are unknown and we do use the optimal mean square nonlinear filter to recast the problem as one with complete observations (this is essentially due to the amount of nonlinearity that is introduced in the problem). However, by using a certain class of filters, we devise in Chapter 6 a separation principle for the case in which θ is assumed to be accessible.
- *Duality* has to be redefined.

Throughout the next chapters, these and other issues will be discussed in detail and, we hope, the distinguishing particularities of MJLS, as well as the differences (and similarities) in relation to the classical case, will be made clear.

1.5 A Few Words On Our Approach

MJLS belongs to a wider class of systems which are known in the specialized literature as systems with switching structure (see, e.g., [17], [105], [179], [207] and references therein). Regarding the Markov switching linear class, three approaches stand out over the last years: the so-called Multiple Model (MM) approach; the one with an operator theoretical bent, whose theory tries to parallel the linear theory, which we shall term here the *Analytical Point of View* (APV), and more recently the so-called *Hidden Markov Model* (HMM) theory.

A very rough attempt to describe and distinguish these three approaches goes as follows. In the MM approach, the idea is to devise strategies which allow us to choose efficiently which mode is running the system, and work with the linear system associated to this mode (see, e.g. [17] and [207], for a comprehensive treatment on this subject). Essentially, the APV, which is the approach in this book, has operator theory as one of its technical underpinnings. In very loose terms the idea is, instead of dealing directly with the state $x(k)$ as in (1.3), which is not Markovian, we couch the systems in the Markov framework via the augmented state $(x(k), \theta(k))$. It is a well known fact that Markov processes are associated to operator theory via semigroup theory. This, in turn, allows us to define operators which play the discrete-time role of versions of the "infinitesimal generator," providing us with a powerful tool to get a dynamical equation for the second moment which is essential to our mean square stability treatment for MJLS. In addition, these operators will give, as a by-product, a criterion for mean square stability in terms of the spectral radius, which is in the spirit of the linear case. Vis-a-vis the MM, perhaps what best differentiates the two approaches, apart from the technical machinery, is the fact that the APV treats the problem as a whole (we do not have to choose between models).

As the very term *hidden* betrays, HMM designate what is known in the control literature as a class of *partially observed* stochastic dynamical system models. Tailored in its simplest version, and perhaps the original idea, the discrete-time HMM refer to a Markov chain $\{x(k), k \in \mathbb{T}\}$ which is not observed directly, but is hidden in a noisy observation process $\{y(k), k \in \mathbb{T}\}$. Roughly, the aim is the estimation of the state of the Markov chain, given the related observations, or the control of the hidden Markov chain (the transition matrix of the chain depends on a control variable u). In [105], an exhaustive study of HMM is carried out, which includes HMMs of increasing complexity. Reference probability method is the very mathematical girder that underpins the study in [105]. Loosely, the idea is to use a discrete-time *change of measure technique* (a discrete-time version of Girsanov's theorem) in such a way that in the new probability space ("the fictitious world"), the problem can now be treated via well-known results for i.i.d. random variables. Besides the methodology, the models and topics considered here differ from those in [105] in many aspects. For instance, in order to go deep in the special structure of MJLS, we analyze the case with complete observations. This allows us, for instance, to devise a mean square stability theory, which parallels that for the linear case. On the other hand, the HMM approach adds to the APV in the sense that the latter has been largely wedded to the complete observation scenario.

1.6 Historical Remarks

There is by now an extensive theory surrounding systems with switching structure. A variety of different approaches emerged over the last 30 years, or so. Yet, no one approach supersedes any others. It would take us too far afield to go into details on all of them (see, e.g., [17], [105], [179], [207] and references therein). Therefore, we confine our historical remarks to the Markov switching class, and particularly to MJLS, with emphasis on the APV approach. The interest in the study of this class of systems can be traced back at least to [155] and [113]. However, [206] and [219] bucked the trend on this scenario. In the first, the jump linear quadratic (JLQ) control problem is considered only in the finite horizon setting, and a stochastic maximum principle approach is used (see also [205]). In the other one, dynamic programming is used and the infinite horizon case is also treated. Although the objective had been carried out successfully, it seemed clear, prima facie, that the stability criteria used in [219] were not fully adequate. The inchoate idea in [219] was to consider the class above as a "natural" extension of the linear class and use as stability criteria the stability for each operation mode of the system plus a certain restrictive assumption which allows us to use fixed point type arguments to treat the coupled Riccati equations. In fact, the Riccati equation results used in [219] come from the seminal paper [220] (the restrictive assumption in [220] is removed in [122]). In the 1970s we can find [27] and [28], where the latter seems to be the first to treat the discrete-time version of the optimal quadratic control problem for the finite-time horizon case (see also [15], for the MM approach). It has been in the last two decades, or so, that a steadily rising level of activity with MJLS has produced a considerable development and flourishing literature on MJLS. For a sample of problems dealing with control (optimal, H_∞, robust, receding horizon), filtering, coupled Riccati equation, MJLS with delay, etc, the readers are referred to: [1], [2], [5], [11], [13], [20], [30], [31], [32], [33], [34], [35], [36], [37], [38], [39], [40], [41], [50], [52], [53], [54], [56], [58], [59], [60], [61], [63], [64], [65], [66], [67], [71], [72], [77], [78], [83], [85], [89], [92], [97], [101], [106], [109], [114], [115], [116], [121],

[123], [125], [127], [128], [132], [137], [143], [144], [145], [146], [147], [163], [166], [167], [169], [170], [171], [172], [173], [176], [177], [178], [188], [191], [192], [197], [198], [199], [210], [211], [215], [223], [226], [227]. In addition, there is by now a growing conviction that MJLS provide a model of wide applicability (see, e.g., [10] and [202]). For instance, it is said in [202] that the results achieved by MJLS, when applied to the synthesis problem of wing deployment of an uncrewed air vehicle, were quite encouraging. The evidence in favor of such a proposition has been amassing rapidly over the last decades. We mention [4], [10], [16], [23], [27], [47], [89], [134], [135], [150], [164], [168], [174], [175], [189], [202], and [208] as works dealing with applications of this class (see also [17], [207], and references therein).

2

Background Material

This chapter consists primarily of some background material, with the selection of topics being dictated by our later needs. Some facts and structural concepts of the linear case have a marked parallel in MJLS, so they are included here in order to facilitate the comparison. In Section 2.1 we introduce the notation, norms, and spaces that are appropriate for our approach. Next, in Section 2.2, we present some important auxiliary results that will be used throughout the book. In Section 2.3 we discuss some issues on the probability space for the underlined model. In Sections 2.4 and 2.5, we recall some basic facts regarding linear systems and linear matrix inequalities.

2.1 Some Basics

We shall use throughout the book some standard definitions and results from operator theory in Banach spaces which can be found, for instance, in [181] or [216]. For $\mathbb{X}$ and $\mathbb{Y}$ complex Banach spaces we set $\mathbb{B}(\mathbb{X}, \mathbb{Y})$ the Banach space of all bounded linear operators of $\mathbb{X}$ into $\mathbb{Y}$, with the uniform induced norm represented by $\|.\|$. For simplicity we set $\mathbb{B}(\mathbb{X}) \triangleq \mathbb{B}(\mathbb{X}, \mathbb{X})$. The spectral radius of an operator $\mathcal{T} \in \mathbb{B}(\mathbb{X})$ is denoted by $r_\sigma(\mathcal{T})$. If $\mathbb{X}$ is a Hilbert space then the inner product is denoted by $\langle .; . \rangle$, and for $\mathcal{T} \in \mathbb{B}(\mathbb{X})$, $\mathcal{T}^*$ denotes the adjoint operator of $\mathcal{T}$. As usual, $\mathcal{T} \geq 0$ ($\mathcal{T} > 0$ respectively) will denote that the operator $\mathcal{T} \in \mathbb{B}(\mathbb{X})$ is positive-semi-definite (positive-definite). In particular, we denote respectively by $\mathbb{R}^n$ and $\mathbb{C}^n$ the n dimensional real and complex Euclidean spaces and $\mathbb{B}(\mathbb{C}^n, \mathbb{C}^m)$ ($\mathbb{B}(\mathbb{R}^n, \mathbb{R}^m)$ respectively) the normed bounded linear space of all $m \times n$ complex (real) matrices, with $\mathbb{B}(\mathbb{C}^n) \triangleq \mathbb{B}(\mathbb{C}^n, \mathbb{C}^n)$ ($\mathbb{B}(\mathbb{R}^n) \triangleq \mathbb{B}(\mathbb{R}^n, \mathbb{R}^n)$). Unless otherwise stated, $\|.\|$ will denote the standard norm in $\mathbb{C}^n$, and for $M \in \mathbb{B}(\mathbb{C}^n, \mathbb{C}^m)$, $\|M\|$ denotes the induced uniform norm in $\mathbb{B}(\mathbb{C}^n, \mathbb{C}^m)$. The superscript * indicates the conjugate transpose of a matrix, while $'$ indicates the transpose. Clearly for real matrices * and $'$ will have the same meaning. The identity operator is denoted by $\mathcal{I}$, and the $n \times n$ identity matrix by I_n (or simply I). Finally, we denote by $\lambda_i(P)$, $i = 1, \ldots, n$ the eigenvalues of a matrix $P \in \mathbb{B}(\mathbb{C}^n)$.

Remark 2.1. We recall that the trace operator $\mathrm{tr}(.) : \mathbb{B}(\mathbb{C}^n) \to \mathbb{C}$ is a linear functional with the following properties:

1. $\mathrm{tr}(KL) = \mathrm{tr}(LK)$. (2.1a)
2. For any $M, P \in \mathbb{B}(\mathbb{C}^n)$ with $M \geq 0, P > 0$,

$$\left(\min_{i=1,\ldots,n} \lambda_i(P) \right) \mathrm{tr}(M) \leq \mathrm{tr}(MP) \leq \left(\max_{i=1,\ldots,n} \lambda_i(P) \right) \mathrm{tr}(M). \quad (2.1b)$$

In this book we shall be dealing with finite dimensional spaces, in which case all norms are equivalent. It is worth recalling that two norms $\|.\|_1$, $\|.\|_2$ in a Banach space $\mathbb{X}$ are equivalent if for some $c_1 > 0$ and $c_2 > 0$, and all $x \in \mathbb{X}$,

$$\|x\|_1 \leq c_2 \|x\|_2 \;,\; \|x\|_2 \leq c_1 \|x\|_1 \;.$$

As we are going to see in the next chapters, to analyze the stochastic model as in (1.3), we will use the indicator function on the jump parameter to *markovianize* the state. This, in turn, will decompose the matrices associated to the second moment and control problems into N matrices. Therefore it comes up naturally that a convenient space to be used is the one we define as $\mathbb{H}^{n,m}$, which is the linear space made up of all N-sequences of complex matrices $V = (V_1, \ldots, V_N)$ with $V_i \in \mathbb{B}(\mathbb{C}^n, \mathbb{C}^m)$, $i \in \mathbb{N}$. For simplicity, we set $\mathbb{H}^n \triangleq \mathbb{H}^{n,n}$. For $V = (V_1, \ldots, V_N) \in \mathbb{H}^{n,m}$, we define the following equivalent norms in the finite dimensional space $\mathbb{H}^{n,m}$:

$$\|V\|_1 \triangleq \sum_{i=1}^{N} \|V_i\|$$

$$\|V\|_2 \triangleq \left(\sum_{i=1}^{N} \mathrm{tr}(V_i^* V_i) \right)^{1/2}$$

$$\|V\|_{\max} \triangleq \max\{\|V_i\| \;;\; i \in \mathbb{N}\}. \quad (2.2)$$

We shall omit the subscripts $1, 2, \max$ whenever the definition of a specific norm does not affect the result being considered. It is easy to verify that $\mathbb{H}^{n,m}$ equipped with any of the above norms is a Banach space and, in fact, $(\|.\|_2, \mathbb{H}^{n,m})$ is a Hilbert space, with the inner product given, for $V = (V_1, \ldots, V_N)$ and $S = (S_1, \ldots, S_N)$ in $\mathbb{H}^{n,m}$, by

$$\langle V; S \rangle \triangleq \sum_{i=1}^{N} \mathrm{tr}(V_i^* S_i). \quad (2.3)$$

It is also convenient to define the following equivalent induced norms $\|.\|_1$ and $\|.\|_2$ in the finite dimensional space $\mathbb{B}(\mathbb{H}^n)$. For $\mathcal{T} \in \mathbb{B}(\mathbb{H}^n)$,

$$\|\mathcal{T}\|_1 \triangleq \sup_{V \in \mathbb{H}^n} \frac{\|\mathcal{T}(V)\|_1}{\|V\|_1} , \quad \|\mathcal{T}\|_2 \triangleq \sup_{V \in \mathbb{H}^n} \frac{\|\mathcal{T}(V)\|_2}{\|V\|_2} .$$

Again, we shall omit the subscripts $1, 2$ whenever the definition of the specific norm does not matter to the problem under consideration. For $V = (V_1, \dots, V_N) \in \mathbb{H}^{n,m}$ we write $V^* = (V_1^*, \dots, V_N^*) \in \mathbb{H}^{m,n}$ and say that $V \in \mathbb{H}^n$ is hermitian if $V = V^*$. We set

$$\mathbb{H}^{n*} \triangleq \{V = (V_1, \dots, V_N) \in \mathbb{H}^n; V_i = V_i^*, \ i \in \mathbb{N}\}$$

and

$$\mathbb{H}^{n+} \triangleq \{V = (V_1, \dots, V_N) \in \mathbb{H}^{n*}; V_i \geq 0, \ i \in \mathbb{N}\}$$

and write, for $V = (V_1, \dots, V_N) \in \mathbb{H}^n$ and $S = (S_1, \dots, S_N) \in \mathbb{H}^n$, that $V \geq S$ if $V - S = (V_1 - S_1, \dots, V_N - S_N) \in \mathbb{H}^{n+}$, and that $V > S$ if $V_i - S_i > 0$ for $i \in \mathbb{N}$. We say that an operator $\mathcal{T} \in \mathbb{B}(\mathbb{H}^n)$ is hermitian if $\mathcal{T}(V) \in \mathbb{H}^{n*}$ whenever $V \in \mathbb{H}^{n*}$, and that it is positive if $\mathcal{T}(V) \in \mathbb{H}^{n+}$ whenever $V \in \mathbb{H}^{n+}$.

We define the operators φ and $\hat{\varphi}$ in the following way: for $V = (V_1, \dots, V_N) \in \mathbb{H}^{n,m}$, with $V_i = \begin{bmatrix} v_{i1} & \cdots & v_{in} \end{bmatrix} \in \mathbb{B}(\mathbb{C}^n, \mathbb{C}^m), v_{ij} \in \mathbb{C}^m$

$$\varphi(V_i) \triangleq \begin{bmatrix} v_{i1} \\ \vdots \\ v_{in} \end{bmatrix} \in \mathbb{C}^{mn} \quad \text{and} \quad \hat{\varphi}(V) \triangleq \begin{bmatrix} \varphi(V_1) \\ \vdots \\ \varphi(V_N) \end{bmatrix} \in \mathbb{C}^{Nmn}.$$

With the Kronecker product $L \otimes K \in \mathbb{B}(\mathbb{C}^{n^2})$ defined in the usual way for any $L, K \in \mathbb{B}(\mathbb{C}^n)$, the following properties hold (see [43]):

$$(L \otimes K)^* = L^* \otimes K^* \tag{2.4a}$$

$$\varphi(LKH) = (H' \otimes L)\varphi(K), H \in \mathbb{B}(\mathbb{C}^n). \tag{2.4b}$$

Remark 2.2. It is easy to verify, through the mapping $\hat{\varphi}$, that the spaces $\mathbb{H}^{n,m}$ and $\mathbb{C}^{Nmn}$ are uniformly homeomorphic (see [181], p. 117) and that any operator $\mathcal{Z}$ in $\mathbb{B}(\mathbb{H}^{n,m})$ can be represented in $\mathbb{B}(\mathbb{C}^{Nmn})$ through the mapping $\hat{\varphi}$. We shall denote this operator by $\hat{\varphi}[\mathcal{Z}]$. Clearly we must have

$$r_\sigma(\mathcal{Z}) = r_\sigma(\hat{\varphi}[\mathcal{Z}]).$$

Remark 2.3. It is well known that if $W \in \mathbb{B}(\mathbb{C}^n)^+$ then there exists a unique $W^{1/2} \in \mathbb{B}(\mathbb{C}^n)^+$ such that $W = (W^{1/2})^2$. The absolute value of $W \in \mathbb{B}(\mathbb{C}^n)$, denoted by $|W|$, is defined as $|W| = (W^*W)^{1/2}$. As shown in [216], p. 170, there exists an orthogonal matrix $U \in \mathbb{B}(\mathbb{C}^n)$ (that is, $U^{-1} = U^*$) such that

$$W = U|W| \ (\text{ or } |W| = U^{-1}W = U^*W), \tag{2.5}$$

and $\|W\| = \| |W| \|$.

Remark 2.4. For any $W \in \mathbb{B}(\mathbb{C}^n)$ there exist W^j, $j = 1, 2, 3, 4$, such that $W^j \geq 0$ and $\left\|W^j\right\| \leq \|W\|$ for $j = 1, 2, 3, 4$, and $W = (W^1 - W^2) + \sqrt{-1}(W^3 - W^4)$. Indeed, we can write

$$W = V^1 + \sqrt{-1}V^2$$

where

$$V^1 = \frac{1}{2}(W^* + W)$$

$$V^2 = \frac{\sqrt{-1}}{2}(W^* - W).$$

Since V^1 and V^2 are self-adjoint (that is, $V^i = V^{i*}, i = 1, 2$), and every self-adjoint element in $\mathbb{B}(\mathbb{C}^n)$ can be decomposed into positive and negative parts (see [181], p. 464), we have that there exist $W^i \in \mathbb{B}(\mathbb{C}^n)^+$, $i = 1, 2, 3, 4$, such that

$$V^1 = W^1 - W^2$$

$$V^2 = W^3 - W^4.$$

Therefore for any $S = (S_1, \ldots, S_N) \in \mathbb{H}^n$, we can find $S^j \in \mathbb{H}^{n+}$, $j = 1, 2, 3, 4$ such that $\left\|S^j\right\|_1 \leq \|S\|_1$ and

$$S = (S^1 - S^2) + \sqrt{-1}(S^3 - S^4).$$

2.2 Auxiliary Results

The next result follows from the decomposition of square matrices into positive semi-definite matrices as seen in Remark 2.4 in conjunction with Lemma 1 and Remark 4 in [156].

Proposition 2.5. *Let* $\mathcal{Z} \in \mathbb{B}(\mathbb{H}^n)$. *The following assertions are equivalent:*

1. $\sum_{k=0}^{\infty} \left\|\mathcal{Z}^k(V)\right\|_1 < \infty$ *for all* $V \in \mathbb{H}^{n+}$.
2. $r_\sigma(\mathcal{Z}) < 1$.
3. $\left\|\mathcal{Z}^k\right\| \leq \beta \zeta^k, k = 0, 1, \ldots$ *for some* $0 < \zeta < 1$ *and* $\beta \geq 1$.
4. $\left\|\mathcal{Z}^k(V)\right\|_1 \to 0$ *as* $k \to \infty$ *for all* $V \in \mathbb{H}^{n+}$.

Proof. From Remark 2.4 for any $S = (S_1, \ldots, S_N) \in \mathbb{H}^n$, we can find $S^j \in \mathbb{H}^{n+}$, $j = 1, 2, 3, 4$ such that $\left\|S^j\right\|_1 \leq \|S\|_1$ and

$$S = (S^1 - S^2) + \sqrt{-1}(S^3 - S^4).$$

Since $\mathcal{Z}$ is a linear operator we get

$$\left\| \mathcal{Z}^k(S) \right\|_1 = \left\| \mathcal{Z}^k(S^1) - \mathcal{Z}^k(S^2) + \sqrt{-1}\left(\mathcal{Z}^k(S^3) - \mathcal{Z}^k(S^4) \right) \right\|_1$$
$$\leq \sum_{i=1}^{4} \left\| \mathcal{Z}^k(S^i) \right\|_1. \tag{2.6}$$

The result now follows easily from Lemma 1 and Remark 4 in [156], after noticing that $\mathbb{H}^n$ is a finite dimensional complex Banach space. ☐

The next result is an immediate adaptation of Lemma 1 in [158].

Proposition 2.6. *Let $\mathcal{Z} \in \mathbb{B}(\mathbb{H}^n)$. If $r_\sigma(\mathcal{Z}) < 1$ then there exists a unique $V \in \mathbb{H}^n$ such that*
$$V = \mathcal{Z}(V) + S$$
for any $S \in \mathbb{H}^n$. Moreover,
$$V = \hat{\varphi}^{-1}\left((I_{Nn^2} - \hat{\varphi}[\mathcal{Z}])^{-1} \hat{\varphi}(S) \right) = \sum_{k=0}^{\infty} \mathcal{Z}^k(S).$$

Furthermore, if $\mathcal{Z}$ is a hermitian operator then
$$S = S^* \Leftrightarrow V = V^*,$$
and if $\mathcal{Z}$ is a positive operator then
$$S \geq 0 \ \Rightarrow V \geq 0$$
$$S > 0 \ \Rightarrow V > 0.$$

The following corollary is an immediate consequence of the previous result.

Corollary 2.7. *Suppose that $\mathcal{Z} \in \mathbb{B}(\mathbb{H}^n)$ is a positive operator with $r_\sigma(\mathcal{Z}) < 1$. If*
$$V = \mathcal{Z}(V) + S$$
$$\tilde{V} = \mathcal{Z}(\tilde{V}) + \tilde{S},$$
with $\tilde{S} \geq S$ ($\tilde{S} > S$) then $\tilde{V} \geq V$ ($\tilde{V} > V$).

Proof. Straightforward from Proposition 2.6. ☐

The following definition and result will be useful in Chapter 3.

Definition 2.8. *We shall say that a Cauchy sequence $\{z(k); k = 0, 1, \ldots\}$ in a complete normed space $\mathbb{Z}$ (in particular, $\mathbb{C}^n$ or $\mathbb{B}(\mathbb{C}^n)$) is Cauchy summable if (cf. [157])*
$$\sum_{k=0}^{\infty} \sup_{\tau \geq 0} \| z(k+\tau) - z(k) \| < \infty.$$

The next proposition was established in Lemma (L1) of [157].

Proposition 2.9. *Let $\{z(k); k = 0, 1, \ldots\}$ be a Cauchy summable sequence in $\mathbb{Z}$ and consider the sequence $\{y(k); k = 0, 1, \ldots\}$ in $\mathbb{Z}$ given by*

$$y(k + 1) = \mathcal{L}y(k) + z(k)$$

where $\mathcal{L} \in \mathbb{B}(\mathbb{Z})$. If $r_\sigma(\mathcal{L}) < 1$, then $\{y(k); k = 0, 1, \ldots\}$ is a Cauchy summable sequence and for any initial condition $y(0) \in \mathbb{Z}$,

$$\lim_{k \to \infty} y(k) = (\mathcal{I} - \mathcal{L})^{-1} \lim_{k \to \infty} z(k).$$

2.3 Probabilistic Space

In this section we present in detail the probabilistic framework we shall consider throughout this book. We shall be dealing with stochastic models as in (1.3) with, at each time k, the jump variable $\theta(k)$ taking values in the set $\mathbb{N} = \{1, \ldots, N\}$, and the remaining input variables taking values in $\tilde{\Omega}_k$. Thus, for the jump variable, we set $\mathfrak{N}$ the σ-field of all subsets of $\mathbb{N}$, and for the remaining input variables, we set $\tilde{\mathfrak{F}}_k$ as the Borel σ-field of $\tilde{\Omega}_k$. To consider all time values, we define

$$\Omega \triangleq \prod_{k \in \mathbb{T}} (\tilde{\Omega}_k \times \mathbb{N}_k)$$

where $\mathbb{N}_k$ are copies of $\mathbb{N}$, $\times$ and $\prod$ denote the product space, and $\mathbb{T}$ represents the discrete-time set, being $\{\ldots, -1, 0, 1, \ldots\}$ when the process starts from $-\infty$, or $\{0, 1, \ldots\}$, when the process starts from 0. Set also $\mathbb{T}_k = \{i \in \mathbb{T}; i \leq k\}$ for each $k \in \mathbb{T}$, and

$$\mathfrak{F} \triangleq \sigma \left\{ \prod_{k \in \mathbb{T}} S_k \times \psi_k; S_k \in \tilde{\mathfrak{F}}_k \text{ and } \psi_k \in \mathfrak{N} \text{ for each } k \in \mathbb{T} \right\}$$

and for each $k \in \mathbb{T}$,

$$\mathfrak{F}_k \triangleq \sigma \left\{ \prod_{l \in \mathbb{T}_k} S_l \times \psi_l \times \prod_{\tau = k+1}^{\infty} \tilde{\Omega}_\tau \times \mathbb{N}_\tau; S_l \in \tilde{\mathfrak{F}}_l \text{ and } \psi_l \in \mathfrak{N} \text{ for } l \in \mathbb{T}_k \right\}$$

so that $\mathfrak{F}_k \subset \mathfrak{F}$. We define then the stochastic basis $(\Omega, \mathfrak{F}, \{\mathfrak{F}_k\}, \mathcal{P})$, where $\mathcal{P}$ is a probability measure such that

$$\mathcal{P}(\theta(k+1) = j \mid \mathfrak{F}_k) = \mathcal{P}(\theta(k+1) = j \mid \theta(k)) = p_{\theta(k)j}$$

with $p_{ij} \geq 0$ for $i, j \in \mathbb{N}$, $\sum_{j=1}^{N} p_{ij} = 1$ for each $i \in \mathbb{N}$, and for each $k \in \mathbb{T}$, $\theta(k)$ is a random variable from Ω to $\mathbb{N}$ defined as $\theta(k)(\omega) = \beta(k)$ with $\omega = \{(\xi(k), \beta(k)); k \in \mathbb{T}\}$, $\xi(k) \in \tilde{\Omega}_k, \beta(k) \in \mathbb{N}$. Clearly $\{\theta(k); k \in \mathbb{T}\}$ is a Markov chain taking values in $\mathbb{N}$ and with transition probability matrix $P = [p_{ij}]$. The initial distribution for $\theta(0)$ is denoted by $v = \{v_1, \ldots, v_N\}$.

We set $C^m = L_2(\Omega, \mathfrak{F}, \mathcal{P}, \mathbb{C}^m)$ the Hilbert space of all second order $\mathbb{C}^m$-valued $\mathfrak{F}$-measurable random variables with inner product given by $\langle x; y \rangle = E(x^*y)$ for all $x, y \in C^m$, where $E(.)$ stands for the expectation of the underlying scalar valued random variables, and norm denoted by $\|.\|_2$. Set $\ell_2(C^m) = \underset{k \in \mathbb{T}}{\oplus} C^m$, the direct sum of countably infinite copies of C^m, which is a Hilbert space made up of $r = \{r(k); k \in \mathbb{T}\}$, with $r(k) \in C^m$ for each $k \in \mathbb{T}$, and such that

$$\|r\|_2^2 \triangleq \sum_{k \in \mathbb{T}} E(\|r(k)\|^2) < \infty.$$

For $r = \{r(k); k \in \mathbb{T}\} \in \ell_2(C^m)$ and $v = \{v(k); k \in \mathbb{T}\} \in \ell_2(C^m)$, the inner product $\langle r; s \rangle$ in $\ell_2(C^m)$ is given by

$$\langle r; s \rangle \triangleq \sum_{k \in \mathbb{T}} E(r^*(k)v(k)) \le \|r\|_2 \|v\|_2.$$

We define $\mathcal{C}^m \subset \ell_2(C^m)$ in the following way: $r = \{r(k); k \in \mathbb{T}\} \in \mathcal{C}^m$ if $r \in \ell_2(C^m)$ and $r(k) \in L_2(\Omega, \mathfrak{F}_k, \mathcal{P}, \mathbb{C}^m)$ for each $k \in \mathbb{T}$. We have that $\mathcal{C}^m$ is a closed linear subspace of $\ell_2(C^m)$ and therefore a Hilbert space. We also define $\mathcal{C}_k^m$ as formed by the elements $r_k = \{r(k); k \in \mathbb{T}_k\}$ such that $r(l) \in L_2(\Omega, \mathfrak{F}_l, \mathcal{P}, \mathbb{C}^m)$ for each $l \in \mathbb{T}_k$. Finally we define Θ_0 as the set of all $\mathfrak{F}_0$-measurable variables taking values in $\mathbb{N}$.

2.4 Linear System Theory

Although MJLS seem, prima facie, a natural extension of the linear class, their subtleties are such that the standard linear theory cannot be directly applied, although it will be most illuminating in the development of the results described in this book. In view of this, it is worth having a brief look at some basic results and properties of the linear time-invariant systems (in short LTI), whose Markov jump counterparts will be considered later.

2.4.1 Stability and the Lyapunov Equation

Consider the following difference equations

$$x(k + 1) = f(x(k)) \tag{2.7}$$

and

$$x(k + 1) = Ax(k) \tag{2.8}$$

with $k \in \{0, 1, 2, \ldots\}$, $x(k) \in \mathbb{C}^n$, $f : \mathbb{C}^n \to \mathbb{C}^n$ and $A \in \mathbb{B}(\mathbb{C}^n)$. A sequence $x(0), x(1), \ldots$ generated according to (2.7) or (2.8) is called a trajectory of the system. The second equation is a particular case of the first one and is of greater interest to us (thus we shall not be concerned on regularity hypotheses over f in (2.7)). It defines what we call a *discrete-time homogeneous linear time-invariant system*. For more information on dynamic systems or proofs of the results presented in this section, the reader may refer to one of the many works on the theme, like [48], [165] and [213].

First we recall that a point $x_e \in \mathbb{C}^n$ is called an *equilibrium point* of System (2.7), if $f(x_e) = x_e$. In particular, $x_e = 0$ is an equilibrium point of System (2.8). The following definitions apply to Systems (2.7) and (2.8).

Definition 2.10 (Lyapunov Stability). *An equilibrium point x_e is said to be stable in the sense of Lyapunov if for each $\epsilon > 0$ there exists $\delta_\epsilon > 0$ such that $\|x(k) - x_e\| \leq \epsilon$ for all $k \geq 0$ whenever $\|x(0) - x_e\| \leq \delta_\epsilon$.*

Definition 2.11 (Asymptotic Stability). *An equilibrium point is said to be asymptotically stable if it is stable in the sense of Lyapunov and there exists $\delta > 0$ such that whenever $\|x(0) - x_e\| \leq \delta$ we have that $x(k) \to x_e$ as k increases. It is globally asymptotically stable if it is asymptotically stable and $x(k) \to x_e$ as k increases for any $x(0)$ in the state space.*

The definition above simply states that the equilibrium point is stable if, given any spherical region surrounding the equilibrium point, we can find another spherical region surrounding the equilibrium point such that trajectories starting inside this second region do not leave the first one. Besides, if the trajectories also converge to this equilibrium point, then it is asymptotically stable.

Definition 2.12 (Lyapunov Function). *Let x_e be an equilibrium point for System (2.7). A positive function $\phi : \Gamma \to \mathbb{R}$, where Γ is such that $x_e \in \Gamma \subseteq \mathbb{C}^n$, is said to be a Lyapunov function for System (2.7) and equilibrium point x_e if*

1. *$\phi(.)$ is continuous,*
2. *$\phi(x_e) < \phi(x)$ for every $x \in \Gamma$ such that $x \neq x_e$,*
3. *$\Delta\phi(x) = \phi(f(x)) - \phi(x) \leq 0$ for all $x \in \Gamma$.*

With this we can proceed to the Lyapunov Theorem. A proof of this result can be found in [165].

Theorem 2.13 (Lyapunov Theorem). *If there exists a Lyapunov function $\phi(x)$ for System (2.7) and x_e, then the equilibrium point is stable in the sense of Lyapunov. Moreover, if $\Delta\phi(x) < 0$ for all $x \neq x_e$, then it is asymptotically stable. Furthermore if ϕ is defined on the entire state space and $\phi(x)$ goes to infinity as any component of x gets arbitrarily large in magnitude then the equilibrium point x_e is globally asymptotically stable.*

The Lyapunov theorem applies to System (2.7) and, of course, to System (2.8) as well. Let us consider a possible Lyapunov function for System (2.8) as follows:

$$\phi(x(k)) = x^*(k)Vx(k) \tag{2.9}$$

with $V > 0$. Then

$$
\begin{aligned}
\Delta\phi(x(k)) &= \phi(x(k+1)) - \phi(x(k)) \\
&= x^*(k+1)Vx(k+1) - x^*(k)Vx(k) \\
&= x^*(k)A^*VAx(k) - x^*(k) * Vx(k) \\
&= x^*(k)(A^*VA - V)x(k).
\end{aligned}
$$

With this we can present the following theorem that establishes the connection between System (2.8), stability, and the so called Lyapunov equation. All assertions are classical applications of the Lyapunov theorem with the Lyapunov function (2.9). The proof can be found, for instance, in [48].

Theorem 2.14. *The following assertions are equivalent.*

1. *$x = 0$ is the only globally asymptotically stable equilibrium point for System (2.8).*
2. *$r_\sigma(A) < 1$.*
3. *For any $S > 0$, there exists a unique $V > 0$ such that*

$$V - A^*VA = S. \tag{2.10}$$

4. *For some $V > 0$, we have*

$$V - A^*VA > 0. \tag{2.11}$$

The above theorem will be extended to the Markov case in Chapter 3 (Theorem 3.9).

Since (2.8) has only one equilibrium point whenever it is stable, we commonly say in this case that System (2.8) is stable.

2.4.2 Controllability and Observability

Let us now consider a non-homogeneous form for System (2.8)

$$x(k+1) = Ax(k) + Bu(k) \tag{2.12}$$

where $B \in \mathbb{B}(\mathbb{C}^m, \mathbb{C}^n)$ and $u(k) \in \mathbb{C}^m$ is a vector of inputs to the system.

The idea behind the concept of controllability is rather simple. It deals with answering the following question: for a certain pair (A, B), is it possible to apply a sequence of $u(k)$ in order to drive the system from any $x(0)$ to a specified final state x_f in a finite time?

The following definition establishes more precisely the concept of controllability. Although not treated here, a concept akin to controllability is the *reachability* of a system. In more general situations these concepts may differ, but in the present case they are equivalent, and therefore we will only use the term controllability.

Definition 2.15 (Controllability). *The pair (A, B) is said to be controllable, if for any $x(0)$ and any given final state x_f, there exists a finite positive integer T and a sequence of inputs $u(0), u(1), \ldots, u(T - 1)$ that, applied to System (2.12), yields $x(T) = x_f$.*

One can establish if a given system is controllable using the following theorem, which also lists some classical results (see [48], p. 288).

Theorem 2.16. *The following assertions are equivalent.*

1. *The pair (A, B) is controllable.*
2. *The following $n \times nm$ matrix (called a* controllability matrix*) has rank n:*

$$\begin{bmatrix} B \ AB \ \cdots \ A^{n-1}B \end{bmatrix}.$$

3. *The* controllability Grammian $S_c \in \mathbb{B}(\mathbb{C}^n)$ *given by*

$$S_c(k) = \sum_{i=0}^{k} A^i BB^*(A^*)^i$$

   is nonsingular for some $k < \infty$.
4. *For A and B real, given any monic real polynomial ψ of degree n, there exists $F \in \mathbb{B}(\mathbb{R}^n, \mathbb{R}^m)$ such that $\det(sI - (A + BF)) = \psi(s)$.*

Moreover, if $r_\sigma(A) < 1$ then the pair (A, B) is controllable if and only if the unique solution S_c of $S = ASA^* + BB^*$ is positive-definite.

The concept of controllability Grammian for MJLS will be presented in Chapter 4, Section 4.4.2.

Item 4 of the theorem above is particularly interesting, since it involves the idea of *state feedback*. Suppose that for some $F \in \mathbb{B}(\mathbb{C}^n, \mathbb{C}^m)$, we apply $u(k) = Fx(k)$ in System (2.12), yielding

$$x(k + 1) = (A + BF)x(k),$$

which is a form similar to (2.8). According to the theorem above, an adequate choice of F (for A, B and F real) would allow us to perform pole placement for the closed loop system $(A + BF)$. For instance we could use state feedback to stabilize an unstable system.

The case in which the state feedback can only change the unstable eigenvalues of the system is of great interest and leads us to the introduction of the concept of *stabilizability*.

Definition 2.17 (Stabilizability). *The pair (A, B) is said to be* stabilizable *if there exists $F \in \mathbb{B}(\mathbb{C}^n, \mathbb{C}^m)$ such that $r_\sigma(A + BF) < 1$.*

This concept will play a crucial role for the MJLS, as will be seen in the next chapters (see Section 3.5). Consider now a system of the form

$$x(k + 1) = Ax(k) \qquad (2.13a)$$
$$y(k) = Lx(k) \qquad (2.13b)$$

where $L \in \mathbb{B}(\mathbb{C}^n, \mathbb{C}^p)$ and $y(k) \in \mathbb{C}^p$ is the vector of outputs of the system. The concepts of controllability and stabilizability just presented, which relate structurally $x(k)$ and the input $u(k)$, have their dual counterparts from the point of view of the output $y(k)$. The following theorem and definitions present them.

Definition 2.18 (Observability). *The pair (L, A) is said to be* observable, *if there exists a finite positive integer T such that knowledge of the outputs $y(0), y(1), \ldots, y(T - 1)$ is sufficient to determine the initial state $x(0)$.*

The concept of observability deals with the following question: is it possible to infer the internal behavior of a system by observing its outputs? This is a fundamental property when it comes to control and filtering issues.

The following theorem is dual to Theorem 2.16, and the proof can be found in [48], p. 282.

Theorem 2.19. *The following assertions are equivalent.*

1. *The pair (L, A) is observable.*
2. *The following $pn \times n$ matrix (called an* observability matrix*) has rank n:*

$$\begin{bmatrix} L \\ LA \\ \vdots \\ LA^{n-1} \end{bmatrix}.$$

3. *The* observability Grammian $S_o \in \mathbb{B}(\mathbb{C}^n)$ *given by*

$$S_o(k) = \sum_{i=0}^{k} (A^*)^i L^* L A^i$$

is nonsingular for some $k < \infty$.

4. *For A and L real, given any monic real polynomial ψ of degree n, there exists $K \in \mathbb{B}(\mathbb{R}^p, \mathbb{R}^n)$ such that $\det(sI - (A + KL)) = \psi(s)$.*

Moreover, if $r_\sigma(A) < 1$ then the pair (L, A) is observable if and only if the unique solution S_o of $S = A^* S A + L^* L$ is positive-definite.

We also define the concept of *detectability*, which is dual to the definition of stabilizability.

Definition 2.20 (Detectability).   *The pair* (L, A) *is said to be detectable if there exists* $K \in \mathbb{B}(\mathbb{C}^p, \mathbb{C}^n)$ *such that* $r_\sigma(A + KL) < 1$.

These are key concepts in linear system theory which will be extended to the Markov jump case in due course throughout this book.

2.4.3 The Algebraic Riccati Equation and the Linear-Quadratic Regulator

Consider again System (2.12)

$$x(k + 1) = Ax(k) + Bu(k).$$

An extensively studied and classical control problem is that of finding a sequence $u(0), u(1), \ldots, u(T - 1)$ that minimizes the cost $\mathfrak{J}_T(x_0, u)$ given by

$$\mathfrak{J}_T(x_0, u) = \sum_{k=0}^{T-1} \left[\|Cx(k)\|^2 + \|Du(k)\|^2 \right] + E(x(T)^* \mathcal{V} x(T)), \qquad (2.14)$$

where $\mathcal{V} \geq 0$ and $D^* D > 0$. The idea of minimizing $\mathfrak{J}_T(x_0, u)$ is to drive the state of the system to the origin without much strain from the control variable which is, in general, a desirable behavior for control systems. This problem is referred to as the *linear-quadratic regulator* (*linear* system + *quadratic* cost) problem. It can be shown (see for instance [48] or [183]) that the solution to this problem is

$$u(k) = F(k)x(k) \qquad (2.15)$$

with $F(k)$ given by

$$F(k) = -(B^* X_T(k + 1)B + D^* D)^{-1} B^* X_T(k + 1)A \qquad (2.16a)$$
$$X_T(k) = C^* C + A^* X_T(k + 1)A - A^* X_T(k + 1)B$$
$$\times (B^* X_T(k + 1)B + D^* D)^{-1} B^* X_T(k + 1)A \qquad (2.16b)$$
$$X_T(T) = \mathcal{V}.$$

Equation (2.16b) is called the *difference Riccati equation*. Another related problem is the infinite horizon linear quadratic regulator problem, in which it is desired to minimize the cost

$$\mathfrak{J}(x_0, u) = \sum_{k=0}^{\infty} \left[\|Cx(k)\|^2 + \|Du(k)\|^2 \right]. \qquad (2.17)$$

Under some conditions, the solution to this problem is

$$u(k) = F(X)x(k), \tag{2.18}$$

where the constant gain $F(X)$ is given by

$$F(X) = -(B^*XB + D^*D)^{-1}B^*XA \tag{2.19}$$

and X is a positive semi-definite solution of

$$W = C^*C + A^*WA - A^*WB(B^*WB + D^*D)^{-1}B^*WA. \tag{2.20}$$

Equation (2.20) is usually referred to as the *algebraic Riccati equation* or in short, ARE . If $r_\sigma(A+BF(X)) < 1$, then X is said to be a stabilizing solution of (2.20). Questions that naturally arise are: under which conditions there is convergence of $X_T(0)$ given by (2.16b) as T goes to infinity to a positive semi-definite solution X of (2.20)? When is there a stabilizing solution for (2.20)? Is it unique? The following theorem, whose proof can be found, for instance, in [48], p. 348, answers these questions.

Theorem 2.21. *Suppose that the pair (A, B) is stabilizable. Then for any $V \geq 0$, $X_T(0)$ converges to a positive semi-definite solution X of (2.20) as T goes to infinity. Moreover if the pair (C, A) is detectable, then there exists a unique positive semi-definite solution X to (2.20), and this solution is the unique stabilizing solution for (2.20).*

Riccati equations like (2.16b) and (2.20) and their variations are employed in a variety of control (as in (2.14) and (2.17)) and filtering problems. As we are going to see in Chapters 4, 5, 6 and 7, they will also play a crucial role for MJLS. For more on Riccati equations and associated problems, see [26], [44], and [195].

2.5 Linear Matrix Inequalities

Some miscellaneous definitions and results involving matrices and matrix equations are presented in this section. These results will be used throughout the book, especially those related with the concept of linear matrix inequalities (or in short LMIs), which will play a very important role in the next chapters.

Definition 2.22 (Generalized Inverse). *The generalized inverse (or Moore–Penrose inverse) of a matrix $A \in \mathbb{B}(\mathbb{C}^n, \mathbb{C}^m)$ is the unique matrix $A^\dagger \in \mathbb{B}(\mathbb{C}^m, \mathbb{C}^n)$ such that*

1. $AA^\dagger A = A$,
2. $A^\dagger AA^\dagger = A^\dagger$,
3. $(AA^\dagger)^* = AA^\dagger$,
4. $(A^\dagger A)^* = A^\dagger A$.

For more on this subject, see [49]. The Schur complements presented below are used to convert quadratic equations into larger dimension linear ones and vice versa.

Lemma 2.23 (Schur complements). *(From [195]). Consider an hermitian matrix Q such that*

$$Q = \begin{bmatrix} Q_{11} & Q_{12} \\ Q_{12}^* & Q_{22} \end{bmatrix}.$$

1. $Q > 0$ if and only if

$$\begin{cases} Q_{22} > 0 \\ Q_{11} - Q_{12}Q_{22}^{-1}Q_{12}^* > 0 \end{cases}$$

or

$$\begin{cases} Q_{11} > 0 \\ Q_{22} - Q_{12}^*Q_{11}^{-1}Q_{12} > 0. \end{cases}$$

2. $Q \geq 0$ if and only if

$$\begin{cases} Q_{22} \geq 0 \\ Q_{12} = Q_{12}Q_{22}^{\dagger}Q_{22} \\ Q_{11} - Q_{12}Q_{22}^{\dagger}Q_{12}^* \geq 0 \end{cases}$$

or

$$\begin{cases} Q_{11} \geq 0 \\ Q_{12} = Q_{11}Q_{11}^{\dagger}Q_{12} \\ Q_{22} - Q_{12}^*Q_{11}^{\dagger}Q_{12} \geq 0. \end{cases}$$

Next we present the definition of LMI.

Definition 2.24. *A linear matrix inequality (LMI) is any constraint that can be written or converted to*

$$F(x) = F_0 + x_1 F_1 + x_2 F_2 + \ldots + x_m F_m < 0, \tag{2.21}$$

where x_i are the variables and the hermitian matrices $F_i \in \mathbb{B}(\mathbb{R}^n)$ for $i = 1, \ldots, m$ are known.

LMI (2.21) is referred to as a strict LMI. Also of interest are the nonstrict LMIs, where $F(x) \leq 0$. From the practical point of view, LMIs are usually presented as

$$f(X_1, \ldots, X_N) < g(X_1, \ldots, X_N), \tag{2.22}$$

where f and g are affine functions of the unknown matrices $X_1, \ldots, X_N$. For example, from the Lyapunov equation, the stability of System (2.8) is equivalent to the existence of a $V > 0$ satisfying the LMI (2.11). Quadratic forms can usually be converted to affine ones using the Schur complements. Therefore we will make no distinctions between (2.21) and (2.22), quadratic and affine forms, or between a set of LMIs or a single one, and will refer to all of them as simply LMIs. For more on LMIs the reader is referred to [7], [42], or any of the many works on the subject.

3

On Stability

Among the requirements in any control system design problem, stability is certainly a mandatory one. This chapter is aimed at developing a set of stability results for MJLS. The main problem is to find necessary and sufficient conditions guaranteeing mean square stability in the spirit of the motivating remarks in Section 1.4. To some extent, the result we derive here is very much in the spirit of the one presented in Theorem 2.14, items 2, 3, and 4 for the linear case, in the sense that mean square stability for MJLS is guaranteed in terms of the spectral radius of an augmented matrix being less than one, or in terms of the existence of a positive-definite solution for a set of coupled Lyapunov equations. We exhibit some examples which uncover some very interesting and peculiar properties of MJLS. Other concepts and issues of stability are also considered in this chapter.

3.1 Outline of the Chapter

The main goal of this chapter is to obtain necessary and sufficient conditions for mean square stability (MSS) for discrete-time MJLS. We start by defining in Section 3.2 some operators that are closely related to the Markovian property of the augmented state $(x(t), \theta(t))$, greatly simplifying the solution for the mean square stability and other problems that will be analyzed in the next chapters. As a consequence, we can adopt an analytical view toward mean square stability, using the operator theory in Banach spaces provided in Chapter 2 as a primary tool. The outcome is a clean and sound theory ready for application. As mentioned in [134], among the advantages of using the MSS concept and the results derived here, are: (1) the fact that it is easy to test for; (2) it implies stability of the expected dynamics; (3) it yields almost sure asymptotic stability of the zero-input state space trajectories.

In order to ease the reading and facilitate the understanding of the main ideas of MSS for MJLS, we have split up the results into two sections (Section 3.3 for the homogeneous and Section 3.4 for the non-homogeneous case). The advantages in doing this (we hope!) is that, for instance, we can state the results for the homogeneous case in a more general setting (without requiring, e.g., that the Markov chain is ergodic) and also it avoids at the beginning the heavy expressions of the non-homogeneous case.

It will be shown in Section 3.3 that MSS is equivalent to the spectral radius of an augmented matrix being less than one, or to the existence of a solution to a set of coupled Lyapunov equations. The first criterion (spectral radius) will show clearly the connection between MSS and the probability of visits to the unstable modes, translating the intuitive idea that unstable operation modes do not necessarily compromise the global stability of the system. In fact, as will be shown through some examples, the stability of all modes of operation is neither necessary nor sufficient for global stability of the system. For the case of one single operation mode (no jumps in the parameters) these criteria reconcile to the well known stability results for discrete-time linear systems. It is also shown that the Lyapunov equations can be written down in four equivalent forms and each of these forms provides an easy to check sufficient condition.

We will also consider the non-homogeneous case in Section 3.4. For the case in which the system is driven by a second order wide-sense stationary random sequence, it is proved that MSS is equivalent to asymptotic wide sense stationary stability, a result that, we believe, gives a rather complete picture for the MSS of discrete-time MJLS. For the case in which the inputs are ℓ_2-stochastic signals, it will be shown that MSS is equivalent to the discrete-time MJLS being a bounded linear operator that maps ℓ_2-stochastic input signals into ℓ_2-stochastic output signals. This result will be particularly useful for the H_∞-control problem, to be studied in Chapter 7.

Some necessary and sufficient conditions for mean square stabilizability and detectability, as well as a study of mean square stabilizability for the case in which the Markov parameter is only partially known, are carried out in Section 3.5.

With relation to almost sure convergence (ASC), we will consider in Section 3.6 the noise free case and obtain sufficient conditions in terms of the norms of some matrices and limit probabilities of a Markov chain constructed from the original one. We also present an application of this result to the Markovian version of the adaptive filtering algorithm proposed in [25], and obtain a very easy to check condition for ASC.

3.2 Main Operators

Throughout the book we fix an underlying stochastic basis $(\Omega, \mathfrak{F}, \{\mathfrak{F}_k\}, \mathcal{P})$ as in Section 2.3 and, unless otherwise stated, we assume that the time set is given by $\mathbb{T} = \{0, 1, \ldots\}$. We recall that $\{\theta(k); k \in \mathbb{T}\}$ represents a Markov chain taking values in $\mathbb{N}$, with transition probability matrix $\mathrm{P} = [p_{ij}]$. The initial state and jump variables satisfy $x_0 \in \mathcal{C}_0^n$, $\theta_0 \in \Theta_0$, with θ_0 having initial distribution $\upsilon = \{\upsilon_i; i \in \mathbb{N}\}$, as defined in Section 2.3. In order to introduce the main operators which will be used throughout this book, it suffices to deal with the homogeneous MJLS, as defined next:

$$\mathcal{G} = \begin{cases} x(k+1) = \Gamma_{\theta(k)}x(k) \\ x(0) = x_0, \theta(0) = \theta_0 \end{cases} \tag{3.1}$$

where $\Gamma = (\Gamma_1, \ldots, \Gamma_N) \in \mathbb{H}^n$. It is easily seen that $\{x(k), k \in \mathbb{T}\}$ is not a Markov process, but the joint process $\{x(k), \theta(k)\}$ is.

For a set $\mathbb{A} \in \mathfrak{F}$ the indicator function $\mathbf{1}_\mathbb{A}$ is defined in the usual way, that is, for any $\omega \in \Omega$,

$$\mathbf{1}_\mathbb{A}(\omega) = \begin{cases} 1 & \text{if } \omega \in \mathbb{A} \\ 0 & \text{otherwise.} \end{cases} \tag{3.2}$$

Notice that for any $i \in \mathbb{N}$, $\mathbf{1}_{\{\theta(k)=i\}}(\omega) = 1$ if $\theta(k)(\omega) = i$, and 0 otherwise. As mentioned in Chapter 2, the gist of the method adopted here is to work with $x(k)\mathbf{1}_{\{\theta(k)=i\}}$ and $x(k)x(k)^*\mathbf{1}_{\{\theta(k)=i\}}$. This allows us to take advantage of the Markovian properties and get difference equations for $E(x(k)\mathbf{1}_{\{\theta(k)=i\}})$ and $E(x(k)x(k)^*\mathbf{1}_{\{\theta(k)=i\}})$. Notice that

$$E(x(k)) = \sum_{i=1}^N E(x(k)\mathbf{1}_{\{\theta(k)=i\}})$$

and

$$E(x(k)x(k)^*) = \sum_{i=1}^N E(x(k)x(k)^*\mathbf{1}_{\{\theta(k)=i\}}).$$

For $k \in \mathbb{T}$, we introduce the following notation:

$$q(k) \triangleq \begin{pmatrix} q_1(k) \\ \vdots \\ q_N(k) \end{pmatrix} \in \mathbb{C}^{Nn}, \tag{3.3a}$$

$$q_i(k) \triangleq E(x(k)\mathbf{1}_{\{\theta(k)=i\}}) \in \mathbb{C}^n, \tag{3.3b}$$

$$Q(k) \triangleq (Q_1(k), \ldots, Q_N(k)) \in \mathbb{H}^{n+}, \tag{3.3c}$$

$$Q_i(k) \triangleq E(x(k)x(k)^*\mathbf{1}_{\{\theta(k)=i\}}) \in \mathcal{B}(\mathbb{C}^n)^+, \tag{3.3d}$$

and

$$\mu(k) \triangleq E(x(k)) = \sum_{i=1}^N q_i(k) \in \mathbb{C}^n, \tag{3.4}$$

$$\mathbb{Q}(k) \triangleq E(x(k)x(k)^*) = \sum_{i=1}^N Q_i(k) \in \mathcal{B}(\mathbb{C}^n)^+. \tag{3.5}$$

We establish next recursive equations for $q_i(k)$ and $Q_i(k)$.

Proposition 3.1. *Consider model (3.1). For every $k \in \mathbb{T}$ and $j \in \mathbb{N}$:*

1. $q_j(k+1) = \sum_{i=1}^{N} p_{ij} \Gamma_i q_i(k)$,
2. $Q_j(k+1) = \sum_{i=1}^{N} p_{ij} \Gamma_i Q_i(k) \Gamma_i^*$,
3. $\|x(k)\|_2^2 = \left\| \Gamma_{\theta(k-1)} \ldots \Gamma_{\theta(0)} x(0) \right\|_2^2 \leq n \|Q(k)\|_1$.

Proof. For (1)

$$q_j(k+1)$$

$$= \sum_{i=1}^{N} E(\Gamma_i x(k) \mathbf{1}_{\{\theta(k+1)=j\}} \mathbf{1}_{\{\theta(k)=i\}})$$

$$= \sum_{i=1}^{N} \Gamma_i E(x(k) \mathbf{1}_{\{\theta(k)=i\}} \mathcal{P}(\theta(k+1) = j|\mathfrak{F}_k))$$

$$= \sum_{i=1}^{N} p_{ij} \Gamma_i q_i(k)$$

showing the first result. For (2), notice that

$$Q_j(k+1)$$

$$= \sum_{i=1}^{N} E(\Gamma_i x(k)(\Gamma_i x(k))^* \mathbf{1}_{\{\theta(k+1)=j\}} \mathbf{1}_{\{\theta(k)=i\}})$$

$$= \sum_{i=1}^{N} p_{ij} \Gamma_i Q_i(k) \Gamma_i^*.$$

Finally notice that

$$\|x(k)\|_2^2 = E(\|x(k)\|^2) = \sum_{i=1}^{N} E(\|x(k)\|^2 \mathbf{1}_{\{\theta(k)=i\}})$$

$$= \sum_{i=1}^{N} E(\operatorname{tr}(x(k)x(k)^* \mathbf{1}_{\{\theta(k)=i\}}))$$

$$= \sum_{i=1}^{N} \operatorname{tr}(E(x(k)x(k)^* \mathbf{1}_{\{\theta(k)=i\}}))$$

$$= \sum_{i=1}^{N} \operatorname{tr}(Q_i(k)) = \operatorname{tr}\left(\sum_{i=1}^{N} Q_i(k)\right)$$

$$\leq n \left\| \sum_{i=1}^{N} Q_i(k) \right\|$$

$$\leq n \sum_{i=1}^{N} \|Q_i(k)\| = n \|Q(k)\|_1,$$

completing the proof.   □

As mentioned before, instead of dealing directly with the state $x(k)$, which is not Markov, we couch the systems in the Markovian framework via the augmented state $(x(k), \theta(k))$, as in Proposition 3.1. In this case the following operators play an important role in our approach. We set

$$\mathcal{E}(.) \triangleq (\mathcal{E}_1(.), \ldots, \mathcal{E}_N(.)) \in \mathbb{B}(\mathbb{H}^n)$$
$$\mathcal{L}(.) \triangleq (\mathcal{L}_1(.), \ldots, \mathcal{L}_N(.)) \in \mathbb{B}(\mathbb{H}^n)$$
$$\mathcal{T}(.) \triangleq (\mathcal{T}_1(.), \ldots, \mathcal{T}_N(.)) \in \mathbb{B}(\mathbb{H}^n)$$
$$\mathcal{J}(.) \triangleq (\mathcal{J}_1(.), \ldots, \mathcal{J}_N(.)) \in \mathbb{B}(\mathbb{H}^n)$$
$$\mathcal{V}(.) \triangleq (\mathcal{V}_1(.), \ldots, \mathcal{V}_N(.)) \in \mathbb{B}(\mathbb{H}^n)$$

as follows. For $V = (V_1, \ldots, V_N) \in \mathbb{H}^n$ and $i, j \in \mathbb{N}$,

$$\mathcal{E}_i(V) \triangleq \sum_{j=1}^{N} p_{ij} V_j \in \mathbb{B}(\mathbb{C}^n) \tag{3.6}$$

$$\mathcal{T}_j(V) \triangleq \sum_{i=1}^{N} p_{ij} \Gamma_i V_i \Gamma_i^* \in \mathbb{B}(\mathbb{C}^n) \tag{3.7}$$

$$\mathcal{L}_i(V) \triangleq \Gamma_i^* \mathcal{E}_i(V) \Gamma_i \in \mathbb{B}(\mathbb{C}^n) \tag{3.8}$$

$$\mathcal{V}_j(V) \triangleq \sum_{i=1}^{N} p_{ij} \Gamma_j V_i \Gamma_j^* \in \mathbb{B}(\mathbb{C}^n) \tag{3.9}$$

$$\mathcal{J}_i(V) \triangleq \sum_{j=1}^{N} p_{ij} \Gamma_j^* V_j \Gamma_j \in \mathbb{B}(\mathbb{C}^n). \tag{3.10}$$

From Proposition 3.1 we have a connection between the operator $\mathcal{T}$ and the second moment in (3.3d) as follows:

$$Q(k+1) = \mathcal{T}(Q(k)). \tag{3.11}$$

It is immediate to check that the operators $\mathcal{E}$, $\mathcal{L}$, $\mathcal{T}$, $\mathcal{V}$, and $\mathcal{J}$ map $\mathbb{H}^{n*}$ into $\mathbb{H}^{n*}$ and $\mathbb{H}^{n+}$ into $\mathbb{H}^{n+}$, that is, they are hermitian and positive operators (see Section 2.1). Also for $V \in \mathbb{H}^n$, $\mathcal{T}(V)^* = \mathcal{T}(V^*)$ and similarly for $\mathcal{L}$, $\mathcal{V}$ and $\mathcal{J}$ replacing $\mathcal{T}$. With the inner product given by (2.3), we have $\mathcal{T} = \mathcal{L}^*$ and $\mathcal{V} = \mathcal{J}^*$, as shown in the next proposition.

Proposition 3.2. $\mathcal{T} = \mathcal{L}^*$ and $\mathcal{V} = \mathcal{J}^*$.

Proof. For any $S, V \in \mathbb{H}^n$,

$$\langle \mathcal{T}(V); S \rangle = \sum_{j=1}^{N} \text{tr}(\mathcal{T}_j(V)^* S_j) = \sum_{j=1}^{N} \text{tr}(\mathcal{T}_j(V^*) S_j)$$

$$= \sum_{j=1}^{N} \sum_{i=1}^{N} p_{ij} \, \text{tr}(\Gamma_i V_i^* \Gamma_i^* S_j) = \sum_{i=1}^{N} \sum_{j=1}^{N} p_{ij} \, \text{tr}(V_i^* \Gamma_i^* S_j \Gamma_i)$$

$$= \sum_{i=1}^{N} \text{tr}\left(V_i^* \Gamma_i^* \left(\sum_{j=1}^{N} p_{ij} S_j \right) \Gamma_i \right) = \sum_{i=1}^{N} \text{tr}(V_i^* \Gamma_i^* \mathcal{E}_i(S_j) \Gamma_i)$$

$$= \sum_{i=1}^{N} \text{tr}(V_i^* \mathcal{L}_i(S)) = \langle V; \mathcal{L}(S) \rangle$$

and similarly,

$$\langle \mathcal{V}(V); S \rangle = \langle V; \mathcal{J}(S) \rangle.$$

□

For $M_i \in \mathbb{B}(\mathbb{C}^n)$, $i \in \mathbb{N}$, we set $\text{diag}[M_i]$ the $Nn \times Nn$ block diagonal matrix formed with M_i in the diagonal and zero elsewhere, that is,

$$\text{diag}[M_i] \triangleq \begin{bmatrix} M_1 & \cdots & 0 \\ \vdots & \ddots & \vdots \\ 0 & \cdots & M_N \end{bmatrix}.$$

In what follows we recall that I_ℓ denotes the identity matrix $\ell \times \ell$ and $\otimes$ the Kronecker product. We define

$$\mathcal{B} \triangleq (\mathbf{P}' \otimes I_n) \, \text{diag}[\Gamma_i] \in \mathbb{B}(\mathbb{C}^{Nn}) \tag{3.12a}$$

$$\mathcal{C} \triangleq (\mathbf{P}' \otimes I_{n^2}) \in \mathbb{B}(\mathbb{C}^{Nn^2}) \tag{3.12b}$$

$$\mathcal{N} \triangleq \text{diag}[\bar{\Gamma}_i \otimes \Gamma_i] \in \mathbb{B}(\mathbb{C}^{Nn^2}) \tag{3.12c}$$

$$\mathcal{A}_1 \triangleq \mathcal{C}\mathcal{N}, \ \mathcal{A}_2 \triangleq \mathcal{N}^* \mathcal{C}^*, \ \mathcal{A}_3 \triangleq \mathcal{N}\mathcal{C}, \ \mathcal{A}_4 \triangleq \mathcal{C}^* \mathcal{N}^*. \tag{3.12d}$$

From Proposition 3.1 we have that matrix $\mathcal{B}$ and the first moment in (3.3b) are related as follows:

$$q(k+1) = \mathcal{B}q(k). \tag{3.13}$$

Remark 3.3. Since $\mathcal{A}_1^* = \mathcal{A}_2$, $\mathcal{A}_3^* = \mathcal{A}_4$, it is immediate that $r_\sigma(\mathcal{A}_1) = r_\sigma(\mathcal{A}_2)$ and $r_\sigma(\mathcal{A}_3) = r_\sigma(\mathcal{A}_4)$. Moreover, since $r_\sigma(\mathcal{C}\mathcal{N}) = r_\sigma(\mathcal{N}\mathcal{C})$, we get $r_\sigma(\mathcal{A}_1) = r_\sigma(\mathcal{A}_3)$. Summing up, we have $r_\sigma(\mathcal{A}_1) = r_\sigma(\mathcal{A}_2) = r_\sigma(\mathcal{A}_3) = r_\sigma(\mathcal{A}_4)$.

The following result is germane to our MSS approach. It gives the connection between the operators $\mathcal{T}, \mathcal{L}, \mathcal{V}, \mathcal{J}$ and $\mathcal{A}_1, \mathcal{A}_2, \mathcal{A}_3, \mathcal{A}_4$, allowing a treatment which reconciles, to some extent, with that of the linear case, i.e., gives stability conditions in terms of the spectral radius of an augmented matrix in the spirit, mutatis mutandis, of the one for the linear case.

Proposition 3.4. *For any $Q \in \mathbb{H}^n$ we have*

$$\hat{\varphi}(\mathcal{T}(Q)) = \mathcal{A}_1 \hat{\varphi}(Q),$$
$$\hat{\varphi}(\mathcal{L}(Q)) = \mathcal{A}_2 \hat{\varphi}(Q),$$
$$\hat{\varphi}(\mathcal{V}(Q)) = \mathcal{A}_3 \hat{\varphi}(Q),$$
$$\hat{\varphi}(\mathcal{J}(Q)) = \mathcal{A}_4 \hat{\varphi}(Q).$$

Proof. Immediate application of the definitions in Section 2.1, (2.4) and (3.12). □

Remark 3.5. From Remark 2.2, Remark 3.3 and Proposition 3.4 it is immediate that

$$r_\sigma(\mathcal{A}_1) = r_\sigma(\mathcal{A}_2) = r_\sigma(\mathcal{A}_3) = r_\sigma(\mathcal{A}_4) = r_\sigma(\mathcal{T}) = r_\sigma(\mathcal{L}) = r_\sigma(\mathcal{V}) = r_\sigma(\mathcal{J}).$$

We conclude this section with the following result, which guarantees that stability of the second moment operator implies stability of the first moment operator.

Proposition 3.6. *If $r_\sigma(\mathcal{A}_1) < 1$ then $r_\sigma(\mathcal{B}) < 1$.*

Proof. Let $\{e_i; i = 1, \ldots, Nn\}$ and $\{v_i; i = 1, \ldots, n\}$ be the canonical orthonormal basis for $\mathbb{C}^{Nn}$ and $\mathbb{C}^n$ respectively. Fix arbitrarily ι, ℓ, $1 \leq \iota \leq n$, $\ell \in \mathbb{N}$, and consider the homogeneous System (3.1) with initial conditions $x(0) = v_\iota$, $v_\ell = 1$. Then $q(0) = e_\kappa$, where $\kappa = \iota + (\ell - 1)n$. From Proposition 3.1,

$$\left\| \mathcal{B}^k e_\kappa \right\|^2 = \|q(k)\|^2 = \sum_{i=1}^N \|q_i(k)\|^2 = \sum_{i=1}^N \left\| E(x(k)\mathbf{1}_{\{\theta(k)=i\}}) \right\|^2$$

$$\leq \sum_{i=1}^N E(\left\| x(k)\mathbf{1}_{\{\theta(k)=i\}} \right\|^2) = \sum_{i=1}^N \mathrm{tr}(Q_i(k))$$

$$\leq n \sum_{i=1}^N \|Q_i(k)\| = n \|Q(k)\|.$$

Now from Proposition 3.4 and Proposition 3.1, it follows that

$$\hat{\varphi}(Q(k)) = \mathcal{A}_1^k \hat{\varphi}(Q(0))$$

with $Q_\ell(0) = v_\ell v_\ell^*$ and $Q_i(0) = 0$ for $i \neq \ell$. Thus, $r_\sigma(\mathcal{A}_1) < 1$ implies that $\|Q(k)\| \to 0$ as $k \to \infty$ and thus $\left\| \mathcal{B}^k e_\kappa \right\| \to 0$ as $k \to \infty$. Since ι and ℓ were arbitrarily chosen, it follows that $\left\| \mathcal{B}^k e_i \right\| \to 0$ as $k \to \infty$ for each $i = 1, \ldots, Nn$. Then $\left\| \mathcal{B}^k q \right\| \to 0$ as $k \to \infty$ for every $q \in \mathbb{C}^{Nn}$. Hence we conclude that $r_\sigma(\mathcal{B}) < 1$. □

Remark 3.7. It can be easily checked that $r_\sigma(\mathcal{B}) < 1$ does not imply that $r_\sigma(\mathcal{A}_1) < 1$. Indeed, consider $n = 1, N = 2, p_{11} = p_{22} = \frac{1}{2}, \Gamma_1 = 0.7, \Gamma_2 = 1.25$. In this case $r_\sigma(\mathcal{B}) = 0.975$ and $r_\sigma(\mathcal{A}_1) = 1.02625$.

3.3 MSS: The Homogeneous Case

In a rather rough way, we say that a MJLS is mean square stable if $E(x(k)x(k)^*)$ converges as k goes to infinity. The difference between the homogeneous and the non-homogeneous case is that for the former the convergence is to the zero matrix. In a formal way, we define mean square stability (MSS) as follows:

Definition 3.8. *We say that the linear system with Markov jump parameter (1.3) with $u(k) = 0$ is mean square stable (MSS) if for any initial condition $x_0 \in C_0^n$, $\theta_0 \in \Theta_0$ there exist $\mu \in \mathbb{C}^n$ and $\mathbb{Q} \in \mathbb{B}(\mathbb{C}^n)^+$ (independent of x_0 and θ_0) such that*

1. $\|\mu(k) - \mu\| \to 0$ *as* $k \to \infty,$
2. $\|\mathbb{Q}(k) - \mathbb{Q}\| \to 0$ *as* $k \to \infty.$

In order to derive some fundamental results in their more general form, without restriction on the Markov chain, and put MSS in a unified basis, we will be considering in this section the homogeneous case. However it is important to bear in mind that a great deal of the main results, to be presented in the next subsection, also hold for the non-homogeneous case, as will be clear in the following sections.

In this section, necessary and sufficient conditions for MSS of the homogeneous System (3.1) are established. It is shown that MSS is equivalent to the spectral radius of the augmented matrix $\mathcal{A}_1$, defined in (3.12d), being less than one or to the existence of a unique solution to a set of coupled Lyapunov equations. Furthermore, it is proved that the Lyapunov equations can be written down in four equivalent forms, each one providing an easy-to-check sufficient condition. The results are derived from the operator theory methods described in Section 2.2 in conjunction with the operators presented in Section 3.2.

3.3.1 Main Result

In order to give a general picture on MSS, we present next the main result of this section. The proof is presented in Subsection 3.3.3.

Theorem 3.9. *The following assertions are equivalent.*

1. *Model (3.1) is MSS.*
2. *$r_\sigma(\mathcal{A}_1) < 1$.*
3. *For any $S \in \mathbb{H}^{n+}$, $S > 0$, there exists a unique $V \in \mathbb{H}^{n+}$, $V > 0$, such that*

$$V - \mathcal{T}(V) = S. \tag{3.14}$$

4. *For some $V \in \mathbb{H}^{n+}$, $V > 0$, we have*

$$V - \mathcal{T}(V) > 0. \tag{3.15}$$

5. *For some $\beta \geq 1$, $0 < \zeta < 1$, we have for all $x_0 \in \mathcal{C}_0^n$ and all $\theta_0 \in \Theta_0$,*

$$E(\|x(k)\|^2) \leq \beta \zeta^k \|x_0\|_2^2, \quad k = 0, 1, \ldots.$$

6. *For all $x_0 \in \mathcal{C}_0^n$ and all $\theta_0 \in \Theta_0$*

$$\sum_{k=0}^{\infty} E(\|x(k)\|^2) < \infty. \tag{3.16}$$

The result also holds replacing $\mathcal{T}$ in (3.14) and (3.15) by $\mathcal{L}, \mathcal{V}$ or $\mathcal{J}$.

Remark 3.10. From Theorem 3.9 it is easy to see that MSS of model (3.1) is equivalent to $\mathbb{Q}(k) \to 0$ and $\mu(k) \to 0$ as $k \to \infty$. Indeed, since $r_\sigma(\mathcal{A}_1) < 1$, we have from Proposition 3.6 that $r_\sigma(\mathcal{B}) < 1$. Thus, from Proposition 3.1 and Proposition 2.9 it follows that $Q(k) \to 0$, $q(k) \to 0$, and $\mathbb{Q}(k) \to 0$, $\mu(k) \to 0$, as $k \to \infty$.

Remark 3.11. It is perhaps worth mentioning here that the equivalence between (1)-(4) in the above theorem remains true for the non-homogeneous case.

Remark 3.12. In the MJLS literature we can also find the concept of stochastic stability (SS), a kind of ℓ_2-stability as in (3.16), which amounts to item 6 of Theorem 3.9. Therefore from Theorem 3.9 we have that MSS for model (3.1) is equivalent to SS. As can be seen in [67], this equivalence does not hold anymore for the case in which the state space of the Markov chain is infinite countable.

Remark 3.13. In Definition 3.8 our set up is in the complex case, and particularly the initial condition x_0 belongs to the complex space $\mathbb{C}^n$. For a discussion on the real set up, see [68].

3.3.2 Examples

In this subsection some examples are given to illustrate applications of Theorem 3.9, unveiling some subtleties of MJLS.

Example 3.14. Consider the following system with two operation modes, defined by matrices $A_1 = \frac{4}{3}$, $A_2 = \frac{1}{3}$. (Note that mode 1 is unstable and mode 2 is stable). The transitions between these modes are given by the transition probability matrix shown below:

$$P = \begin{bmatrix} \frac{1}{2} & \frac{1}{2} \\ \frac{1}{2} & \frac{1}{2} \end{bmatrix}.$$

It is easy to verify that for this transition probability matrix we have

$$\mathcal{A}_1 = \frac{1}{2} \begin{bmatrix} \frac{16}{9} & \frac{1}{9} \\ \frac{16}{9} & \frac{1}{9} \end{bmatrix}, \quad r_\sigma(\mathcal{A}_1) = \frac{17}{18} < 1$$

and so the system is MSS. Suppose now that we have a different transition probability matrix, say

$$\bar{P} = \begin{bmatrix} 0.9 & 0.1 \\ 0.9 & 0.1 \end{bmatrix},$$

so that the system will most likely stay longer in mode 1, which is unstable. Then

$$\mathcal{A}_1 = \begin{bmatrix} \frac{144}{90} & \frac{1}{10} \\ \frac{16}{9} & \frac{1}{9} \end{bmatrix}, \quad r_\sigma(\mathcal{A}_1) = 1.61 > 1$$

and the system is no longer MSS. This evinces a connection between MSS and the probability of visits to the unstable modes, which is translated in the expression for $\mathcal{A}_1$, given in (3.12d).

Example 3.15. Consider the above example again with probability transition matrix given by

$$P = \begin{bmatrix} \frac{1}{2} & \frac{1}{2} \\ \frac{1}{2} & \frac{1}{2} \end{bmatrix}.$$

Applying Theorem 3.9 (item 3) with $S_1 = S_2 = \frac{1}{18}$ we obtain

$$V_j - \frac{1}{2}(V_1\frac{16}{9} + V_2\frac{1}{9}) = \frac{1}{18}, \; j = 1, 2,$$

which has the positive-definite solution $V_1 = V_2 = 1$. If we use the equivalent forms, $V - \mathcal{L}(V) = S$ or $V - \mathcal{V}(V) = S$, for the Lyapunov equations with $S_1 = \frac{1}{9}$ and $S_2 = \frac{4}{9}$, we get

$$V_1 - \frac{1}{2}(V_1 + V_2)\frac{16}{9} = \frac{1}{9}, \quad V_2 - \frac{1}{2}(V_1 + V_2)\frac{1}{9} = \frac{4}{9}$$

which have positive-definite solutions $V_1 = 9$, $V_2 = 1$.

Example 3.16. For the system of Example 1.1, we have $r_\sigma(\mathcal{A}_1) = 1.0482 > 1$, so the system is not MSS.

The next examples, borrowed from [146], illustrate how sometimes the switching between operation modes can play tricks with our intuition. They show that System (3.1) carries a great deal of subtleties which distinguish it from the linear case and provide us with a very rich structure. For instance, the next example deals with the fact that a system composed only of unstable modes can be MSS (and also the opposite).

Example 3.17 (A Non MSS System with Stable Modes). Consider a system with two operation modes, defined by matrices

$$A_1 = \begin{bmatrix} 0 & 2 \\ 0 & 0.5 \end{bmatrix} \text{ and } A_2 = \begin{bmatrix} 0.5 & 0 \\ 2 & 0 \end{bmatrix}$$

and the transition probability matrix

$$P = \begin{bmatrix} 0.5 & 0.5 \\ 0.5 & 0.5 \end{bmatrix}.$$

Note that both modes are stable. Curiously, we have that $r_\sigma(A_1) = 2.125 > 1$, which means that the system is not MSS. A brief analysis of the trajectories for each mode is useful to clarify the matter. First consider only trajectories for mode 1. For initial conditions given by

$$x(0) = \begin{bmatrix} x_{10} \\ x_{20} \end{bmatrix},$$

the trajectories are given by

$$x(k) = \begin{bmatrix} x_1(k) \\ x_2(k) \end{bmatrix} = \begin{bmatrix} 2(0.5)^{k-1}x_{20} \\ 0.5(0.5)^{k-1}x_{20} \end{bmatrix} \text{ for } k = 1, 2, \ldots.$$

So, with the exception of the point given by $x(0)$, the whole trajectory lies along the line given by $x_1(k) = 4x_2(k)$ for any initial condition. This means that if, in a given time, the state is not in this line, mode 1 dynamics will transfer it to the line in one time step and it will remain there thereafter. For mode 2, it can be easily shown that the trajectories are given by

$$x(k) = \begin{bmatrix} x_1(k) \\ x_2(k) \end{bmatrix} = \begin{bmatrix} 0.5(0.5)^{k-1}x_{10} \\ 2(0.5)^{k-1}x_{10} \end{bmatrix} \text{ for } k = 1, 2, \ldots.$$

Similarly to mode 1, if the state is not in the line $x_1(k) = x_2(k)/4$, mode 2 dynamics will transfer it to the line in one time step. The equations for the trajectories also show that the transitions make the state switch between these two lines. Notice that transitions from mode 1 to mode 2 cause the state to move away from the origin in the direction of component x_2, while transitions from mode 2 to mode 1 do the same with respect to component x_1. Figure 3.1.a shows the trajectory of the system with mode 1 dynamics only, for a given initial condition while Figure 3.1.b shows the same for mode 2. Figure 3.2 shows the trajectory for a possible sequence of switchings between the two modes, evincing unstability of the system.

Example 3.18 (A MSS System with Unstable Modes). Consider the following system:

$$A_1 = \begin{bmatrix} 2 & -1 \\ 0 & 0 \end{bmatrix} \text{ and } A_2 = \begin{bmatrix} 0 & 1 \\ 0 & 2 \end{bmatrix}$$

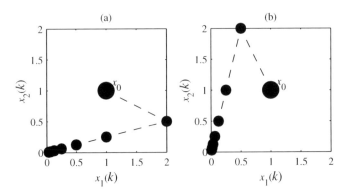

Fig. 3.1. Trajectories for operation modes 1 (a) and 2 (b)

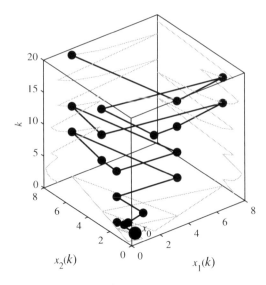

Fig. 3.2. One of the possible trajectories for the Markov system. Note that the trajectory tends to move away from the origin. The time $k = 1, \ldots, 20$ is presented on the z-axis. The gray lines are trajectory projections

and the transition probability matrix

$$P = \begin{bmatrix} 0.1 & 0.9 \\ 0.9 & 0.1 \end{bmatrix}.$$

Note that both modes are unstable, but we have that $r_\sigma(\mathcal{A}_1) = 0.4 < 1$.

The general conclusion one extracts from these examples is that stability of each operation mode is neither a necessary nor sufficient condition for mean square stability of the system. MSS depends upon a balance between transition probability of the Markov chain and the operation modes.

3.3.3 Proof of Theorem 3.9

We first establish our main results related to the coupled Lyapunov equations, which give us four equivalent forms of writing them.

Theorem 3.19. *If there exists $V \in \mathbb{H}^{n+}$, $V > 0$, such that*

$$V = \mathcal{T}(V) + S \tag{3.17}$$

for some $S \in \mathbb{H}^{n+}$, $S > 0$, then $r_\sigma(\mathcal{A}_1) < 1$. The result also holds if we replace $\mathcal{T}$ by $\mathcal{L}, \mathcal{V}$ or $\mathcal{J}$.

Proof. Consider the homogeneous system

$$Y(k+1) = \mathcal{L}(Y(k)), \ Y(0) \in \mathbb{H}^{n+}. \tag{3.18}$$

$Y(k) \in \mathbb{H}^{n+}$ for $k = 0, 1, \ldots$. Define the function ϕ from $\mathbb{H}^{n+}$ to $\mathbb{R}$ as: for $Y \in \mathbb{H}^{n+}$,

$$\phi(Y) \triangleq \langle V; Y \rangle = \sum_{j=1}^{N} \mathrm{tr}(V_j Y_j)$$

$$= \sum_{j=1}^{N} \mathrm{tr}(V_j^{1/2} Y_j V_j^{1/2}) \geq 0.$$

To obtain global asymptotic stability of the origin we need to show that ϕ is a Lyapunov function for (3.18) satisfying (see the Lyapunov Theorem 2.13):

1. $\phi(Y) \to \infty$ whenever $\|Y\|_2 \to \infty$ and $Y \in \mathbb{H}^{n+}$,
2. $\phi(.)$ is continuous,
3. $\phi(0) = 0$,
4. $\phi(Y) > 0$ for all $Y \neq 0, Y \in \mathbb{H}^{n+}$,
5. $\phi(Y(k+1)) - \phi(Y(k)) < 0$ whenever $Y(k) \neq 0$.

For any $P = (P_1, \ldots, P_N) \in \mathbb{H}^{n+}$ let $\lambda_i(P_j) \geq 0$ be the ith eigenvalue of P_j and

$$c_0(P) \triangleq \min_{\substack{1 \leq i \leq n \\ 1 \leq j \leq N}} \lambda_i(P_j) \geq 0,$$

$$c_1(P) \triangleq \max_{\substack{1 \leq i \leq n \\ 1 \leq j \leq N}} \lambda_i(P_j) \geq 0.$$

Since $V_j > 0$ and $S_j > 0$ for each $j = 1, \ldots, N$, it follows that $c_0(V) > 0, c_1(V) > 0$ and $c_0(S) > 0, c_1(S) > 0$. From (2.1b) it follows that

$$c_0(V)\left(\sum_{j=1}^{N}\sum_{i=1}^{n}\lambda_i(Y_j)\right) \le \phi(y) \le c_1(V)\left(\sum_{j=1}^{N}\sum_{i=1}^{n}\lambda_i(Y_j)\right). \qquad (3.19)$$

Note that

$$\|Y\|_2^2 = \sum_{j=1}^{N} \mathrm{tr}\left(Y_j^2\right) = \sum_{j=1}^{N}\sum_{i=1}^{n}\lambda_i(Y_j)^2.$$

From the fact that $\lambda_i(Y_j)$ is nonnegative, we know that $\|Y\|_2 \to \infty$ if and only if $\sum_{j=1}^{N}\sum_{i=1}^{n}\lambda_i(Y_j) \to \infty$, and $Y = 0$ if and only if $\lambda_i(Y_j) = 0$, $i = 1,\ldots,n$, $j = 1,\ldots,N$. Therefore (1), (3) and (4) follow from these assertions and (3.19). Continuity of ϕ is easily established and thus it only remains to show (5). We have from Proposition 3.2, (2.1b), and (3.17) that

$$
\begin{aligned}
\phi(Y(k+1)) - \phi(Y(k)) &= \langle V; \mathcal{L}(Y(k))\rangle - \langle V; Y(k)\rangle \\
&= \langle \mathcal{T}(V); Y(k)\rangle - \langle V; Y(k)\rangle \\
&= \langle \mathcal{T}(V) - V; Y(k)\rangle \\
&= -\langle S; Y(k)\rangle \\
&\le -c_0(S)\left(\sum_{j=1}^{N}\sum_{i=1}^{n}\lambda_i(Y_j(k))\right) < 0
\end{aligned}
$$

whenever $Y(k) \ne 0$. Therefore we have shown that (3.18) is asymptotically stable and thus $\|\mathcal{L}^k(Y)\| \to 0$ as $k \to \infty$ for all $Y \in \mathbb{H}^{n+}$, which yields, from Proposition 2.5, and Remark 3.5, that $r_\sigma(\mathcal{L}) = r_\sigma(\mathcal{A}_1) < 1$. The result also holds replacing $\mathcal{T}$ by $\mathcal{L}, \mathcal{V}$ or $\mathcal{J}$, bearing in mind Proposition 3.2, Remark 3.5, and their relations to $\mathcal{A}_\ell$ as in Proposition 3.4.   □

The following result is an immediate consequence of Proposition 2.6 (recalling that $r_\sigma(\mathcal{A}_1) = r_\sigma(\mathcal{T})$).

Proposition 3.20. *If $r_\sigma(\mathcal{A}_1) < 1$ then there exists a unique $V \in \mathbb{H}^n$, such that*

$$V = \mathcal{T}(V) + S$$

for any $S \in \mathbb{H}^n$. Moreover,

1. $V = \hat{\varphi}^{-1}((I_{Nn^2} - \mathcal{A}_1)^{-1}\hat{\varphi}(S)) = \sum_{k=0}^{\infty}\mathcal{T}^k(S)$,
2. $S = S^* \Leftrightarrow V = V^*$,
3. $S \ge 0 \Rightarrow V \ge 0$,
4. $S > 0 \Rightarrow V > 0$.

The result also holds replacing $\mathcal{T}$ by $\mathcal{L}, \mathcal{V}$ or $\mathcal{J}$.

From Theorem 3.19 and Proposition 3.20 we get the following corollary.

Corollary 3.21. *The following assertions are equivalent.*

1. $r_\sigma(\mathcal{A}_1) < 1$.
2. *For any given* $S \in \mathbb{H}^{n+}$, $S > 0$, *there exists a unique* $V \in \mathbb{H}^{n+}$, $V > 0$, *such that* $V = \mathcal{T}(V) + S$. *Moreover*

$$V = \hat{\varphi}^{-1}((I_{Nn^2} - \mathcal{A}_1)^{-1}\hat{\varphi}(S)) = \sum_{k=0}^{\infty} \mathcal{T}^k(S).$$

3. *For some* $V \in \mathbb{H}^{n+}$, $V > 0$, *we have* $V - \mathcal{T}(V) > 0$.

The result above also holds when $\mathcal{T}$ *is replaced by* $\mathcal{L}, \mathcal{V}$ *or* $\mathcal{J}$.

Proof. Clearly (2) implies (3). From Theorem 3.19 we get that (3) implies (1). Finally, from Proposition 3.20, we get that (1) implies (2).   □

In order to connect the Lyapunov equations, the spectral radius of $\mathcal{A}_\ell$, and MSS, we need the following results. First recall from Proposition 3.1 that $Q(k + 1) = \mathcal{T}(Q(k))$, and thus $Q(k) = \mathcal{T}^k(Q(0))$ and

$$\mathbb{Q}(k) = \sum_{i=1}^{N} Q_i(k) = \sum_{i=1}^{N} \mathcal{T}_i^k(Q(0)). \tag{3.20}$$

Proposition 3.22. *If* $r_\sigma(\mathcal{A}_1) < 1$ *then System (3.1) is MSS according to Definition 3.8.*

Proof. Since from Remark 3.5, $r_\sigma(\mathcal{T}) = r_\sigma(\mathcal{A}_1) < 1$, it follows from Proposition 2.9 and (3.20) that $\mathbb{Q}(k) = \sum_{i=1}^{N} \mathcal{T}_i^k(Q(0)) \to 0$ as $k \to \infty$.   □

Proposition 3.23. *If model (3.1) is MSS according to Definition 3.8 then* $r_\sigma(\mathcal{A}_1) < 1$.

Proof. By hypothesis, for some $\mathbb{Q} \geq 0$, $\mathbb{Q}(k) \to \mathbb{Q}$ as $k \to \infty$ for any $\mathbb{Q}(0) = E(x_0 x_0^*)$. By taking $x_0 = 0$ it follows that $\mathbb{Q} = 0$. Therefore from (3.20), for any initial conditions x_0 and v,

$$0 \leq \sum_{i=1}^{N} \mathcal{T}_i^k(Q(0)) \longrightarrow 0 \text{ as } k \longrightarrow \infty$$

which shows that $\mathcal{T}^k(Q(0)) \to 0$ as $k \to \infty$. By choosing suitable initial conditions x_0 and v, we have that any element in $\mathbb{H}^{n+}$ can be written as $Q(0)$ so that, from Proposition 2.5, $r_\sigma(\mathcal{A}_1) = r_\sigma(\mathcal{T}) < 1$.   □

The next result shows that SS (see Remark 3.12) implies MSS.

Proposition 3.24. *If for all* $x_0 \in \mathcal{C}_0^n$ *and all* $\theta_0 \in \Theta_0$ *we have that (3.16) is satisfied then System (3.1) is MSS.*

Proof. From (3.16) we have that

$$0 \leq E(\operatorname{tr}(x(k)x(k)^*)) = \operatorname{tr}(\mathbb{Q}(k)) = E(x(k)^*x(k))$$

goes to zero as $k \to \infty$, and thus $\mathbb{Q}(k)$ converges to 0 for any initial condition x_0 and θ_0. This implies MSS of System (3.1). □

Finally we have the following proposition.

Proposition 3.25. *System (3.1) is MSS if and only if for some $\beta \geq 1$, $0 < \zeta < 1$, we have*

$$E(\|x(k)\|^2) \leq \beta \zeta^k \|x_0\|_2^2, \quad k = 0, 1, \dots. \tag{3.21}$$

Proof. If System (3.1) is MSS then from Proposition 3.23, $r_\sigma(\mathcal{T}) < 1$ and therefore, according to Proposition 2.5, for some $\bar\beta \geq 1$ and $0 < \zeta < 1$, we have $\left\|\mathcal{T}^k\right\|_1 \leq \bar\beta \zeta^k$ for all $k = 0, 1, \dots$. Therefore, recalling from Proposition 3.1 that $Q(k+1) = \mathcal{T}(Q(k))$, where $Q(k) = (Q_1(k), \dots, Q_N(k)) \in \mathbb{H}^{n+}$ and $Q_i(k) = E(x(k)x(k)^* \mathbf{1}_{\{\theta(k)=i\}})$, we get

$$E(\|x(k)\|^2) = E(\operatorname{tr}(x(k)x(k)^*))$$
$$= \sum_{i=1}^N \operatorname{tr} E\left(x(k)x(k)^* \mathbf{1}_{\{\theta(k)=i\}}\right) = \sum_{i=1}^N \operatorname{tr}(Q_i(k))$$
$$\leq n \sum_{i=1}^N \|Q_i(k)\| = n \|Q(k)\|_1 \leq n \left\|\mathcal{T}^k\right\|_1 \|Q(0)\|_1$$
$$\leq n\bar\beta \zeta^k \|x_0\|_2^2,$$

since

$$\|Q(0)\|_1 = \sum_{i=1}^N \|Q_i(0)\| = \sum_{i=1}^N \left\|E(x(0)x(0)^* \mathbf{1}_{\{\theta(0)=i\}})\right\|$$
$$\leq \sum_{i=1}^N E(\|x_0\|^2 \mathbf{1}_{\{\theta(0)=i\}}) = \|x_0\|_2^2,$$

showing (3.21). On the other hand, if (3.21) is satisfied, then clearly (3.16) holds for all $x_0 \in \mathcal{C}_0^n$ and all $\theta_0 \in \Theta_0$, and from Proposition 3.24 we have that System (3.1) is MSS, completing the proof of the proposition. □

We present now the proof of Theorem 3.9.

Proof. The equivalence among (1), (2), (3), (4), (5), and (6) follows from Corollary 3.21, and Propositions 3.22, 3.23, 3.24, and 3.25. □

3.3.4 Easy to Check Conditions for Mean Square Stability

For the case in which $p_{ij} = p_j$ for every $i, j \in \mathbb{N}$, we have the following necessary and sufficient condition for MSS:

Corollary 3.26. *If $p_{ij} = p_j$ for every $i, j \in \mathbb{N}$, then the following assertions are equivalent:*

1. *Model (3.1) is MSS according to Definition 3.8.*
2. *For some $V_1 > 0$, $V_1 \in \mathbb{B}(\mathbb{C}^n)$, $V_1 - \sum_{j=1}^{N} p_j \Gamma_j^* V_1 \Gamma_j > 0$.*
3. *For any $S_1 > 0$, $S_1 \in \mathbb{B}(\mathbb{C}^n)$, there exists a unique $V_1 > 0$, $V_1 \in \mathbb{B}(\mathbb{C}^n)$, such that $V_1 - \sum_{j=1}^{N} p_j \Gamma_j^* V_1 \Gamma_j = S_1$.*

Proof. Using Theorem 3.9 with the operator $\mathcal{J}$, it follows that

$$V_1 - \sum_{j=1}^{N} p_j \Gamma_j^* V_1 \Gamma_j > 0$$

implies MSS, so that (2) implies (1). On the other hand, MSS implies that $r_\sigma(\mathcal{A}_1) < 1$, which implies in turn, from Theorem 3.9 and Proposition 3.20, that there exists a unique $V \in \mathbb{H}^{n+}$ satisfying

$$V = \mathcal{J}(V) + S \tag{3.22}$$

where

$$V = \sum_{k=0}^{\infty} \mathcal{J}^k(S). \tag{3.23}$$

Notice that whenever $U = (U_1, \ldots, U_1) \geq 0$, we have that

$$\mathcal{J}_i(U) = \mathcal{J}_1(U) = \sum_{j=1}^{N} p_j \Gamma_j^* U_1 \Gamma_j \tag{3.24}$$

and therefore from (3.24),

$$\mathcal{J}_i^k(U) = \mathcal{J}_1^k(U) \tag{3.25}$$

for each $i \in \mathbb{N}$ and all $k = 0, 1, \ldots$. Choosing $S = (S_1, \ldots, S_1) > 0$, we have from (3.25) and (3.23) that $V = (V_1, \ldots, V_1)$ and therefore, from (3.22),

$$V_1 - \sum_{j=1}^{N} p_j \Gamma_j^* V_1 \Gamma_j = S_1,$$

showing that (1) implies (3). Clearly (3) implies (2). □

From Theorem 3.9 we can derive some sufficient conditions easier to check for MSS.

Corollary 3.27. *Conditions (1) to (4) below are equivalent:*

1. *There exist $\alpha_j > 0$ such that $\alpha_j - \sum_{i=1}^{N} p_{ij}\alpha_i r_\sigma(\Gamma_i \Gamma_i^*) > 0$, for each $j \in \mathbb{N}$.*
2. *There exist $\alpha_i > 0$ such that $\alpha_i - \sum_{j=1}^{N} p_{ij}\alpha_j r_\sigma(\Gamma_i \Gamma_i^*) > 0$, for each $i \in \mathbb{N}$.*
3. *There exist $\alpha_j > 0$ such that $\alpha_j - \sum_{i=1}^{N} p_{ij}\alpha_i r_\sigma(\Gamma_j \Gamma_j^*) > 0$, for each $j \in \mathbb{N}$.*
4. *There exist $\alpha_i > 0$ such that $\alpha_i - \sum_{j=1}^{N} p_{ij}\alpha_j r_\sigma(\Gamma_j \Gamma_j^*) > 0$, for each $i \in \mathbb{N}$.*

Moreover if one of the above conditions is satisfied then model (3.1) is MSS.

Proof. Consider the homogeneous system

$$\hat{x}(k+1) = \hat{a}_{\theta(k)}\hat{x}(k), \ k = 0, 1, \dots$$

where $\hat{a}_j = r_\sigma(\Gamma_j \Gamma_j^*)^{1/2}, \ j \in \mathbb{N}$. Applying Theorem 3.9 to this system, we get that conditions (1) to (4) above are equivalent.

Suppose that condition (3) is satisfied and make $V_j = \alpha_j I > 0$, $j \in \mathbb{N}$, $V = (V_1, \dots, V_N)$. Since

$$V_j - \mathcal{V}_j(V) = \alpha_j I - \sum_{i=1}^{N} p_{ij}\alpha_i \Gamma_j \Gamma_j^* \geq (\alpha_j - \sum_{i=1}^{N} p_{ij}\alpha_i r_\sigma(\Gamma_j \Gamma_j^*))I > 0$$

we have from Theorem 3.9 again that model (3.1) is MSS.   □

Corollary 3.28. *If for real numbers $\delta_i > 0, i \in \mathbb{N}$, one of the conditions below is satisfied*

1. $r_\sigma\left(\sum_{i=1}^{N} p_{ij}\delta_i \Gamma_i \Gamma_i^*\right) < \delta_j, \ j = 1, \dots, N,$
2. $r_\sigma\left(\sum_{j=1}^{N} p_{ij}\delta_j \Gamma_j^* \Gamma_j\right) < \delta_i, \ i = 1, \dots, N,$

then model (3.1) is MSS. Besides, these conditions are stronger than those in Corollary 3.27.

Proof. Suppose that condition (1) is satisfied and make $V_i = \alpha_i I > 0$, $i \in \mathbb{N}$, $V = (V_1, \dots, V_N)$. Following the same arguments as in the proof of the previous corollary, we have

$$V_j - \mathcal{T}_j(V) = \delta_j I - \sum_{i=1}^{N} p_{ij}\delta_i \Gamma_i \Gamma_i^* \geq (\delta_j - \sum_{i=1}^{N} p_{ij}\alpha_i r_\sigma(\Gamma_i \Gamma_i^*))I > 0$$

and the model (3.1) is MSS. The proof of condition (2) is similar. Note that if the conditions of Corollary 3.27 are satisfied then for each $j \in \mathbb{N}$,

$$r_\sigma \left(\sum_{i=1}^{N} p_{ij} \alpha_i \Gamma_i \Gamma_i^* \right) = \left\| \sum_{i=1}^{N} p_{ij} \alpha_i \Gamma_i \Gamma_i^* \right\|$$

$$\leq \sum_{i=1}^{N} p_{ij} \alpha_i \left\| \Gamma_i \Gamma_i^* \right\|$$

$$= \sum_{i=1}^{N} p_{ij} \alpha_i r_\sigma (\Gamma_i \Gamma_i^*)$$

$$< \alpha_j$$

which implies condition (1). Similarly we can show that condition (4) of Corollary 3.27 implies condition (2) above.   □

Let us illustrate these conditions through some examples.

Example 3.29. This example shows that conditions (1) and (2) of Corollary 3.28 are not equivalent. Indeed, consider $n = 2$, $p_{ij} = p_j \in (0,1)$, $i, j = 1, 2$, and

$$\Gamma_1 = \begin{bmatrix} 0 & 0 \\ \zeta_1 & 0 \end{bmatrix} \text{ and } \Gamma_2 = \begin{bmatrix} \zeta_2 & 0 \\ 0 & 0 \end{bmatrix},$$

with ζ_1, ζ_2 real numbers. Condition (1) will be satisfied if, for some $\delta_i > 0$, $i = 1, 2$, we have

$$r_\sigma \left(\begin{bmatrix} \delta_2 \zeta_2^2 & 0 \\ 0 & \delta_1 \zeta_1^2 \end{bmatrix} \right) p_j = \max\{\delta_1 \zeta_1^2, \delta_2 \zeta_2^2\} p_j < \delta_j, \ j = 1, 2$$

or, equivalently, if $p_1 \zeta_1^2 < 1$, $p_2 \zeta_2^2 < 1$. On the other hand, condition (2) will be satisfied if for some $\alpha_i > 0$, $i = 1, 2$, we have

$$r_\sigma \left(\begin{bmatrix} \alpha_1 p_1 \zeta_1^2 + \alpha_2 p_2 \zeta_2^2 & 0 \\ 0 & 0 \end{bmatrix} \right) = \alpha_1 p_1 \zeta_1^2 + \alpha_2 p_2 \zeta_2^2 < \alpha_i, \ i = 1, 2. \quad (3.26)$$

Note that if (3.26) is satisfied then $p_1 \zeta_1^2 < 1$ and $p_2 \zeta_2^2 < 1$. Consider $p_1 = p_2 = \frac{1}{2}$, $\zeta_1^2 = \zeta_2^2 = \frac{4}{3}$. For this choice, $p_1 \zeta_1^2 < 1$, $p_2 \zeta_2^2 < 1$, but (3.26) will not be satisfied for any $\alpha_i > 0$, $i = 1, 2$. Therefore conditions (1) and (2) of Corollary 3.28 are not equivalent. Note that in this example, condition (4) of Corollary 3.27 will be the same as (3.26), and there will be a positive-definite solution for

$$V - p_1 \Gamma_1^* V \Gamma_1 - p_2 \Gamma_2^* V \Gamma_2 = I$$

if and only if $\zeta_2^2 p_2 < 1$, and thus from Corollary 3.26 this is a necessary and sufficient condition for MSS.

Example 3.30. Consider Example 3.14 with a different Γ_1, that is,

$$\Gamma_1 = \begin{bmatrix} 0 & \zeta_1 \\ 0 & 0 \end{bmatrix}.$$

Repeating the same arguments above, we get that condition (2) of Corollary 3.28 will be satisfied if $p_1 a_1^2 < 1$ and $p_2 a_2^2 < 1$, while condition (1) will be the same as (3.26). This shows that we cannot say that any of the conditions of Corollary 3.28 will be stronger than the other. Again in this problem we have that MSS is equivalent to $\zeta_2^2 p_2 < 1$.

3.4 MSS: The Non-homogeneous Case

3.4.1 Main Results

We consider in this section the class of non-homogeneous dynamical systems modeled by the following equation:

$$\mathcal{G}_{nh} = \begin{cases} x(k+1) = \Gamma_{\theta(k)}x(k) + G_{\theta(k)}w(k) \\ x(0) = x_0 \in \mathcal{C}_0^n, \theta(0) = \theta_0 \in \Theta_0, \end{cases} \qquad (3.27)$$

where $\Gamma = (\Gamma_1, \ldots, \Gamma_N) \in \mathbb{H}^n$, $G = (G_1, \ldots, G_N) \in \mathbb{H}^{r,n}$. We set

$$\pi_i(k) = \mathcal{P}(\theta(k) = i), \qquad (3.28)$$

and regarding the additive disturbance $w = \{w(k); k \in \mathbb{T}\}$ we analyze two cases:

Case (1):   a second order independent wide sense stationary sequence of random variables with $E(w(k)) = \gamma \in \mathbb{C}^r$ and $E(w(k)w(k)^*) = W \in \mathbb{B}(\mathbb{C}^r)^+$, $k \in \mathbb{T}$, with x_0 and $\{\theta(k); k \in \mathbb{T}\}$ independent of $\{w(k); k \in \mathbb{T}\}$.

Case (2):   a sequence in $\mathcal{C}^r$.

Case (1): For this case the Markov chain $\{\theta(k), k \in \mathbb{T}\}$ is assumed to be ergodic, referred to from now on as Assumption 3.31:

Assumption 3.31 (Ergodic Assumption) $\{\theta(k); k = 0, 1, \ldots\}$ *is an ergodic Markov chain.*

We recall that under Assumption 3.31, for any given initial distribution $\{v_i; i \in \mathbb{N}\}$, there exists a limit probability distribution $\{\pi_i; \pi_i > 0, i \in \mathbb{N}\}$ (which doesn't depend on v) such that for each $j \in \mathbb{N}$,

$$\sum_{i=1}^{N} p_{ij}\pi_i = \pi_j, \quad \sum_{i=1}^{N} \pi_i = 1$$

and

$$|\pi_i(k) - \pi_i| \leq \eta\xi^k$$

for some $\eta \geq 0$ and $0 < \xi < 1$ (cf. [24], [194]).

We introduce the following notation:

$$U(k,l) \triangleq (U_1(k,l), \ldots, U_N(k,l)) \in \mathbb{H}^n, \tag{3.29}$$

$$U_j(k,l) \triangleq E(x(k+l)x(k)^* \mathbf{1}_{\{\theta(k+l)=j\}}) \in \mathbb{B}(\mathbb{C}^n),$$

$$\mathbb{U}(k,l) \triangleq E(x(k+l)x(k)^*) = \sum_{i=1}^{N} U_i(k,l) \in \mathbb{B}(\mathbb{C}^n)$$

and shall be interested in the following stability concept, which generalizes Definition 3.8:

Definition 3.32. *We say that the linear system with Markov jump parameter (3.27) and input sequence as in Case (1) is asymptotically wide sense stationary (AWSS) if for any initial condition $x_0 \in \mathcal{C}_0^n, \theta_0 \in \Theta_0$ there exist $\mu \in \mathbb{C}^n$, and $\{\mathbb{U}(l); l = 0, 1, \ldots\}$, $\mathbb{U}(l) \in \mathbb{B}(\mathbb{C}^n)$ (independent of x_0 and θ_0) such that*

1. $\|\mu(k) - \mu\| \to 0$ *as* $k \to \infty$,
2. $\|\mathbb{U}(k,l) - \mathbb{U}(l)\| \to 0$ *as* $k \to \infty$.

The main result for this case, which will be proved in Section 3.4.2, is as follows.

Theorem 3.33. *The following assertions are equivalent:*

1. *System (3.1) is MSS.*
2. *System (3.27) is MSS.*
3. *System (3.27) is AWSS.*

Case (2): For this situation we will be interested in showing that the state sequence $x = \{x(k); k \in \mathbb{T}\} \in \mathcal{C}^n$, which is the usual scenario for the H_∞ approach to be seen in Chapter 7. In Subsection 3.4.3 we prove the following result.

Theorem 3.34. $r_\sigma(\mathcal{T}) < 1$ *if and only if for $x = \{x(k); k \in \mathbb{T}\}$ as in (3.27), we have that $x \in \mathcal{C}^n$ for every $w = \{w(k); k \in \mathbb{T}\} \in \mathcal{C}^r$, $x_0 \in \mathcal{C}_0^n$ and $\theta_0 \in \Theta_0$.*

Proof. See Subsection 3.4.3. $\square$

3.4.2 Wide Sense Stationary Input Sequence

We consider here the first scenario regarding additive disturbance: the one characterized by a second order independent wide sense stationary sequence of random variables. We will prove Theorem 3.33 in this subsection, unifying the results for the homogeneous and non-homogeneous cases. We make the following definitions. For

$$q = \begin{pmatrix} q_1 \\ \vdots \\ q_N \end{pmatrix} \in \mathbb{C}^{Nn}$$

set the following auxiliary sequences:

$$\psi(k) \triangleq \begin{pmatrix} \psi_1(k) \\ \vdots \\ \psi_N(k) \end{pmatrix} \in \mathbb{C}^{Nn}, \tag{3.30}$$

where

$$\psi_j(k) \triangleq \sum_{i=1}^{N} p_{ij} G_i \gamma \pi_i(k) \in \mathbb{C}^n, \tag{3.31}$$

and

$$R(k, q) \triangleq (R_1(k, q), \dots, R_N(k, q)) \in \mathbb{H}^n, \tag{3.32}$$

where

$$R_j(k, q) \triangleq \sum_{i=1}^{N} p_{ij}(G_i W G_i^* \pi_i(k) + \Gamma_i q_i \gamma^* G_i^* + G_i \gamma q_i^* \Gamma_i^*) \in \mathbb{B}(\mathbb{C}^n). \tag{3.33}$$

In what follows recall the definition of $q(k)$, $Q(k)$, and $U(k)$ in (3.3) and (3.29).

Proposition 3.35. *For every* $k = 0, 1, \dots, \iota = 0, 1, \dots$ *and* $j \in \mathbb{N}$,

1. $q(k+1) = \mathcal{B}q(k) + \psi(k)$,
2. $Q(k+1) = \mathcal{T}(Q(k)) + R(k, q(k))$,
3. $U_j(k, \iota) = \sum_{i=1}^{N} p_{ij} \Gamma_i U_i(k, \iota - 1) + \sum_{i=1}^{N} p_{ij} G_i \gamma \Gamma \left(\sum_{l=1}^{N} p_{li}^{\iota-1} q_l^*(k) \right)$

where $p_{li}^{\iota-1} = \mathcal{P}(\theta(\iota - 1) = i | \theta(0) = l)$.

Proof. From the hypothesis that x_0 and $\{\theta(k); k \in \mathbb{T}\}$ are independent of $\{w(k); k \in \mathbb{T}\}$, it easily follows that $x(k)$ is independent of $w(l)$ for every $l \geq k$. Therefore

$$q_j(k+1)$$

$$= \sum_{i=1}^{N} E((\Gamma_i x(k) + G_i w(k)) \mathbf{1}_{\{\theta(k+1)=j\}} \mathbf{1}_{\{\theta(k)=i\}})$$

$$= \sum_{i=1}^{N} \Gamma_i E(x(k) \mathbf{1}_{\{\theta(k)=i\}} \mathcal{P}(\theta(k+1) = j | \mathfrak{F}_k))$$

$$+ \sum_{i=1}^{N} G_i \gamma E(\mathbf{1}_{\{\theta(k)=i\}} \mathcal{P}(\theta(k+1) = j | \mathfrak{F}_k))$$

$$= \sum_{i=1}^{N} p_{ij} \Gamma_i q_i(k) + \sum_{i=1}^{N} p_{ij} G_i \gamma \pi_i(k)$$

showing (1). For (2), notice that

$$Q_j(k+1)$$
$$= \sum_{i=1}^{N} E((\Gamma_i x(k) + G_i w(k))(\Gamma_i x(k) + G_i w(k))^* \mathbf{1}_{\{\theta(k+1)=j\}} \mathbf{1}_{\{\theta(k)=i\}})$$
$$= \sum_{i=1}^{N} p_{ij} \Gamma_i Q_i(k) \Gamma_i^*$$
$$+ \sum_{i=1}^{N} p_{ij} \left(G_i \gamma q_i(k)^* \Gamma_i^* + \Gamma_i q_i(k) \gamma^* G_i^* + G_i W G_i^* \pi_i(k) \right)$$
$$= \mathcal{T}_j(Q(k)) + R_j(k, q(k)).$$

For (3) we have that

$$U_j(k, \iota)$$
$$= \sum_{i=1}^{N} E((\Gamma_i x(k + \iota - 1) + G_i w(k + \iota - 1)) x(k)^* \mathbf{1}_{\{\theta(k+\iota)=j\}} \mathbf{1}_{\{\theta(k+\iota-1)=i\}})$$
$$= \sum_{i=1}^{N} p_{ij} \Gamma_i U_i(k, \iota - 1) + \sum_{i=1}^{N} p_{ij} G_i \gamma E(x(k)^* \mathbf{1}_{\{\theta(k+\iota-1)=i\}})$$
$$= \sum_{i=1}^{N} p_{ij} \Gamma_i U_i(k, \iota - 1) + \sum_{i=1}^{N} p_{ij} G_i \gamma \left(\sum_{l=1}^{N} p_{li}^{\iota-1} q_l^*(k) \right).$$

$\square$

Define now

$$\psi \triangleq \begin{pmatrix} \psi_1 \\ \vdots \\ \psi_N \end{pmatrix} \in \mathbb{C}^{Nn} \qquad (3.34)$$

where

$$\psi_j \triangleq \sum_{i=1}^{N} p_{ij} G_i \gamma \pi_i \in \mathbb{C}^n, \qquad (3.35)$$

and

$$R(q) \triangleq (R_1(q), \dots, R_N(q)) \in \mathbb{H}^n, \qquad (3.36)$$

where

$$R_j(q) \triangleq \sum_{i=1}^{N} p_{ij} (G_i W G_i^* \pi_i + \Gamma_i q_i \gamma^* G_i^* + G_i \gamma q_i^* \Gamma_i^*) \in \mathbb{B}(\mathbb{C}^n). \qquad (3.37)$$

Set the following recursive equations,

$$v(k+1) = \mathcal{B}v(k) + \psi(k), \tag{3.38}$$

$$Z(k+1) = \mathcal{T}(Z(k)) + R(k, v(k)). \tag{3.39}$$

We have now the following result (recall the definition of $\psi(k),R(k,q)$ in (3.30), (3.31), (3.32),(3.33), and $\psi, R(q)$ in (3.34), (3.35), (3.36), (3.37)).

Proposition 3.36. *If $r_\sigma(\mathcal{T}) < 1$ then there exists a unique solution $q \in \mathbb{C}^n$, $Q = (Q_1, \ldots, Q_N) \in \mathbb{H}^n$ for the system of equations in $v \in \mathbb{C}^n$ and $Z \in \mathbb{H}^n$ below*

$$v = \mathcal{B}v + \psi, \tag{3.40}$$

$$Z = \mathcal{T}(Z) + R(v). \tag{3.41}$$

Moreover,

$$q = (I - \mathcal{B})^{-1}\psi, \tag{3.42}$$

$$Q = (\mathcal{I} - \mathcal{T})^{-1}R(q)$$
$$= \hat{\varphi}^{-1}((I - \mathcal{A}_1)^{-1}\hat{\varphi}(R(q))) \tag{3.43}$$

with $Q \in \mathbb{H}^{n+}$. Furthermore for any $v(0) \in \mathbb{C}^n$, $Z(0) = (Z_1(0), \ldots, Z_N(0)) \in \mathbb{H}^n$, we have that $v(k)$ and $Z(k)$ given by (3.38) and (3.39) satisfy,

$$v(k) \to q \text{ as } k \to \infty,$$

$$Z(k) \to Q \text{ as } k \to \infty.$$

Proof. Since $r_\sigma(\mathcal{T}) = r_\sigma(\mathcal{A}_1) < 1$, we have from Proposition 3.6 that $r_\sigma(\mathcal{B}) < 1$. Let us show that $v(k)$ in (3.38) is a Cauchy summable sequence. Recall that by hypothesis the Markov chain is ergodic (see Assumption 3.31) and therefore for some $\eta \geq 1$ and $0 < \xi < 1$,

$$|\pi_i - \pi_i(k)| \leq \eta\xi^k. \tag{3.44}$$

From (3.31), (3.35), and (3.44) we have that

$$\sum_{k=0}^{\infty} \sup_{\tau \geq 0} \|\psi(k+\tau) - \psi(k)\|$$

$$\leq \sum_{k=0}^{\infty} \sup_{\tau \geq 0} \{\|\psi(k+\tau) - \psi\| + \|\psi - \psi(k)\|\}$$

$$\leq \max_{1 \leq j \leq N} \|G_j\gamma\| \sum_{k=0}^{\infty} \sup_{\tau \geq 0} \max_{1 \leq i \leq N} \{|\pi_i - \pi_i(k+\tau)| + |\pi_i - \pi_i(k)|\}$$

$$\leq \max_{1 \leq j \leq N} \|G_j\gamma\| \sum_{k=0}^{\infty} \sup_{\tau \geq 0}(\eta\xi^k(1+\xi^\tau)) = \max_{1 \leq j \leq N} \|G_j\gamma\| \frac{2\eta}{1-\xi}.$$

From Proposition 2.9 it follows that the sequence $\{v(k); k = 0, 1, \ldots\}$ is Cauchy summable, and $v(k) \to q$ with q as in (3.42). Moreover, from (3.33) and the fact that $\{v(k); k = 0, 1, \ldots\}$ is a Cauchy summable sequence it follows, from arguments similar to those just presented above, that the sequence $\{\hat{\varphi}(R(k, v(k))); k = 0, 1, \ldots\}$ is also Cauchy summable. Since

$$\hat{\varphi}(R(k, v(k))) \longrightarrow \hat{\varphi}(R(q)) \text{ as } k \longrightarrow \infty$$

it follows from Proposition 2.9 that

$$Z(k) \to Q = \hat{\varphi}^{-1}((I - \mathcal{A}_1)^{-1} \hat{\varphi}(R(q)).$$

From (3.41) it follows that Q can be written also as

$$Q = (\mathcal{I} - \mathcal{T})^{-1} R(q),$$

showing (3.43). Finally notice that by making $v(0) = q(0)$ and $Z(0) = Q(0)$ in (3.38) and (3.39) respectively, we have from Proposition 3.35 that $v(k) = q(k)$ and $Z(k) = Q(k)$ for all $k = 0, 1, \ldots$. Since $Q(k) \in \mathbb{H}^{n+}$ for all k, and $Q(k)$ converges to Q, it follows that $Q \in \mathbb{H}^{n+}$.   □

Proposition 3.37. *If $r_\sigma(\mathcal{T}) < 1$ then System (3.27) is MSS and AWSS according to Definitions 3.8 and 3.32 respectively, with*

$$\mu = \sum_{i=1}^{N} q_i, \tag{3.45}$$

where

$$q = \begin{bmatrix} q_1 \\ \vdots \\ q_N \end{bmatrix} = (I - \mathcal{B})^{-1} \begin{bmatrix} \psi_1 \\ \vdots \\ \psi_N \end{bmatrix} = (I - \mathcal{B})^{-1} \psi, \tag{3.46}$$

$$\mathbb{Q} = \sum_{i=1}^{N} Q_i \tag{3.47}$$

$$Q = (\mathcal{I} - \mathcal{T})^{-1}(R(q))$$
$$= \hat{\varphi}^{-1}((I - \mathcal{A}_1)^{-1} \hat{\varphi}(R(q))), \tag{3.48}$$

and

$$\mathbb{U}(\iota) = \sum_{i=1}^{N} U_i(\iota), \tag{3.49}$$

$$U_j(\iota + 1) = \sum_{i=1}^{N} p_{ij} \Gamma_i U_i(\iota) + \sum_{i=1}^{N} p_{ij} G_i \gamma \left(\sum_{l=1}^{N} p_{li}^{\iota-1} q_l^* \right),$$

$$U(0) = Q. \tag{3.50}$$

Proof. From Proposition 3.36 it is immediate to obtain convergence of $\mu(k)$ to μ and $\mathbb{Q}(k)$ to $\mathbb{Q}$ as in (3.45), (3.46), (3.47), and (3.48). Let us now show by induction on ι that $U(k,\iota) \to U(\iota)$ as $k \to \infty$, for some $U(\iota) \in \mathbb{H}^n$. Indeed since $q(k) \to q$ and $Q(k) \to Q$ as $k \to \infty$ it is immediate from Proposition 3.35 (3), with $\iota = 1$, that

$$U_j(k,1) \longrightarrow \sum_{i=1}^{N} p_{ij}\Gamma_i Q_i + \sum_{i=1}^{N} p_{ij} G_i \gamma q_i^* \text{ as } k \longrightarrow \infty.$$

Suppose that the induction hypothesis holds for ι, that is, $U(k,\iota) \to U(i)$ as $k \to \infty$, for some $U(\iota) \in \mathbb{H}^n$. Then, similarly,

$$U_j(k,\iota+1) \longrightarrow \sum_{i=1}^{N} p_{ij}\Gamma_i U_i(\iota) + \sum_{i=1}^{N} p_{ij} G_i \gamma \left(\sum_{l=1}^{N} p_{li}^\iota q_l^* \right) \text{ as } k \longrightarrow \infty$$

showing (3.49) and (3.50).   □

Proposition 3.38. *If System (3.27) is MSS according to Definition 3.8 then $r_\sigma(\mathcal{T}) < 1$.*

Proof. Recall that

$$Q_i(k) = E\left(x(k)x(k)^* \mathbf{1}_{\{\theta(k)=i\}}\right),$$
$$Q(k+1) = \mathcal{T}(Q(k)) + R(q(k)),$$

and

$$\mathbb{Q}(k) = E(x(k)x(k)^*) = \sum_{i=1}^{N} Q_i(k).$$

Thus,

$$\mathbb{Q}(k) = \sum_{i=1}^{N} \mathcal{T}_i^k(Q(0)) + \sum_{i=1}^{N} \left(\sum_{j=0}^{k-1} \mathcal{T}_i^{k-j-1}(R(q(j))) \right). \tag{3.51}$$

By hypothesis, there exists $\mathbb{Q} \in \mathbb{H}^{n+}$ (depending only on γ and W) such that $\mathbb{Q}(k) \to \mathbb{Q}$ as $k \to \infty$ for any $\mathbb{Q}(0) = E(x_0 x_0^*)$. For $x_0 = 0$, the second term on the right hand side of (3.51) (which does not depend on x_0) converges to $\mathbb{Q}$ as $k \to \infty$. Therefore for any initial conditions x_0 and v,

$$0 \le \sum_{i=1}^{N} \mathcal{T}_i^k(Q(0)) \longrightarrow 0 \text{ as } k \longrightarrow \infty.$$

This shows that $\mathcal{T}^k(Q(0)) \to 0$ as $k \to \infty$. By choosing suitable initial conditions x_0 and v, we have that any element in $\mathbb{H}^{n+}$ can be written as $Q(0)$ so that, from Proposition 2.6, $r_\sigma(\mathcal{T}) < 1$.   □

Finally we have the following proposition, completing the proof of Theorem 3.33.

Proposition 3.39. *System (3.27) is MSS if and only if it is AWSS.*

Proof. This follows from Propositions 3.37, 3.38, and the fact that AWSS clearly implies MSS.   □

3.4.3 The ℓ_2-disturbance Case

We consider in this subsection the second scenario regarding additive disturbance: the one characterized by an ℓ_2-sequence. It is shown that for model (3.27), with $w = (w(0), w(1), \ldots) \in \mathcal{C}^r$, MSS is equivalent to the MJLS being a bounded linear operator from $\mathcal{C}^r$ into $\mathcal{C}^n$.

Consider the class of dynamical systems modeled by (3.27) where the additive disturbance $w = \{w(k); k \in \mathbb{T}\}$ is a sequence in $\mathcal{C}^r$. We prove next Theorem 3.34.

Proof (of Theorem 3.34). To prove necessity, all we have to show is that $\|x\|_2 < \infty$ since clearly $x_k = (x(0), \ldots, x(k)) \in \mathcal{C}_k^n$ for every $k = 0, 1, \ldots$. We have

$$x(k) = \Gamma_{\theta(k-1)} \ldots \Gamma_{\theta(0)} x(0) + \sum_{\iota=0}^{k-1} \Gamma_{\theta(k-1)} \ldots \Gamma_{\theta(\iota+1)} G_{\theta(\iota)} w(\iota)$$

and by the triangular inequality in $\mathcal{C}^n$,

$$\|x(k)\|_2 \leq \left\| \Gamma_{\theta(k-1)} \ldots \Gamma_{\theta(0)} x(0) \right\|_2 + \sum_{\iota=0}^{k-1} \left\| \Gamma_{\theta(k-1)} \ldots \Gamma_{\theta(\iota+1)} G_{\theta(\iota)} w(\iota) \right\|_2.$$

Set

$$W_i(\iota) \triangleq E(w(\iota)w(\iota)^* \mathbf{1}_{\{\theta(\iota)=i\}}),$$
$$W(\iota) \triangleq (W_1(\iota), \ldots, W_N(\iota)) \in \mathbb{H}^{r+},$$
$$GW(\iota)G^* \triangleq (G_1 W_1(\iota) G_1^*, \ldots, G_N W_N(\iota) G_N^*) \in \mathbb{H}^{n+}.$$

From Proposition 3.1, we get

$$\left\| \Gamma_{\theta(k-1)} \ldots \Gamma_{\theta(\iota+1)} G_{\theta(\iota)} w(\iota) \right\|_2^2 \leq n \left\| \mathcal{T}^{k-\iota-1}(GW(\iota)G^*) \right\|_1$$
$$\leq n \left\| \mathcal{T}^{k-\iota-1} \right\|_1 \left\| GW(\iota)G^* \right\|_1$$
$$\leq n \|G\|_{\max}^2 \left\| \mathcal{T}^{k-\iota-1} \right\|_1 \|w(\iota)\|_2^2$$

since

$$\|W(\iota)\|_1 = \sum_{i=1}^{N} \|W_i(\iota)\|$$

$$\leq \sum_{i=1}^{N} E(\|w(\iota)\|^2 \mathbf{1}_{\{\theta(\iota)=i\}})$$

$$= E(\|w(\iota)\|^2)$$

$$= \|w(\iota)\|_2^2 .$$

Similarly,

$$\left\| \Gamma_{\theta(k-1)} \ldots \Gamma_{\theta(0)} x(0) \right\|_2^2 \leq n \left\| \mathcal{T}^k \right\|_1 \|x(0)\|_2^2 .$$

From Theorem 3.9, there exists $0 < \zeta < 1$ and $\beta \geq 1$ such that $\left\| \mathcal{T}^k \right\|_1 \leq \beta \zeta^k$, and therefore,

$$\|x(k)\|_2 \leq \sum_{\iota=0}^{k} \zeta_{k-\iota} \beta_\iota ,$$

where $\zeta_{k-\iota} \triangleq (\zeta^{1/2})^{(k-\iota)}$, $\beta_0 \triangleq (n\beta)^{1/2} \|x(0)\|_2$, and

$$\beta_\iota \triangleq (n\beta)^{1/2} \|G\|_{\max} \|w(\iota - 1)\|_2 , \ \iota \geq 1.$$

Set $a \triangleq (\zeta_0, \zeta_1, \ldots)$ and $b \triangleq (\beta_0, \beta_1, \ldots)$. Since $a \in \ell_1$ (that is, $\sum_{\iota=0}^{\infty} |\zeta_\iota| < \infty$) and $b \in \ell_2$ (that is, $\sum_{\iota=0}^{\infty} |\beta_\iota|^2 < \infty$) it follows that the convolution $c \triangleq a*b = (c_0, c_1, \ldots)$, $c_k \triangleq \sum_{\iota=0}^{k} \zeta_{k-\iota} \beta_\iota$, lies itself in ℓ_2 with $\|c\|_2 \leq \|a\|_1 \|b\|_2$ (cf. [103], p. 529). Hence,

$$\|x\|_2 = \left\{ \sum_{k=0}^{\infty} E(\|x(k)\|^2) \right\}^{1/2} \leq \left\{ \sum_{\iota=0}^{k} c_\iota^2 \right\}^{1/2} = \|c\|_2 < \infty.$$

To prove sufficiency, first notice that for any $V = (V_1, \ldots, V_N) \in \mathbb{H}^{n+}$, we can define $x_0 \in \mathcal{C}_0^n$ and $\theta_0 \in \Theta_0$ such that $V_i = E(x(0)x(0)^* \mathbf{1}_{\{\theta_0=i\}})$, $i \in \mathbb{N}$. Making $w(k) = 0$, $k = 0, 1, \ldots$ in (3.27) and recalling that $Q_i(k) = E(x(k)x(k)^* \mathbf{1}_{\{\theta(k)=i\}})$, $Q(k) = (Q_1(k), \ldots, Q_N(k))$, $Q(0) = V$, we have from Proposition 3.1 that

$$\left\| \mathcal{T}^k(V) \right\| = \|Q(k)\| = \sum_{i=1}^{N} \|Q_i(k)\| = \sum_{i=1}^{N} \left\| E(x(k)x(k)^* \mathbf{1}_{\{\theta(k)=i\}}) \right\|$$

$$\leq \sum_{i=1}^{N} E(\|x(k)x(k)^*\| \mathbf{1}_{\{\theta(k)=i\}})$$

$$= \sum_{i=1}^{N} E(\|x(k)\|^2 \mathbf{1}_{\{\theta(k)=i\}})$$

$$= E(\|x(k)\|^2)$$

and thus

$$\sum_{k=0}^{\infty} \left\|\mathcal{T}^k(V)\right\| \leq \sum_{k=0}^{\infty} E(\|x(k)\|^2) < \infty$$

for every $V \in \mathbb{H}^{n+}$ which is equivalent, according to Proposition 2.5, to $r_\sigma(\mathcal{T}) < 1$. □

3.5 Mean Square Stabilizability and Detectability

3.5.1 Definitions and Tests

This section deals with the concepts of mean square detectability and mean square stabilizability. As far as control for Markov systems is concerned, these are central concepts, which will be further explored in Chapter 4. We shall consider in this section another version of System (3.1)

$$\mathcal{G} = \begin{cases} x(k+1) = A_{\theta(k)}x(k) + B_{\theta(k)}u(k) \\ y(k) = C_{\theta(k)}x(k) \\ x(0) = x_0, \theta(0) = \theta_0. \end{cases} \tag{3.52}$$

In the following we will present the definition of these properties and also tests, based on convex programming, to establish both stabilizability and detectability for a given system. The reader should keep in mind that these concepts in the Markovian context are akin to their deterministic equivalents, and that stabilizability and detectability for the deterministic case could be considered as special cases of the definitions given below (see Definitions 2.17 and 2.20).

Definition 3.40 (Mean Square Stabilizability). *Let $A = (A_1, \ldots, A_N) \in \mathbb{H}^n$, $B = (B_1, \ldots, B_N) \in \mathbb{H}^{m,n}$. We say that the pair (A, B) is mean square stabilizable if there is $F = (F_1, \ldots, F_N) \in \mathbb{H}^{n,m}$ such that when $u(k) = F_{\theta(k)}x(k)$, System (3.52) is MSS. In this case, F is said to stabilize the pair (A, B).*

Definition 3.41 (Mean Square Detectability). *Let $C = (C_1, \ldots, C_N) \in \mathbb{H}^{n,s}$. We say that the pair (C, A) is mean square detectable if there is $H = (H_1, \ldots, H_N) \in \mathbb{H}^{s,n}$ such that $r_\sigma(\mathcal{T}) < 1$ as in (3.7) with $\Gamma_i = A_i + H_iC_i$ for $i \in \mathbb{N}$.*

We now introduce tests to establish detectability and stabilizability for a given system. These tests are based on the resolution of feasibility problems, with restrictions given in terms of LMIs.

Proposition 3.42 (Mean square Stabilizability Test). *The pair (A, B) is mean square stabilizable if and only if there are $W_1 = (W_{11}, \ldots, W_{N1}) \in \mathbb{H}^{n+}$,*

$W_2 = (W_{12}, \ldots, W_{N2}) \in \mathbb{H}^{m,n}$, $W_3 = (W_{13}, \ldots, W_{N3}) \in \mathbb{H}^{m+}$ *such that for all* $j \in \mathbb{N}$,

$$\sum_{i=1}^{N} p_{ij}(A_i W_{i1} A_i^* + B_i W_{i2}^* A_i^* + A_i W_{i2} B_i^* + B_i W_{i3} B_i^*) - W_{j1} < 0, \quad (3.53a)$$

$$\begin{bmatrix} W_{j1} & W_{j2} \\ W_{j2}^* & W_{j3} \end{bmatrix} \geq 0, \quad (3.53b)$$

$$W_{j1} > 0. \quad (3.53c)$$

Proof. Consider the set below:

$$\Phi \triangleq \{ W_1 \in \mathbb{H}^{n+}, W_2 \in \mathbb{H}^{m,n}, W_3 \in \mathbb{H}^{m+};$$
$$W_{j1}, W_{j2}, W_{j3} \text{ satisfy } (3.53) \text{ for all } j \in \mathbb{N}\}.$$

For necessity, assume that the pair (A, B) is mean square stabilizable. Then there is $F \in \mathbb{H}^{n,m}$ such that System (3.52) with $\Gamma_i = A_i + B_i F_i$ for $i \in \mathbb{N}$ is MSS. From Theorem 3.9 there exists $X = (X_1, \ldots, X_N) > 0$ in $\mathbb{H}^{n+}$ such that

$$-X_j + \sum_{i=1}^{N} p_{ij}(A_i + B_i F_i) X_i (A_i + B_i F_i)^* < 0, \quad (3.54)$$

for all $j \in \mathbb{N}$. Taking $W_{j1} = X_j$, $W_{j2} = W_{j1} F_j^*$ and $W_{j3} = W_{j2}^* W_{j1}^{-1} W_{j2}$ for $j \in \mathbb{N}$, it is easy to verify from (3.54) that (3.53) are satisfied and therefore $\Phi \neq \emptyset$. Note that (3.53b) is equivalent to $W_{j3} \geq W_{j2}^* W_{j1}^{-1} W_{j2}$ by the Schur complement (see Lemma 2.23).

For sufficiency, assume that $\Phi \neq \emptyset$, so that there are $W_1 \in \mathbb{H}^{n+}$, $W_2 \in \mathbb{H}^{m,n}$ and $W_3 \in \mathbb{H}^{m+}$ satisfying (3.53). Let $F = (F_1, \ldots, F_N) \in \mathbb{H}^{n,m}$ be given by $F_j = W_{j2}^* W_{j1}^{-1}$ for $j \in \mathbb{N}$, so

$$-W_{j1} + \sum_{i=1}^{N} p_{ij}(A_i + B_i F_i) W_{i1} (A_i + B_i F_i)^*$$

$$= -W_{j1} + \sum_{i=1}^{N} p_{ij}(A_i + B_i W_{i2}^* W_{i1}^{-1}) W_{i1} (A_i + B_i W_{i2}^* W_{i1}^{-1})^*$$

$$= -W_{j1} + \sum_{i=1}^{N} p_{ij}(A_i W_{i1} A_i^* + B_i W_{i2}^* A_i^*$$
$$+ A_i W_{i2} B_i^* + B_i W_{i2}^* W_{i1}^{-1} W_{i2} B_i^*)$$

$$\leq -W_{j1} + \sum_{i=1}^{N} p_{ij}(A_i W_{i1} A_i^* + B_i W_{i2}^* A_i^* + A_i W_{i2} B_i^* + B_i W_{i3} B_i^*) < 0$$

for all $j \in \mathbb{N}$. From Theorem 3.9 we have that F stabilizes (A, B) in the mean square sense (see Definition 3.40). So it is clear that (A, B) is mean square stabilizable. □

Proposition 3.43 (Mean Square Detectability Test). *The pair* (C, A) *is mean square detectable if and only if there are* $W_1 = (W_{11}, \ldots, W_{N1}) \in \mathbb{H}^{n+}$, $Z = (Z_1, \ldots, Z_N) \in \mathbb{H}^{n+}$, $W_2 = (W_{12}, \ldots, W_{N2}) \in \mathbb{H}^{s,n}$, $W_3 = (W_{13}, \ldots, W_{N3}) \in \mathbb{H}^{s+}$ *such that for all* $i \in \mathbb{N}$,

$$A_i^* Z_i A_i + C_i^* W_{i2}^* A_i + A_i^* W_{i2} C_i + C_i^* W_{i3} C_i - W_{i1} < 0,$$

$$\begin{bmatrix} Z_i & W_{i2} \\ W_{i2}^* & W_{i3} \end{bmatrix} \geq 0,$$

$$Z_i \geq \mathcal{E}_i(W_1), \; W_{i1} > 0, \; Z_i > 0.$$

Proof. Analogous to the proof of Proposition 3.42. $\square$

3.5.2 Stabilizability with Markov Parameter Partially Known

In the previous subsection we have considered the problem of the existence of a mean square stabilizing controller for (A, B) assuming that the parameter $\theta(k)$ is available at time k. In practical situations it may not be the case and it is possible that only an estimate $\hat{\theta}(k)$ of $\theta(k)$ may be available. In this case a possible approach would be to use, for $F = (F_1, \ldots, F_N) \in \mathbb{H}^{n,m}$ that stabilizes (A, B) in the mean square sense, the controller

$$u(k) = F_{\hat{\theta}(k)} x(k)$$

that is, replace $\theta(k)$ by its available estimate $\hat{\theta}(k)$ at time k. The feedback system takes the form

$$x(k + 1) = (A_{\theta(k)} + B_{\theta(k)} F_{\hat{\theta}(k)}) x(k) \tag{3.55}$$

and a question that immediately arises is whether mean square stability will be preserved. To answer this question we have to make some assumptions about the relationship between $\theta(k)$ and $\hat{\theta}(k)$. We assume that $\{\hat{\theta}(k); k = 0, 1, \ldots\}$ is a sequence of $\tilde{\mathfrak{F}}_k$-adapted random variables taking values in $\mathbb{N}$ and such that

$$\mathcal{P}(\hat{\theta}(k) = s \mid \tilde{\mathfrak{F}}_k) = \mathcal{P}(\hat{\theta}(k) = s \mid \theta(k)) = \rho_{\theta(k)s}, s \in \mathbb{N}$$

with

$$\rho_{is} \geq 0 \text{ and } \sum_{s=1}^{N} \rho_{is} = 1 \text{ for all } i, s \in \mathbb{N}$$

where $\tilde{\mathfrak{F}}_0 = \mathfrak{F}_0$ and, for $k = 1, \ldots,$ $\tilde{\mathfrak{F}}_k$ is the σ-field generated by the random variables $\{x(0), \theta(0), x(1), \theta(1), \hat{\theta}(0), \ldots, x(k), \theta(k), \hat{\theta}(k-1)\}$. Therefore the probability that $\hat{\theta}(k) = s$, given $\tilde{\mathfrak{F}}_k$, depends only on the present value of $\theta(k)$. We define $p = \min_{i \in \mathbb{N}} \rho_{ii}$ so that

$$\mathcal{P}(\hat{\theta}(k) = i \mid \theta(k) = i) \geq p \quad \text{for } i = 1, \ldots, N,$$

and

$$P(\hat{\theta}(k) \neq i \mid \theta(k) = i) = \sum_{s=1, s \neq i}^{N} \rho_{is} = 1 - \rho_{ii} \leq 1 - p, \text{ for } i \in \mathbb{N}.$$

Set $\mathcal{T}(.)$ as in (3.7) with $\Gamma_i = (A_i + B_i F_i)$ and define

$$c_0 \triangleq \max\{\|(A_i + B_i F_s)\|^2 ; i, s \in \mathbb{N}, i \neq s, \rho_{is} \neq 0\}.$$

From the fact that F stabilizes (A, B) we have that there exists $\beta \geq 1$ and $0 < \zeta < 1$ such that (see Theorem 3.9) $\|\mathcal{T}\|_1 \leq \beta \zeta^k$, $k = 0, 1, \ldots$. The following lemma provides a condition on the lower bound p of the probability of correct estimation of $\theta(k)$ which guarantees mean square stability of (3.55).

Lemma 3.44. *Assume that F stabilizes (A, B). If $p > 1 - (1 - \zeta)/c_0\beta$, with ζ, β and c_0 as above, then (3.55) is mean square stable.*

Proof. Set $Q(k) = (Q_1(k), \ldots, Q_N(k)) \in \mathbb{H}^{n+}$ where (see (3.3d))

$$Q_i(k) = E(x(k)x(k)^* \mathbf{1}_{\{\theta(k)=i\}}).$$

We get

$$Q_j(k+1) = E\Big((A_{\theta(k)} + B_{\theta(k)} F_{\hat{\theta}(k)}) x(k) x(k)^*$$

$$\times (A_{\theta(k)} + B_{\theta(k)} F_{\hat{\theta}(k)})^* \mathbf{1}_{\{\theta(k+1)=j\}} \Big)$$

$$= \sum_{i=1}^{N} E\Big(E\big((A_{\theta(k)} + B_{\theta(k)} F_{\hat{\theta}(k)}) x(k) x(k)^*$$

$$\times (A_{\theta(k)} + B_{\theta(k)} F_{\hat{\theta}(k)})^* \mathbf{1}_{\{\theta(k)=i\}} \mathbf{1}_{\{\theta(k+1)=j\}} \mid \mathfrak{F}_k\big) \Big)$$

$$= \sum_{i=1}^{N} E\Big(P\big(\theta(k+1) = j \mid \mathfrak{F}_k\big)(A_i + B_i F_{\hat{\theta}(k)})(x(k)x(k)^* \mathbf{1}_{\{\theta(k)=i\}})$$

$$\times (A_i + B_i F_{\hat{\theta}(k)})^* \Big)$$

$$= \sum_{i=1}^{N} p_{ij} \sum_{s=1}^{N} E\Big((A_i + B_i F_s) \big(x(k)x(k)^* \mathbf{1}_{\{\theta(k)=i\}} \mathbf{1}_{\{\hat{\theta}(k)=s\}} \big) (A_i + B_i F_s)^* \Big)$$

and since $\mathbf{1}_{\{\hat{\theta}(k)=i\}} \leq 1$, it follows that

$$Q_j(k+1) \leq \sum_{i=1}^{N} p_{ij} \ (A_i + B_i F_i) Q_i(k)(A_i + B_i F_i)^*$$

$$+ \sum_{i=1}^{N} p_{ij} \sum_{s=1,s\neq i}^{N} (A_i + B_i F_s) E \left(x(k)x(k)^* \mathbf{1}_{\{\theta(k)=i\}} \mathbf{1}_{\{\hat{\theta}(k)=s\}} \right) (A_i + B_i F_s)^*$$

$$= \mathcal{T}_j(Q(k)) + \sum_{i=1}^{N} p_{ij} \sum_{s=1,s\neq i}^{N} (A_i + B_i F_s) E \Big(E \Big(x(k)x(k)^* \mathbf{1}_{\{\theta(k)=i\}}$$

$$\times \mathbf{1}_{\{\hat{\theta}(k)=s\}} \mid \tilde{\mathfrak{F}}_k \Big) \Big) (A_i + B_i F_s)^*$$

$$= \mathcal{T}_j(Q(k)) + \sum_{i=1}^{N} p_{ij} \sum_{s=1,s\neq i}^{N} (A_i + B_i F_s) E \Big(\mathcal{P}(\hat{\theta}(k) = s \mid \tilde{\mathfrak{F}}_k)$$

$$\times E(x(k)x(k)^* \mathbf{1}_{\{\theta(k)=i\}}) \Big) (A_i + B_i F_s)^*$$

$$= \mathcal{T}_j(Q(k)) + \sum_{i=1}^{N} p_{ij} \sum_{s=1,s\neq i}^{N} \rho_{is}(A_i + B_i F_s) Q_i(k)(A_i + B_i F_s)^*$$

$$= \mathcal{T}_j(Q(k)) + \mathcal{S}_j(Q(k)), \tag{3.56}$$

where $\mathcal{S} \in \mathbb{B}(\mathbb{H}^n)$ is defined for $V = (V_1,\ldots,V_N) \in \mathbb{H}^n$ as $\mathcal{S}(V) \triangleq (\mathcal{S}_1(V),\ldots,\mathcal{S}_N(V))$ with

$$\mathcal{S}_j(V) \triangleq \sum_{i=1}^{N} p_{ij} \sum_{s=1,s\neq i}^{N} \rho_{is}(A_i + B_i F_s)V_i(A_i + B_i F_s)^*.$$

Then

$$\|\mathcal{S}\|_1 = \sup_{\|V\|_1=1} \{\|\mathcal{S}(V)\|_1\}$$

$$= \sup_{\|V\|_1=1} \left\{ \sum_{j=1}^{N} \left\| \sum_{i=1}^{N} p_{ij} \sum_{s=1,s\neq i}^{N} \rho_{is}(A_i + B_i F_s)V_i(A_i + B_i F_s)^* \right\| \right\}$$

$$\leq \sup_{\|V\|_1=1} \left\{ \sum_{j=1}^{N} \sum_{i=1}^{N} p_{ij} \left\{ \sum_{s=1,s\neq i}^{N} \rho_{is} \|A_i + B_i F_s\|^2 \right\} \|V_i\| \right\}$$

$$\leq c_0 \left\{ \sup_{\|V\|_1=1} \left\{ \sum_{s-1,s\neq i}^{N} \rho_{is} \right\} \left\{ \sum_{j=1}^{N} p_{ij} \|V_i\| \right\} \right\}$$

$$\leq c_0(1-p) \sup_{\|V\|_1=1} \sum_{i=1}^{N} \|V_i\| = c_0(1-p) \sup_{\|V\|_1=1} \|V\|_1$$

$$= c_0(1-p). \tag{3.57}$$

For any $V(0) = (V_1(0),\dots,V_N(0)) \in \mathbb{H}^n$ set for $k = 0,1,\dots,$ $V(k) = (V_1(k),\dots,V_N(k)) \in \mathbb{H}^n$ as

$$V(k+1) = \mathcal{T}(V(k)) + \mathcal{S}(V(k)).$$

Then

$$V(k) = \mathcal{T}^k(V(0)) + \sum_{\kappa=0}^{k-1} \mathcal{T}^{k-1-\kappa}(\mathcal{S}(V(\kappa))),$$

so that

$$\|V(k)\|_1 \leq \|\mathcal{T}^k\|_1 \|V(0)\|_1 + \sum_{\kappa=0}^{k-1} \|\mathcal{T}^{k-1-\kappa}\|_1 \|\mathcal{S}\|_1 \|V(\kappa)\|_1$$

$$\leq \beta\{\zeta^k \|V(0)\|_1 + \sum_{\kappa=0}^{k-1} \zeta^{k-1-\kappa} \|\mathcal{S}\|_1 \|V(\kappa)\|_1\}.$$

We claim now that $\|V(k)\|_1 \leq \beta(\zeta + \beta\|\mathcal{S}\|_1)^k \|V(0)\|_1$. Indeed, set $\psi(0) = \beta\|V(0)\|_1$ and

$$\psi(k) = \zeta^k \psi(0) + \sum_{\iota=0}^{k-1} \zeta^{k-1-\iota} \beta\|\mathcal{S}\|_1 \|V(\iota)\|_1,$$

so that $\psi(k) \geq \|V(k)\|_1$ for all $k = 0,1,\dots.$ Moreover,

$$\psi(k+1) = \zeta\psi(k) + \beta\|\mathcal{S}\|_1 \|V(k)\|_1 \leq \zeta\psi(k) + \beta\|\mathcal{S}\|_1 \psi(k)$$
$$= (\zeta + \beta\|\mathcal{S}\|_1)\psi(k)$$

and thus,

$$\|V(k)\|_1 \leq \psi(k) \leq (\zeta + \beta\|\mathcal{S}\|_1)^k \psi(0) = \beta(\zeta + \beta\|\mathcal{S}\|_1)^k \|V(0)\|_1,$$

proving our claim. From the hypothesis that $p > 1 - (1-\zeta)/c_0\beta$, and bearing in mind (3.57), we get $(\zeta + \beta\|\mathcal{S}\|_1) \leq (\zeta + \beta c_0(1-p)) < 1$, and it follows that

$$\|(\mathcal{T} + \mathcal{S})^k(V(0))\|_1 = \|V(k)\|_1 \leq \beta(\zeta + \beta c_0(1-p))^k \|V(k)\|_1$$

for any $V(0) \in \mathbb{H}^n$, that is, $\|(\mathcal{T} + \mathcal{S})^k\|_1 \leq \beta(\zeta + \beta c_0(1-p))^k$ for all $k = 0,1,\dots.$ Moreover if we pick $V(0) = Q(0)$, from (3.56) and use induction on k it is easy to verify that $Q(k) \leq V(k)$ for all $k = 0,1,\dots$ so that

$$E(\|x(k)\|^2) = \mathrm{tr}\left(\sum_{i=1}^N Q_i(k)\right) \leq \mathrm{tr}\left(\sum_{i=1}^N V_i(k)\right)$$
$$\leq n\|V(k)\|_1 \leq n\beta(\zeta + \beta c_0(1-p))^k \|V(0)\|_1$$

which shows that (3.55) is MSS. $\square$

Example 3.45. Consider

$$n = 1, \ \theta(k) \in \{1, 2\}, \ A_1 = 2, \ A_2 = 4, \ B_1 = B_2 = 1,$$
$$p_{11} = p_{22} = 0.5, \ F_1 = -1.5, \ F_2 = -3.5.$$

In this case, $\mathcal{T}_1(V) = 0.5^3 V_1 + 0.5^3 V_2 = \mathcal{T}_2(V)$ so that $\beta = 1$, $\zeta = 0.5^2$ and $c_0 = \max\{(2-3.5)^2, (4-1.5)^2\} = 2.5^2$. Thus, if $p > 1-(1-0.5^2)/(2.5)^2 = 0.88$, then model (3.55) is MSS.

Therefore Lemma 3.44 says that if we have a "nice detector", then mean square stability still holds if we use a "certainty equivalent" approach. Furthermore notice that, in fact, Lemma 3.44 holds for

$$p > 1 - \max_{\zeta, \beta}(1 - \zeta)/c_0\beta.$$

3.6 Stability With Probability One

3.6.1 Main Results

In this section we return to the homogeneous version of the MJLS (3.1), reproduced here for convenience:

$$\mathcal{G} = \begin{cases} x(k+1) = \Gamma_{\theta(k)} x(k) \\ x(0) = x_0, \theta(0) = \theta_0. \end{cases} \tag{3.58}$$

Our goal is to derive sufficient conditions for almost sure convergence (ASC) of $x(k)$ to 0. Initially we show that MSS implies ASC.

Corollary 3.46. *If System (3.58) is MSS then $x(k) \to 0$ with probability one (w.p.1).*

Proof. From Theorem 3.9, $E(\|x(k)\|^2) \leq \beta\zeta^k \|x_0\|_2^2$, $k = 0, 1, \ldots$. Therefore,

$$\sum_{k=0}^{\infty} E(\|x(k)\|^2) \leq \frac{\beta}{1-\zeta} \|x_0\|_2^2$$

and the result follows from the Borel–Cantelli Lemma ([24]). □

Throughout the remainder of this section we shall assume that the Markov chain $\{\theta(k); k = 0, 1, \ldots\}$ is ergodic, as in Assumption 3.31. For ν positive integer, define the stochastic process $\tilde{\theta}_\nu(k)$, taking values in $\mathbb{N}^\nu$ (the ν-fold product space of $\mathbb{N}$) as

$$\tilde{\theta}_\nu(k) \triangleq (\theta(k\nu + \nu - 1), \ldots, \theta(k\nu)) \in \mathbb{N}^\nu, k = 0, 1, \ldots.$$

It follows that $\tilde{\theta}_\nu(k)$ is a Markov chain and for $\tilde{\iota}_\nu, \tilde{\tau}_\nu \in \mathbb{N}^\nu$, $\tilde{\iota}_\nu = (i_{\nu-1}, \ldots, i_0)$, $\tilde{\tau}_\nu = (j_{\nu-1}, \ldots, j_0)$, the transition probabilities for $\tilde{\theta}_\nu(k)$ are

$$\tilde{p}_{\tilde{\iota}_\nu \tilde{\tau}_\nu} \triangleq \mathcal{P}(\tilde{\theta}_\nu(k+1) = \tilde{\tau}_\nu \mid \tilde{\theta}_\nu(k) = \tilde{\iota}_\nu) = p_{i_{\nu-1} j_0} p_{j_0 j_1} \cdots p_{j_{\nu-2} j_{\nu-1}}.$$

From Assumption 3.31, there exists a limit probability distribution $\{\pi_i; i \in \mathbb{N}\}$ (which does not depend on the initial probability distribution $\{v_i; i \in \mathbb{N}\}$) such that $\pi_i(k) \to \pi_i$ for each $i \in \mathbb{N}$ as $k \to \infty$, and therefore, as $k \to \infty$,

$$\mathcal{P}(\tilde{\theta}_\nu(k) = \tilde{\tau}_\nu) = \mathcal{P}(\theta(k\nu) = j_0) p_{j_0 j_1} \cdots p_{j_{\nu-2} j_{\nu-1}}$$

$$\to \tilde{\pi}_{\tilde{\tau}_\nu} \triangleq \pi_{j_0} p_{j_0 j_1} \cdots p_{j_{\nu-2} j_{\nu-1}}.$$

Define for every $\tilde{\iota}_\nu \in \mathbb{N}^\nu$,

$$\tilde{\Gamma}_{\tilde{\iota}_\nu} \triangleq \Gamma_{i_{\nu-1}} \cdots \Gamma_{i_0}$$

and

$$\tilde{x}_\nu(k) \triangleq x(k\nu).$$

Thus,

$$\tilde{x}_\nu(k+1) = x((k+1)\nu) = \Gamma_{\theta(k\nu+\nu-1)} \cdots \Gamma_{\theta(k\nu)} x(k\nu) = \tilde{\Gamma}_{\tilde{\theta}_\nu(k)} \tilde{x}_\nu(k). \quad (3.59)$$

In what follows, $\|.\|$ will denote any norm in $\mathbb{C}^n$, and for $M \in \mathbb{B}(\mathbb{C}^n, \mathbb{C}^m)$, $\|M\|$ denotes the induced uniform norm in $\mathbb{B}(\mathbb{C}^n, \mathbb{C}^m)$. For instance, for $P \in \mathbb{B}(\mathbb{C}^n)$, $P > 0$, we define $\|x\|$ in Example 3.49 below as $\|x\|^2 = x^* P x$. From (3.59),

$$\|\tilde{x}_\nu(k+1)\| \le \prod_{i=0}^{k} \left\| \tilde{\Gamma}_{\tilde{\theta}_\nu(i)} \right\| \|x(0)\|. \quad (3.60)$$

Theorem 3.47. *If for some positive integer ν we have*

$$\prod_{\tilde{\iota}_\nu \in \mathbb{N}^\nu} \left\| \tilde{\Gamma}_{\tilde{\iota}_\nu} \right\|^{\tilde{\pi}_{\tilde{\iota}_\nu}} < 1, \quad (3.61)$$

then $x(k) \to 0$ as $k \to \infty$ w.p.1.

Proof. From (3.60), it is easy to see that for $x(0) \ne 0$,

$$\frac{1}{k+1} \ln \left(\frac{\|\tilde{x}_\nu(k+1)\|}{\|x(0)\|} \right) \le \frac{1}{k+1} \sum_{i=0}^{k} \ln \left(\left\| \tilde{\Gamma}_{\tilde{\theta}_\nu(i)} \right\| \right).$$

From the hypothesis that the Markov chain $\{\theta(k)\}$ is ergodic it follows that $\tilde{\theta}_\nu(k)$ is a regenerative process, and from Theorem 5.10 of [194], we have

$$\frac{1}{k+1} \sum_{i=0}^{k} \ln \left(\left\| \tilde{\Gamma}_{\tilde{\theta}_\nu(i)} \right\| \right) \longrightarrow \sum_{\tilde{\iota}_\nu \in \mathbb{N}^\nu} \ln \left(\left\| \tilde{\Gamma}_{\tilde{\iota}_\nu} \right\| \right) \tilde{\pi}_{\tilde{\iota}_\nu}$$

$$= \ln \left(\prod_{\tilde{\iota}_\nu \in \mathbb{N}^\nu} \left\| \tilde{\Gamma}_{\tilde{\iota}_\nu} \right\|^{\tilde{\pi}_{\tilde{\iota}_\nu}} \right) \quad \text{w.p.1}$$

as $k \to \infty$. If (3.61) is satisfied, then

$$\frac{\|\tilde{x}_\nu(k+1)\|}{\|x(0)\|} \longrightarrow 0 \text{ as } k \longrightarrow \infty \text{ w.p.1.}$$

Note that for any positive integer s, we can find k and $\ell \in \{1,\dots,\nu\}$ such that $s = k\nu + \ell$, and

$$x(s) = x(k\nu + \ell) = \Gamma_{\theta(k\nu+\ell-1)} \dots \Gamma_{\theta(k\nu)} x(k\nu)$$

so that

$$\|x(s)\| = \|x(k\nu + \ell)\| \leq \max\left\{\|\Gamma\|_{\max}^\nu, 1\right\} \|\tilde{x}_\nu(k)\|$$

which shows that $\|x(s)\| \to 0$ as $s \to \infty$ w.p.1. $\qquad\square$

Example 3.48. We can see that the system in Example 3.29 is always ASC. Indeed, $\Gamma_2\Gamma_1 = 0$ and for $\nu = 2$, $\tilde{\iota}_2 = (2,1)$ we have $\tilde{\pi}_{\tilde{\iota}_2} = p_2 p_1 > 0$ so that the product in (3.61) is equal to zero.

Example 3.49. By making an appropriate choice of the norm in $\mathbb{C}^n$ we can force the inequality in (3.61) to hold. Consider $n = 2, N = 2, \pi_1 = \pi_2 = \frac{1}{2}$, and

$$\Gamma_1 = \begin{bmatrix} 1 & 1 \\ 0 & 1 \end{bmatrix}, \ \Gamma_2 = \begin{bmatrix} 1/2 & 0 \\ 0 & 2/3 \end{bmatrix}.$$

Using the Euclidean norm we obtain $\|\Gamma_1\| = 1.618$, $\|\Gamma_2\| = \frac{2}{3}$, so that $\|\Gamma_1\| \|\Gamma_2\| = 1.078 > 1$. Defining

$$P = \begin{bmatrix} \frac{1}{25} & 0 \\ 0 & 1 \end{bmatrix}$$

and considering the norm $\|.\|_P$ in $\mathbb{B}(\mathbb{C}^n)$ induced by $\|x\|_P = (x^* P x)^{1/2}$ in $\mathbb{C}^n$, we obtain $\|\Gamma_1\|_P = 1.105$, $\|\Gamma_2\|_P = \frac{2}{3}$ and thus $\|\Gamma_1\|_P \|\Gamma_2\|_P = 0.737$, and the product in (3.61) with $\nu = 1$ is less than 1.

We have also the following corollary. For $P > 0$ in $\mathbb{B}(\mathbb{C}^n)$ and $A \in \mathbb{B}(\mathbb{C}^n)$, define

$$\zeta_P(A) \triangleq \min_{x \in \mathbb{C}^n, x \neq 0} \frac{x^* A^* P A x}{x^* P x}.$$

Corollary 3.50. *If for some positive integer ν and $P > 0$ in $\mathbb{B}(\mathbb{C}^n)$ we have*

$$\prod_{\tilde{\iota}_\nu \in \mathbb{N}^\nu} \zeta_P(\tilde{\Gamma}_{\tilde{\iota}_\nu})^{\tilde{\pi}_{\tilde{\iota}_\nu}} > 1$$

then $\|x(k)\| \to \infty$ as $k \to \infty$ w.p.1.

Proof. We have

$$\tilde{x}_\nu^*(k+1)P\tilde{x}_\nu(k+1) = \tilde{x}_\nu^*(k)\tilde{\Gamma}_{\hat{\theta}_\nu(k)}^* P\tilde{\Gamma}_{\hat{\theta}_\nu(k)}\tilde{x}_\nu(k)$$

$$\geq \zeta_P(\tilde{\Gamma}_{\hat{\theta}_\nu(k)})\tilde{x}_\nu^*(k)P\tilde{x}_\nu(k)$$

and thus for $x(0) \neq 0$,

$$\frac{\tilde{x}_\nu^*(k+1)P\tilde{x}_\nu(k+1)}{\tilde{x}_\nu^*(0)P\tilde{x}_\nu(0)} \geq \prod_{i=0}^{k} \zeta_P(\tilde{\Gamma}_{\hat{\theta}_\nu(i)}).$$

Repeating the same reasoning of the previous theorem, we conclude that $\|\tilde{x}_\nu(k)\| \to \infty$ as $k \to \infty$ w.p.1. Note that for $\ell \in \{1,\ldots,\nu\}$,

$$x(k\nu+\nu) = \Gamma_{\theta(k\nu+\nu-1)}\ldots\Gamma_{\theta(k\nu+\nu-\ell)}x(k\nu+\nu-\ell)$$

so that

$$\|\tilde{x}_\nu(k+1)\| \leq \|\Gamma\|_{\max}^\ell \|x(k\nu+\nu-\ell)\|,$$

which shows that $\|x(k)\| \to \infty$ as $k \to \infty$ w.p.1.   □

3.6.2 An Application of Almost Sure Convergence Results

In this subsection we deal with a problem studied in [25]. Consider a sequence of n-dimensional random vectors $u_{\theta(k)}$, where $\theta(k)$ is a Markov chain taking values in $\mathbb{N}$. $\{u_1,\ldots,u_N\}$, $u_i \in \mathbb{R}^n$, is the set of possible values of $u_{\theta(k)}$. Suppose that $u_i \neq 0$ for all $i \in \mathbb{N}$. This sequence represents the input of a system which has output $y(k)$, given by

$$y(k) = u_{\theta(k)}^* w, \ w \in \mathbb{R}^n.$$

The recursive estimator for w is in the following form:

$$w(k+1) = w(k) + \mu\frac{u_{\theta(k)}}{u_{\theta(k)}^* u_{\theta(k)}}(y(k) - \hat{y}(k))$$

$$\hat{y}(k) = u_{\theta(k)}^* w(k)$$

for $\mu \in (0,2)$. Writing $e(k) = w(k) - w$, we get

$$e(k+1) = \left(I - \mu\frac{u_{\theta(k)}u_{\theta(k)}^*}{u_{\theta(k)}^* u_{\theta(k)}}\right)e(k) = \Gamma_{\theta(k)}e(k),$$

where

$$\Gamma_i = I - \mu\frac{u_i u_i^*}{u_i^* u_i}, \quad \text{for } i \in \mathbb{N}.$$

It is easy to verify that $\|\Gamma_i\| = 1$ for $i \in \mathbb{N}$. For any positive integer κ, consider $i_p \in \mathbb{N}$, $p = 1,\ldots,\kappa$. We have the following series of lemmas which will give sufficient conditions for ASC.

Lemma 3.51. *Suppose that $\{u_{i_1}, \ldots, u_{i_\kappa}\}$ generates $\mathbb{R}^n$. Then for any $x \neq 0$ in $\mathbb{R}^n$, we can find $p \in \{1, \ldots, \kappa\}$ such that $u_{i_p}^* \Gamma_{i_{p+1}} \ldots \Gamma_{i_\kappa} x \neq 0$ ($u_{i_p}^* x \neq 0$ for $p = \kappa$).*

Proof. Clearly $\kappa \geq n$. Suppose there exists $x \neq 0$ in $\mathbb{R}^n$ such that

$$u_{i_\kappa}^* x = 0, u_{i_{\kappa-1}}^* \Gamma_{i_\kappa} x = 0, \ldots, u_{i_1}^* \Gamma_{i_2} \ldots \Gamma_{i_\kappa} x = 0.$$

Notice that

$$\Gamma_{i_\kappa} x = x - \mu \frac{u_{i_\kappa} u_{i_\kappa}^* x}{u_{i_\kappa}^* u_{i_\kappa}} = x$$

so that

$$u_{i_{\kappa-1}}^* \Gamma_{i_\kappa} x = u_{i_{\kappa-1}}^* x = 0.$$

Repeating the same arguments we have $u_{i_j}^* x = 0$ for all $j = 1, \ldots, \kappa$, which is a contradiction, since $\{u_{i_1}, \ldots, u_{i_\kappa}\}$ generates $\mathbb{R}^n$ by hypothesis. □

Lemma 3.52. *Suppose that $\{u_{i_1}, \ldots, u_{i_\kappa}\}$ generates $\mathbb{R}^n$. Then*

$$\|\Gamma_{i_1} \ldots \Gamma_{i_\kappa}\| < 1.$$

Proof. For $x \in \mathbb{R}^n$ arbitrary, $x \neq 0$, we have from Lemma 3.51 that there exists $p \in \{1, \ldots, \kappa\}$ such that $u_{i_p}^* \Gamma_{i_{p+1}} \ldots \Gamma_{i_\kappa} x \neq 0$. Consider an orthonormal basis $\{v_j; j = 1, \ldots, n\}$ of $\mathbb{R}^n$ such that $v_1 = \frac{u_{i_p}}{\|u_{i_p}\|}$, and denote $y = \Gamma_{i_{p+1}} \ldots \Gamma_{i_\kappa} x$ ($y = x$ if $p = \kappa$). Since

$$\Gamma_{i_p} y = y - \mu \frac{u_{i_p} u_{i_p}^*}{u_{i_p}^* u_{i_p}} y$$

and $v_j^* u_{i_p} = 0$ for $j = 2, \ldots, n$, we have that

$$v_j^* \Gamma_{i_p} y = v_j^* \left(y - \mu \frac{u_{i_p} u_{i_p}^*}{u_{i_p}^* u_{i_p}} y \right) = v_j^* y$$

and

$$v_1^* \Gamma_{i_p} y = \frac{u_{i_p}^*}{\|u_{i_p}\|} \left(y - \mu \frac{u_{i_p} u_{i_p}^*}{u_{i_p}^* u_{i_p}} y \right) = \frac{(1 - \mu) u_{i_p}^* y}{\|u_{i_p}\|}.$$

Then

$$\|\Gamma_{i_p} y\|^2 = \sum_{j=1}^n (v_j^* \Gamma_{i_p} y)^2$$

$$= \left(\frac{u_{i_p}^* y (1 - \mu)}{\|u_{i_p}\|} \right)^2 + \sum_{j=2}^n (v_j^* y)^2$$

$$< \left(\frac{u_{i_p}^* y}{\|u_{i_p}\|} \right)^2 + \sum_{j=2}^n (v_j^* y)^2 = \|y\|^2, \tag{3.62}$$

since $u_{i_p}^* y \neq 0$ and $\mu \in (0,2)$. From (3.62) and recalling that $\|\Gamma_i\| = 1$, for $i \in \mathbb{N}$, we have

$$
\begin{aligned}
\|\Gamma_{i_1} \ldots \Gamma_{i_{p-1}} \Gamma_{i_p} \Gamma_{i_{p+1}} \ldots \Gamma_{i_\kappa} x\| &\leq \prod_{j=1}^{p-1} \|\Gamma_{i_j}\| \, \|\Gamma_{i_p} y\| \\
&= \|\Gamma_{i_p} y\| < \|y\| \\
&= \|\Gamma_{i_{p+1}} \ldots \Gamma_{i_\kappa} x\| \\
&\leq \prod_{j=p+1}^{\kappa} \|\Gamma_{i_j}\| \, \|x\| \\
&= \|x\|
\end{aligned}
$$

and we conclude that $\|\Gamma_{i_1} \ldots \Gamma_{i_\kappa} x\| < \|x\|$ for all $x \neq 0$ in $\mathbb{R}^n$. Recalling that the maximum of a continuous function in a compact set is reached at some point of this set, we get

$$
\|\Gamma_{i_1} \ldots \Gamma_{i_\kappa}\| = \max_{\|x\|=1, x \in \mathbb{R}^n} (\|\Gamma_{i_1} \ldots \Gamma_{i_\kappa} x\|) < 1.
$$

$\square$

Finally the following lemma presents sufficient conditions for ASC.

Lemma 3.53. *If $\{u_1, \ldots, u_N\}$ generates $\mathbb{R}^n$ then $\|e(k)\| \to 0$ as $k \to \infty$ w.p.1.*

Proof. Consider $\{i_1, \ldots, i_n\}$ such that $\{u_{i_1}, \ldots, u_{i_n}\}$ generates $\mathbb{R}^n$. From the hypothesis that the Markov chain $\{\theta(k)\}$ is ergodic (and thus irreducible), we can find for each i_j, $j = 1, \ldots, n-1$, a positive integer ν_j and $i_j^\ell \in \mathbb{N}$, $\ell = 1, \ldots, \nu_j$ such that $i_j^1 = j$, $i_j^{\nu_j+1} = i_{j+1}$ and

$$
p_{i_j i_j^2} p_{i_j^2 i_j^3} \ldots p_{i_j^{\nu_j} i_{j+1}} > 0.
$$

Clearly the set $\{u_{i_1}, \ldots, u_{i_1^{\nu_1}}, \ldots, u_{i_{n-1}}, \ldots, u_{i_{n-1}^{\nu_{n-1}}}, u_n\}$ generates $\mathbb{R}^n$ so that from Lemma 3.52,

$$
\|\Gamma_{i_1} \ldots \Gamma_{i_1^{\nu_1}} \ldots \Gamma_{i_{n-1}} \ldots \Gamma_{i_{n-1}^{\nu_{n-1}}} \Gamma_n\| < 1. \tag{3.63}
$$

Moreover,

$$
\pi_{i_1} p_{i_1 i_1^2} \ldots p_{i_1^{\nu_1} i_2} \ldots p_{i_{n-1} i_{n-1}^2} \ldots p_{i_{n-1}^{\nu_{n-1}} i_n} > 0. \tag{3.64}
$$

Defining $\nu \triangleq \nu_1 + \ldots + \nu_{n-1} + 1$ and recalling that $\|\Gamma_{\tilde{\iota}_\nu}\| \leq 1$ for any $\tilde{\iota}_\nu \in \mathbb{N}^\nu$, we obtain from (3.63) and (3.64) that (3.61) is satisfied.

$\square$

3.7 Historical Remarks

Stability is certainly one of the fundamental themes of control theory. Since the influential works of eminent mathematicians such as C. Hermite, J.E. Routh, A.M. Lyapunov and A. Hurwitz, which have laid the foundation of stability theory in the twentieth century, stabilty stands firmly as a topic of intense research, despite the notable abundance of relevant works and reference materials on this subject. For the stochastic scenario, the interested reader is referred, for instance, to the classical [139] and [159] for an authoritative account (see also [154]). In particular, stability questions for linear systems with *random parameters* have been treated by several authors over the last few years. As a small sample of some of these results, we can mention [18], [20], [25], [29], [35], [36], [41], [54], [55], [56], [61], [66], [67], [93], [107], [108], [109], [118], [119], [120], [121], [122], [143], [145], [146], [147], [160], [161], [166], [169], [170], [171], [173], [176], [177], [178], [191], [215], [219], and references therein. These include the case in which the random mechanism is not necessarily modeled by a Markov chain (see also [163]).

Regarding MJLS, it is perhaps noteworthy here that the *assumption* mentioned in Section 1.6, which was used in [219] to treat the infinite horizon case, was seen as an inconvenient technical hypothesis. To some extent, it indicated that the notion used there for stabilizability (observability and detectability) was not fully adequate. Although it seemed logical by that time to consider the MJLS class as a "natural" extension of the linear class and use as stability criteria the stability for each operation mode of the system, it came up that MJLS carry a great deal of subtleties which distinguish them from their linear counterparts and provide us with a very rich structure. The examples exhibited here illustrate how sometimes the switching between operation modes can surprise us and run counter to our intuition. Furthermore, it portrays how a balance between the modes and the transition probability is essential in stability issues for MJLS. These nuances came through with a different point of view regarding stability for MJLS. Instead of mimicking the structural criteria of stability from the classical linear system theory, the idea was to devise structural criteria right from stability concepts based on the state of the system as, for instance, mean square stability.

The desideratum to clear up adequately stability questions for MJLS (discrete and continuous-time) has given rise to a number of papers on this subject. It was, initially, the introduction of the concept of *mean square stability* (stochastic stability, or ℓ_2-stability), mean square stabilizability (stochastic stabilizability) and mean square detectability (stochastic stabilizability) that has carved out an appropriate framework to study stability for MJLS. We mention here, for instance, [66], [109], [122], [143], [145], [146], and [147], as key works in the unfolding of mean square stability theory for MJLS, which has by now a fairly complete body of results. Most of the results in [66] are described in this book (the continuous-time version is in [120]). For the case in which the state space of the Markov chain is *countably infinite* it is shown in [67] and [119] that mean square and stochastic stability (ℓ_2-stability) are no longer equivalent (see also [118]). A new fresh look at detectability questions for discrete-time MJLS was given recently in [54] and [56] (see [57] for the infinite countable case and [55] for the continuous-time case).

Almost sure stability for MJLS is examined, for instance, in [66], [93], [107], [108], [161] and [169]. A historical account of earlier works can be found in [107]. This is a topic which certainly deserves further study. Regarding other issues such as robust stability (including the case with delay) the readers are referred, for instance, to [20], [35], [36], [61], [166], [191] and [215].

4

Optimal Control

This chapter is devoted to the study of the quadratic optimal control problems of Markov Jump Linear Systems (MJLS) for the case in which the jump variable $\theta(k)$ as well as the state variable $x(k)$ are available to the controller. Both the finite horizon and the infinite horizon cases are considered, and the optimal controllers derived from a set of coupled difference Riccati equations for the former problem, and the mean square stabilizing solution of a set of coupled algebraic Riccati equations (CARE) for the latter problem.

4.1 Outline of the Chapter

The objective of this chapter is to study the quadratic optimal control and some other related problems for MJLS. We will assume throughout this chapter that the jump variable $\theta(k)$ and the state variable $x(k)$ are available to the controller. The case in which the state variable is not available, that is, the case with partial information, will be considered in Chapter 6. We start by analyzing in Section 4.2 the quadratic optimal control problem. For the finite horizon cost criteria, we show that the solution of the problem can be derived from a set of coupled Riccati difference equations. For the infinite horizon quadratic cost case, we present in Subsection 4.3.1 the different cost functionals and additive stochastic inputs that will be considered throughout the chapter. The case with no additive stochastic inputs, which we shall term here the Markov jump linear quadratic regulator problem, is solved in Subsection 4.3.2 through the mean square stabilizing solution of a set of coupled algebraic Riccati equations (CARE). In fact all infinite horizon quadratic control problems presented in this chapter will rely on the mean square stabilizing solution of the CARE.

Subsection 4.3.3 and Sections 4.4 and 4.5 consider infinite horizon quadratic problems for a MJLS subject to additive stochastic inputs. In Subsection 4.3.3 we consider an additive wide sense white noise sequence input with a long run average cost function, a problem which is very much similar to the linear regulator problem considered in Subsection 4.3.2. The same stochastic input is considered in Section 4.4, but with the cost function being the H_2-norm of the system. This problem is called the H_2-control problem and some of its aspects, especially the connection with the CARE and a certain convex programming problem, are addressed in Section 4.4. We generalize in Section 4.5 the linear quadratic regulator problem presented in Subsection 4.3.2 by considering a MJLS subject to a ℓ_2-stochastic input $w = \{w(k); k = 0, 1, \ldots\} \in \mathcal{C}^r$. In this case it is shown that the optimal control law at time k has a feedback term given by the stabilizing solution of the CARE as well as a term characterized by the projection of the ℓ_2-stochastic inputs and Markov chain into the filtration at time k. We illustrate the application of these results through an example for the optimal control of a manufacturing system subject to random breakdowns. We analyze the simple case of just one item in production, although the technique could be easily extended to the more complex case of several items being manufactured. Finally we mention that when restricted to the case with no jumps, all the results presented here coincide with those known in the general literature (see, for instance, [151], [204]).

As aforementioned, all the quadratic optimal control problems presented in this chapter are related to a set of coupled algebraic (difference for the finite horizon problem) Riccati equations. These equations are analyzed in Appendix A where, among other results, necessary and sufficient conditions for existence and uniqueness of the mean square stabilizing solution for the CARE are provided. These results can be seen as a generalization of those analyzed in [67], [143], [144], [146], [147], [176], [177] and [178] using different approaches.

4.2 The Finite Horizon Quadratic Optimal Control Problem

4.2.1 Problem Statement

We consider in this section the finite horizon quadratic optimal control problem for MJLS when the state variable $x(k)$ and jump variable $\theta(k)$ are available to the controller. The case in which the state variable is not available, that is, the case with partial information, will be presented in Chapter 6, and will rely on the results presented in this chapter. On the stochastic basis $(\Omega, \mathfrak{F}, \{\mathfrak{F}_k\}, \mathcal{P})$ with $\theta(0)$ having initial distribution $\{v(i); i \in \mathbb{N}\}$, consider the following MJLS $\mathcal{G}$:

$$\mathcal{G} = \begin{cases} x(k+1) = A_{\theta(k)}(k)x(k) + B_{\theta(k)}(k)u(k) + M_{\theta(k)}(k)\nu(k) \\ z(k) = C_{\theta(k)}(k)x(k) + D_{\theta(k)}(k)u(k) \\ x(0) = x_0 \in \mathcal{C}_0^n, \theta(0) = \theta_0 \in \Theta_0 \end{cases} \qquad (4.1)$$

where $A(k) = (A_1(k), \ldots, A_N(k)) \in \mathbb{H}^n$, $B(k) = (B_1(k), \ldots, B_N(k)) \in \mathbb{H}^{m,n}$, $M(k) = (M_1(k), \ldots, M_N(k)) \in \mathbb{H}^{r,n}$, $C(k) = (C_1(k), \ldots, C_N(k)) \in \mathbb{H}^{n,q}$, $D(k) = (D_1(k), \ldots, D_N(k)) \in \mathbb{H}^{m,q}$, and $\nu = \{\nu(k); k \in \{0, \ldots, T-1\}\}$ represents a noise sequence satisfying

$$E(\nu(k)\nu(k)^*\mathbf{1}_{\{\theta(k)=i\}}) = \Xi_i(k), \quad E(\nu(0)x(0)^*\mathbf{1}_{\{\theta(0)=i\}}) = 0 \tag{4.2}$$

for $k \in \{0, \ldots, T-1\}$ and all $i \in \mathbb{N}$ (recall the definition of the indicator function $\mathbf{1}_{\{\theta(k)=i\}}$ in (3.2)).

In this section we assume that the Markov chain $\{\theta(k)\}$ has time-varying transition probabilities $p_{ij}(k)$. The operator $\mathcal{E}$ defined in (3.6) for the time-invariant case, is redefined here as follows: for $V = (V_1, \ldots, V_N) \in \mathbb{H}^n$, we set $\mathcal{E}(V, k) = (\mathcal{E}_1(V, k), \ldots, \mathcal{E}_N(V, k))$ as

$$\mathcal{E}_i(V, k) = \sum_{j=1}^{N} p_{ij}(k)V_j. \tag{4.3}$$

We assume that the state variable and operation modes ($x(k)$ and $\theta(k)$ respectively) are known at each time k, and as in (3.3),

$$E(x(0)x(0)^*\mathbf{1}_{\{\theta(k)=i\}}) = Q_i(0) \tag{4.4}$$

for each $i \in \mathbb{N}$. We also assume that

$$C_i(k)^* D_i(k) = 0 \tag{4.5}$$
$$D_i(k)^* D_i(k) > 0 \tag{4.6}$$

for all $i \in \mathbb{N}, k \in \{0, \ldots, T-1\}$. Notice that (4.6) means that all components of the control variable will be penalized in the cost function. As pointed out in Remark 4.1 below, there is no loss of generality in assuming (4.5).

We assume that for any measurable functions f and g,

$$E(f(\nu(k))g(\theta(k+1))|\mathfrak{G}_k) = E(f(\nu(k))|\mathfrak{G}_k) \sum_{j=1}^{N} p_{\theta(k)j}(k)g(j) \tag{4.7}$$

where $\mathfrak{G}_k$ is the σ-field generated by the random variables $\{x(t), \theta(t); t = 0, \ldots, k\}$, so that $\mathfrak{G}_k \subset \mathfrak{G}_{k+1}$ and $\mathfrak{G}_k \subset \mathfrak{F}_k \subset \mathfrak{F}$ for each $k = 0, 1, \ldots$.

The set of admissible controllers, denoted by $\mathcal{U}_T$, is given by the sequence of control laws $u = (u(0), \ldots, u(T-1))$ such that for each k, $u(k)$ is $\mathfrak{G}_k$-measurable,

$$E(\nu(k)x(k)^*\mathbf{1}_{\{\theta(k)=i\}}) = 0, \tag{4.8}$$

and

$$E(\nu(k)u(k)^*\mathbf{1}_{\{\theta(k)=i\}}) = 0. \tag{4.9}$$

Notice that we don't assume here the usual wide sense white noise sequence assumption for $\{\nu(k); k = 0, \ldots, T\}$ since it will not be suitable for the partial observation case, to be analyzed in Chapter 6. What will be in fact required for the partial observation case, and also suitable for the complete observation case, are conditions (4.7), (4.8) and (4.9).

The quadratic cost associated to system $\mathcal{G}$ with an admissible control law $u = (u(0), \ldots, u(T-1))$ and initial conditions (x_0, θ_0) is denoted by $\mathfrak{J}(\theta_0, x_0, u)$, and is given by

$$\mathfrak{J}(\theta_0, x_0, u) \triangleq \sum_{k=0}^{T-1} E(\|z(k)\|^2) + E(x(T)^* V_{\theta(T)} x(T)) \tag{4.10}$$

where $\mathcal{V} = (\mathcal{V}_1, \ldots, \mathcal{V}_N) \in \mathbb{H}^{n+}$. Therefore, the finite horizon optimal quadratic control problem we want to study is: find $u \in \mathcal{U}_T$ that produces minimal cost $\mathfrak{J}(\theta_0, x_0, u)$. This minimal is denoted by $\tilde{\mathfrak{J}}(\theta_0, x_0)$.

Remark 4.1. There is no loss of generality in considering $C_i(k)^* D_i(k) = 0$, $i \in \mathbb{N}$ as in (4.5). Indeed, if this is not the case, setting

$$u(k) = -(D_{\theta(k)}(k)^* D_{\theta(k)}(k))^{-1} D_{\theta(k)}(k)^* C_{\theta(k)}(k) x(k) + v(k)$$

we would get

$$\begin{aligned}
\|z(k)\|^2 =& \|C_{\theta(k)}(k) x(k) + D_{\theta(k)}(k) u(k)\|^2 \\
=& \| \left(I - D_{\theta(k)}(k)(D_{\theta(k)}(k)^* D_{\theta(k)}(k))^{-1} D_{\theta(k)}(k)^*\right) C_{\theta(k)}(k) x(k) \\
& + D_{\theta(k)}(k) v(k)\|^2 \\
=& \| \left(I - D_{\theta(k)}(k)(D_{\theta(k)}(k)^* D_{\theta(k)}(k))^{-1} D_{\theta(k)}(k)^*\right) C_{\theta(k)}(k) x(k)\|^2 \\
& + \|D_{\theta(k)}(k) v(k)\|^2
\end{aligned}$$

and thus, by redefining $A_i(k)$ and $C_i(k)$ as

$$A_i(k) - B_i(k)(D_i(k)^* D_i(k))^{-1} D_i(k)^* C_i(k) \tag{4.11}$$

and

$$(I - D_i(k)(D_i(k)^* D_i(k))^{-1} D_i(k)^*) C_i(k) \tag{4.12}$$

respectively we would get that (4.5) would be satisfied.

4.2.2 The Optimal Control Law

For the case in which the state variable $x(k)$ is available to the controller the solution of the quadratic optimal control problem has already been solved in the literature (see for instance, [115]). In this section we will adapt the results presented in [115] that will be useful for the partial observation case.

For a fixed integer $\kappa \in \{0, \ldots, T-1\}, \kappa \neq T$ we consider, bearing in mind the classical principle of optimality, the intermediate problem of minimizing

$$\mathfrak{J}(\theta(\kappa), x(\kappa), \kappa, u_\kappa) \triangleq \sum_{k=\kappa}^{T-1} E(\|z(k)\|^2 | \mathfrak{G}_\kappa) + E(x(T)^* \mathcal{V}_{\theta(T)} x(T) | \mathfrak{G}_\kappa) \quad (4.13)$$

where the control law $u_\kappa = (u(\kappa), \ldots, u(T-1))$ and respective state sequence $(x(\kappa), \ldots, x(T))$ are such that, for each $\kappa \leq k \leq T-1$, (4.8) and (4.9) hold, and $u(k)$ is $\mathfrak{G}_k$-measurable. We denote this set of admissible controllers by $\mathcal{U}_T(\kappa)$, and the optimal cost, usually known in the literature as the value function, by $\bar{\mathfrak{J}}(\theta_\kappa, x_\kappa, \kappa)$. The intermediate problem is thus to optimize the performance of the system over the last $T - \kappa$ stages starting at the point $x(\kappa) = x$ and mode $\theta(\kappa) = i$. As in the stochastic linear regulator problem (see, for instance, [80]) we apply dynamic programming to obtain, by induction on the intermediate problems, the solution of minimizing $\mathfrak{J}(\theta_0, x_0, u)$. For $i \in \mathbb{N}$ and $k = T-1, \ldots, 0$, define the following recursive coupled Riccati difference equations $X(k) = (X_1(k), \ldots, X_N(k)) \in \mathbb{H}^{n+}$,

$$X_i(k) \triangleq A_i(k)^* \mathcal{E}_i(X(k+1), k) A_i(k) - A_i(k)^* \mathcal{E}_i(X(k+1), k) B_i(k)$$
$$\times \left(D_i(k)^* D_i(k) + B_i(k)^* \mathcal{E}_i(X(k+1), k) B_i(k) \right)^{-1}$$
$$\times B_i(k)^* \mathcal{E}_i(X(k+1), k) A_i(k) + C_i(k)^* C_i(k) \quad (4.14)$$

where $X_i(T) \triangleq \mathcal{V}_i$. Set also

$$R_i(k) \triangleq D_i(k)^* D_i(k) + B_i(k)^* \mathcal{E}_i(X(k+1), k) B_i(k) > 0$$

and

$$F_i(k) \triangleq -R_i(k)^{-1} B_i(k)^* \mathcal{E}_i(X(k+1), k) A_i(k). \quad (4.15)$$

Define now for $i \in \mathbb{N}$ and $k = T-1, \ldots, 0$

$$W(i, x, k) \triangleq x^* X_i(k) x + \alpha(k),$$

where

$$\alpha(k) \triangleq \sum_{t=k}^{T-1} \delta(t), \quad \alpha(T) \triangleq 0$$

$$\delta(t) \triangleq \sum_{i=1}^{N} \text{tr}(M_i(t) \Xi_i(t) M_i(t)^* \mathcal{E}_i(X(t+1), t)).$$

The following theorem, which is based on the results in [115] taking into account (4.2), (4.7), (4.8), (4.9), and bearing in mind the Markov property for $\{\theta(k)\}$, shows that $\bar{\mathfrak{J}}(i, x, k) = W(i, x, k)$.

Theorem 4.2. *For the intermediate stochastic control problems described in (4.13) the optimal control law* $\bar{u}_\kappa = (\bar{u}(\kappa), \ldots, \bar{u}(T-1))$ *is given by*

$$\bar{u}(k) = F_{\theta(k)}(k)x(k) \qquad (4.16)$$

and the value function $\bar{\mathfrak{J}}(i, x, k) = W(i, x, k)$. In particular we have that the optimal control law $\bar{u}$ for the problem described in (4.1) and (4.10) is $\bar{u} = \bar{u}_0 = (\bar{u}(0), \dots, \bar{u}(T-1))$ with optimal cost

$$\bar{\mathfrak{J}}(\theta_0, x_0) = E(\bar{\mathfrak{J}}(\theta(0), x(0), 0)) = E(W(\theta(0), x(0), 0)),$$

that is,

$$\bar{\mathfrak{J}}(\theta_0, x_0) = \sum_{i=1}^{N}\left[\mathrm{tr}(\pi_i(0)Q_i(0)X_i(0)) \right.$$

$$\left. + \sum_{k=0}^{T-1} \pi_i(k)\, \mathrm{tr}(M_i(k)\Xi_i(k)M_i(k)^*\mathcal{E}_i(X(k+1), k)) \right]. \qquad (4.17)$$

Proof. Notice that for any $u_\kappa \in \mathcal{U}_T(\kappa)$ we have from the Markov property, (4.7), (4.14) and (4.15) that

$$E(W(\theta(k+1), x(k+1), k+1)|\mathfrak{G}_k) - W(\theta(k), x(k), k) =$$
$$- \|z(k)\|^2 + \|R_{\theta(k)}^{1/2}(k)(u(k) - F_{\theta k}(k)x(k))\|^2 - \delta(k)$$
$$+ \mathrm{tr}(M_{\theta(k)}(k)E(\nu(k)\nu(k)^*|\mathfrak{G}_k)M_{\theta(k)}(k)^*\mathcal{E}_{\theta(k)}(X(k+1), k))$$
$$+ 2\,\mathrm{tr}(M_{\theta(k)}(k)E(\nu(k)x(k)^*|\mathfrak{G}_k)A_{\theta(k)}(k)^*\mathcal{E}_{\theta(k)}(X(k+1), k))$$
$$+ 2\,\mathrm{tr}(M_{\theta(k)}(k)E(\nu(k)u(k)^*|\mathfrak{G}_k)B_{\theta(k)}(k)^*\mathcal{E}_{\theta(k)}(X(k+1), k)). \qquad (4.18)$$

From (4.2),

$$E(\mathrm{tr}(M_{\theta(k)}(k)E(\nu(k)\nu(k)^*|\mathfrak{G}_k)M_{\theta(k)}(k)^*\mathcal{E}_{\theta(k)}(X(k+1), k)))$$
$$= E(\mathrm{tr}(M_{\theta(k)}(k)\nu(k)\nu(k)^*M_{\theta(k)}(k)^*\mathcal{E}_{\theta(k)}(X(k+1), k)))$$
$$= \sum_{i=1}^{N} \mathrm{tr}(M_i(k)E(\nu(k)\nu(k)^*\mathbf{1}_{\{\theta(k)=i\}})M_i(k)^*\mathcal{E}_i(X(k+1), k))$$
$$= \sum_{i=1}^{N} \mathrm{tr}(M_i(k)\Xi_i(k)M_i(k)^*\mathcal{E}_i(X(k+1), k))$$
$$= \delta(k) \qquad (4.19)$$

and similarly from (4.8)

$$E(\mathrm{tr}(M_{\theta(k)}(k)E(\nu(k)x(k)^*|\mathfrak{G}_k)A_{\theta(k)}(k)^*\mathcal{E}_{\theta(k)}(X(k+1), k)))$$
$$= E(\mathrm{tr}(M_{\theta(k)}(k)\nu(k)x(k)^*A_{\theta(k)}(k)^*\mathcal{E}_{\theta(k)}(X(k+1), k)))$$
$$= \sum_{i=1}^{N} \mathrm{tr}(M_i(k)E(\nu(k)x(k)^*\mathbf{1}_{\{\theta(k)=i\}})A_i(k)^*\mathcal{E}_i(X(k+1), k))$$
$$= 0. \qquad (4.20)$$

The same arguments using now (4.9) yield

$$E(\text{tr}(M_{\theta(k)}(k)E(\nu(k)u(k)^*|\mathfrak{G}_k)B_{\theta(k)}(k)^*\mathcal{E}_{\theta(k)}(X(k+1),k)))$$
$$= E(\text{tr}(M_{\theta(k)}(k)\nu(k)u(k)^*B_{\theta(k)}(k)^*\mathcal{E}_{\theta(k)}(X(k+1),k)))$$
$$= \sum_{i=1}^{N} \text{tr}(M_i(k)E(\nu(k)u(k)^*\mathbf{1}_{\{\theta(k)=i\}})A_i(k)^*\mathcal{E}_i(X(k+1),k))$$
$$= 0. \tag{4.21}$$

From (4.18)–(4.21) we have that

$$E(W(\theta(k+1),x(k+1),k+1)|\mathfrak{G}_k) - W(\theta(k),x(k),k)$$
$$= -\|z(k)\|^2 + \|R_{\theta(k)}^{1/2}(k)(u(k) - F_{\theta k}(k)x(k))\|^2.$$

Taking the sum from $k = \kappa$ to $T - 1$, and recalling that $W(\theta(T),x(T),T) = x(T)^*\mathcal{V}_{\theta(T)}x(T)$ we get that

$$E(x(T)^*\mathcal{V}_{\theta(T)}x(T)|\mathfrak{G}_\kappa) - W(\theta(\kappa),x(\kappa),\kappa)$$
$$= E(W(\theta(T),x(T),T)|\mathfrak{G}_\kappa) - W(\theta(\kappa),x(\kappa),\kappa)$$
$$= \sum_{k=\kappa}^{T-1} E\left(E(W(\theta(k+1),x(k+1),k+1)|\mathfrak{G}_k) - W(\theta(k),x(k),k)|\mathfrak{G}_\kappa\right)$$
$$= \sum_{k=\kappa}^{T-1} E(-\|z(k)\|^2 + \|R_{\theta(k)}^{1/2}(k)(u(k) - F_{\theta k}(k)x(k))\|^2|\mathfrak{G}_\kappa)$$

that is,

$$\mathfrak{J}(\theta(\kappa),x(\kappa),\kappa,u_\kappa) = E(\sum_{k=\kappa}^{T-1} \|z(k)\|^2 + x(T)^*\mathcal{V}_{\theta(T)}x(T)|\mathfrak{G}_\kappa)$$
$$= W(\theta(\kappa),x(\kappa),\kappa)+$$
$$E(\sum_{k=\kappa}^{T-1} \|R_{\theta(k)}^{1/2}(k)(u(k) - F_{\theta k}(k)x(k))\|^2|\mathfrak{G}_\kappa). \tag{4.22}$$

Taking the minimum over $u_\kappa \in \mathcal{U}_T(\kappa)$ in (4.22) we obtain that the optimal control law is given by (4.16), so that the second term on the right hand side of (4.22) equals to zero and $\bar{\mathfrak{J}}(\theta_\kappa,x_\kappa,\kappa) = \mathfrak{J}(\theta(\kappa),x(\kappa),\kappa,\bar{u}_\kappa) = W(\theta(\kappa),x(\kappa),\kappa)$. In particular for $\kappa = 0$ we obtain that

$$\bar{\mathfrak{J}}(\theta_0,x_0) = E(x(0)^*X_{\theta(0)}(0)x(0)) + \alpha(0)$$

which, after some manipulation, leads to (4.17), proving the desired result.   □

Remark 4.3. For the case in which $A_i, B_i, C_i, D_i, p_{ij}$ in (4.1) are time-invariant the control coupled Riccati difference equations (4.14) lead to the following control coupled algebraic Riccati equations (CARE)

$$X_i = A_i^* \mathcal{E}_i(X) A_i - A_i^* \mathcal{E}_i(X) B_i (D_i^* D_i + B_i^* \mathcal{E}_i(X) B_i)^{-1} B_i^* \mathcal{E}_i(X) A_i$$
$$+ C_i^* C_i, \tag{4.23}$$

and respective gains:

$$\mathcal{F}_i(X) \triangleq -(D_i^* D_i + B_i^* \mathcal{E}_i(X) B_i)^{-1} B_i^* \mathcal{E}_i(X) A_i. \tag{4.24}$$

It is natural to ask under which condition there will be a convergence of $X(k)$ derived from the control coupled Riccati difference equations (4.14) to a solution X of the CARE (4.23). In Appendix A we present sufficient conditions for the existence of a unique solution $X = (X_1, \ldots, X_N) \in \mathbb{H}^{n+}$ for (4.23), and convergence of $X_i(k)$ to X_i. Thus a time-invariant approximation for the optimal control problem would be to replace $F_i(k)$ in (4.16) by $\mathcal{F}_i(X)$ as in (4.24), thus just requiring to keep in memory the gains $\mathcal{F}(X) = (\mathcal{F}_1(X), \ldots, \mathcal{F}_N(X))$.

4.3 Infinite Horizon Quadratic Optimal Control Problems

4.3.1 Definition of the Problems

For the infinite horizon quadratic optimal control problems to be analyzed in this chapter we consider a time-invariant version of the model (4.1), given by

$$\mathcal{G} = \begin{cases} x(k+1) = A_{\theta(k)} x(k) + B_{\theta(k)} u(k) + G_{\theta(k)} w(k) \\ z(k) = C_{\theta(k)} x(k) + D_{\theta(k)} u(k) \\ x(0) = x_0 \in \mathcal{C}_0^n, \theta(0) = \theta_0 \in \Theta_0 \end{cases} \tag{4.25}$$

with $A = (A_1, \ldots, A_N) \in \mathbb{H}^n$, $B = (B_1, \ldots, B_N) \in \mathbb{H}^{m,n}$, $G = (G_1, \ldots, G_N) \in \mathbb{H}^{r,n}$, $C = (C_1, \ldots, C_N) \in \mathbb{H}^{n,q}$, $D = (D_1, \ldots, D_N) \in \mathbb{H}^{m,q}$. Here $\mathbb{T} = \{0, 1, \ldots, \}$, and $w = \{w(k); k \in \mathbb{T}\}$ represents a noise sequence to be specified for the different problems that will be considered. The initial condition (x_0, θ_0) satisfies (4.4).

Matrices $C_{\theta(k)}$ and $D_{\theta(k)}$ are used to assign different weights to the influence of the components of vectors $x(k)$ and $u(k)$ in the cost functions. As in (4.5) we assume without loss of generality (see Remark 4.1) that

$$C_i^* D_i = 0 \tag{4.26}$$

and, to penalize all control components in the quadratic cost as in (4.6),

$$D_i^* D_i > 0 \tag{4.27}$$

for all $i \in \mathbb{N}$. The control variable $u = (u(0), \dots)$ is such that $u(k)$ is $\mathfrak{G}_k$-measurable for each $k = 0, 1, \dots$ and

$$\lim_{k \to \infty} E(\|x(k)\|^2) = 0 \tag{4.28}$$

whenever $w(k) = 0$ for all $k = 0, 1, \dots$. We shall denote by $\mathcal{U}$ the set of u satisfying these properties.

For the case in which there is no additive input noise, that is, $w(k) = 0$ for all $k = 0, 1, \dots$, we will be interested in the problem of minimizing the infinite horizon cost function given by

$$\mathfrak{J}(\theta_0, x_0, u) \triangleq \sum_{k=0}^{\infty} E(\|z(k)\|^2) \tag{4.29}$$

subject to (4.25) and $u \in \mathcal{U}$. The optimal cost is defined as

$$\bar{\mathfrak{J}}(\theta_0, x_0) \triangleq \inf_{u \in \mathcal{U}} \mathfrak{J}(\theta_0, x_0, u). \tag{4.30}$$

This is the Markov jump version of the well known LQR (linear quadratic regulator), a classical control problem, discussed in Chapter 2. The cost functional (4.29) will also be considered in Section 4.5 for the more general case in which the state equation includes an additive ℓ_2-stochastic input $w = \{w(k); k \in \mathbb{T}\} \in \mathcal{C}^r$. These results will be used in the min-max problem associated to the H_∞-control problem, to be presented in Chapter 7.

For the case in which the state equation includes an additive wide sense white noise sequence (that is, when $w = \{w(k); k \in \mathbb{T}\}$ is such that $E(w(k)) = 0$, $E(w(k)w(k)^*) = I$, $E(w(k)w(l)^*) = 0$) independent of the Markov chain $\{\theta(k)\}$ and initial state value x_0, a cost function as in (4.29) might lead to an unbounded value. So in this case two cost functions will be considered. The first one is the long run average cost function $\mathfrak{J}_{av}(\theta_0, x_0, u)$ defined for arbitrary $u \in \mathcal{U}$ as

$$\mathfrak{J}_{av}(\theta_0, x_0, u) \triangleq \limsup_{t \to \infty} \frac{1}{t} \left\{ \sum_{k=0}^{t-1} E\left(\|z(k)\|^2\right) \right\}, \tag{4.31}$$

and the optimal cost is defined as

$$\bar{\mathfrak{J}}_{av}(\theta_0, x_0) \triangleq \inf_{u \in \mathcal{U}} \mathfrak{J}_{av}(\theta_0, x_0, u). \tag{4.32}$$

The second one, which is somewhat equivalent to the long run average cost criterion, is the H_2-norm of system $\mathcal{G}$. For this criterion we consider control laws in the form of a state feedback $u = F_{\theta(k)} x(k)$. To make sure that (4.28) is satisfied, we define (see Definition 3.40)

$$\mathbb{F} \triangleq \{F = (F_1, \dots, F_N) \in \mathbb{H}^{m,n};$$
$$F \text{ stabilizes } (A, B) \text{ in the mean square sense}\}$$

which contains all controllers that stabilize system $\mathcal{G}$ in the mean square sense. When the input u applied to system $\mathcal{G}$ is given by $u(k) = F_{\theta(k)}x(k)$, we will denote it by $\mathcal{G}_F$, and define the objective function as

$$\|\mathcal{G}_F\|_2^2 \triangleq \lim_{k \to \infty} E(\|z(k)\|^2). \tag{4.33}$$

The optimal cost is defined as

$$\|\bar{\mathcal{G}}\|_2^2 \triangleq \inf_{F \in \mathbb{F}} \|\mathcal{G}_F\|_2^2. \tag{4.34}$$

These problems will be considered in Subsection 4.3.3 and Section 4.4 respectively.

As will be made clear in the next sections, the solutions of all the control problems posed above are closely related to the mean square stabilizing solution for the CARE, defined next.

Definition 4.4 (Control CARE). *We say that $X = (X_1, \ldots, X_N) \in \mathbb{H}^{n+}$ is the mean square stabilizing solution for the control CARE (4.35) if it satisfies for $i \in \mathbb{N}$*

$$X_i = A_i^* \mathcal{E}_i(X) A_i - A_i^* \mathcal{E}_i(X) B_i (D_i^* D_i + B_i^* \mathcal{E}_i(X) B_i)^{-1} B_i^* \mathcal{E}_i(X) A_i$$
$$+ C_i^* C_i \tag{4.35}$$

and $r_\sigma(\mathcal{T}) < 1$ where $\mathcal{T}(.) = (\mathcal{T}_1(.), \ldots, \mathcal{T}_N(.))$ is defined as in (3.7) with $\Gamma_i = A_i + B_i \mathcal{F}_i(X)$ and $\mathcal{F}_i(X)$ as in

$$\mathcal{F}_i(X) \triangleq -(D_i^* D_i + B_i^* \mathcal{E}_i(X) B_i)^{-1} B_i^* \mathcal{E}_i(X) A_i \tag{4.36}$$

for $i \in \mathbb{N}$.

It will be convenient to define, for $V = (V_1, \ldots, V_N) \in \mathbb{H}^{n+}$, $\mathcal{R}(V) = (\mathcal{R}_1(V), \ldots, \mathcal{R}_N(V))$ as follows:

$$\mathcal{R}_i(V) \triangleq D_i^* D_i + B_i^* \mathcal{E}_i(V) B_i > 0. \tag{4.37}$$

Necessary and sufficient conditions for the existence of the mean square stabilizing solution for the control CARE and other related results are presented in Appendix A.

4.3.2 The Markov Jump Linear Quadratic Regulator Problem

We consider in this subsection model (4.25) without additive noise (that is, $w(k) = 0$ for all $k = 0, 1, \ldots$). We show next that if there is the mean square stabilizing solution $X \in \mathbb{H}^{n+}$ for the CARE (4.35) then the control law given by

$$\bar{u}(k) = \mathcal{F}_{\theta(k)}(X) x(k) \tag{4.38}$$

is an optimal solution for the quadratic optimal control problem posed in Subsection 4.3.1, with cost function given by (4.29).

Theorem 4.5. *Consider model (4.25) with $w(k) = 0$ for all $k = 0, 1, \ldots$. Suppose that there exists the mean square stabilizing solution $X \in \mathbb{H}^n$ for the CARE (4.35). Then the control law (4.38) belongs to $\mathcal{U}$ and minimizes the cost (4.29). Moreover the optimal cost (4.30) is given by $\bar{\mathfrak{J}}(\theta_0, x_0) = E\left(x(0)^* X_{\theta(0)} x(0)\right).$*

Proof. Let $\bar{x}(k)$ be the sequence generated by (4.25) when we apply control law (4.38). From the fact that the solution X is mean square stabilizing, we have that (4.28) is satisfied and thus $\bar{u} \in \mathcal{U}$. For any $u = \{u(k); k \in \mathbb{T}\} \in \mathcal{U}$ we have from the CARE (4.35) that

$$E(x(k+1)^* X_{\theta(k+1)} x(k+1) - x(k)^* X_{\theta(k)} x(k) + u(k)^* D_{\theta(k)}^* D_{\theta(k)} u(k) \mid \mathfrak{G}_k)$$
$$= (u(k) - \mathcal{F}_{\theta(k)}(X) x(k))^* \mathcal{R}_{\theta(k)}(X)(u(k) - \mathcal{F}_{\theta(k)}(X) x(k))$$
$$\quad - x(k)^* C_{\theta(k)}^* C_{\theta(k)} x(k)$$

and thus

$$E\Bigg(x(k)^* X_{\theta(k)} x(k) - x(k+1)^* X_{\theta(k+1)} x(k+1)$$
$$+ \|\mathcal{R}_{\theta(k)}(X)^{1/2}(u(k) - \mathcal{F}_{\theta(k)}(X) x(k))\|^2 \Bigg)$$
$$= E\left(x(k)^* C_{\theta(k)}^* C_{\theta(k)} x(k) + u(k)^* D_{\theta(k)}^* D_{\theta(k)} u(k) \right).$$

Taking the sum from 0 to ∞ and noticing that $E(x(k)^* X_{\theta(k)} x(k))$ goes to 0 as k goes to infinity from the hypothesis that $u \in \mathcal{U}$, it follows that

$$\mathfrak{J}(\theta_0, x_0, u) = \sum_{k=0}^{\infty} E\left(\|\mathcal{R}_{\theta(k)}(X)^{1/2}(u(k) - \mathcal{F}_{\theta(k)}(X) x(k))\|^2 \right)$$
$$+ E\left(x(0)^* X_{\theta(0)} x(0) \right)$$

completing the proof of the theorem.   □

4.3.3 The Long Run Average Cost

For this case we consider model (4.25) with an additive wide sense white noise sequence (that is, when $w = (w(0), w(1), \ldots)$ is such that $E(w(k)) = 0$, $E(w(k)w(k)^*) = I$, $E(w(k)w(l)^*) = 0$ for $k \neq l$) independent of the Markov chain $\{\theta(k)\}$ and initial state value x_0, and the long run cost function as in (4.31). We need here the extra assumption 3.31, that is, the Markov chain $\{\theta(k)\}$ is ergodic with limit probability π_i. We have the following theorem.

Theorem 4.6. *Consider model (4.25) with $\{w(k); k = 0, 1, \ldots\}$ a wide sense white noise sequence. Suppose that there exists the mean square stabilizing solution $X \in \mathbb{H}^n$ for the CARE (4.35). Then the control law (4.38) belongs to $\mathcal{U}$ and minimizes the cost (4.31). Moreover the optimal cost (4.32) is given by $\bar{\mathfrak{J}}_{av} = \sum_{i=1}^{N} \pi_i tr\{G_i^* \mathcal{E}_i(X) G_i\}.$*

Proof. From the fact that the solution X is mean square stabilizing, we have that (4.28) is satisfied and thus $\bar{u} \in \mathcal{U}$. Consider (4.10) with $V = X$. From (4.17) in Theorem 4.2 we have for any $u = (u(0), \ldots) \in \mathcal{U}$ that

$$\sum_{k=0}^{T-1} E(\|z(k)\|^2) + E(x(T)^* X_{\theta(T)} x(T)) =$$

$$\sum_{i=1}^{N} \left[\operatorname{tr}(\pi_i(0) Q_i(0) X_i) + \sum_{k=0}^{T-1} \pi_i(k) \operatorname{tr}(G_i G_i^* \mathcal{E}_i(X)) \right], \qquad (4.39)$$

since in this case $X(k) = X$ for all $k = 0, 1, \ldots$. Notice that we have equality in (4.39) when the control law is given by $\bar{u}(k)$ as in (4.38). Dividing (4.39) by T and taking the limit as T goes to ∞, we obtain from the ergodic assumption 3.31 for the Markov chain that

$$\mathfrak{J}_{av}(\theta_0, x_0, u) = \limsup_{T \to \infty} \frac{1}{T} \left\{ \sum_{k=0}^{T-1} E\left(\|z(k)\|^2 \right) \right\}$$

$$\geq \sum_{i=1}^{N} \limsup_{T \to \infty} \frac{1}{T} \left\{ \sum_{k=0}^{T-1} \pi_i(k) tr\{G_i^* \mathcal{E}_i(X) G_i\} \right\}$$

$$= \sum_{i=1}^{N} \pi_i tr\{G_i^* \mathcal{E}_i(X) G_i\}$$

with equality when $\bar{u}(k)$ is as in (4.38), completing the proof of the theorem.
□

4.4 The H_2-control Problem

4.4.1 Preliminaries and the H_2-norm

This section deals with the infinite horizon quadratic optimal control problem when the state equation includes an additive wide sense white noise sequence, as in model (4.25) discussed in Subsection 4.3.1, and it is desired to minimize the limit value of the expected value of a quadratic cost among all closed loop mean square stabilizing dynamic Markov jump controllers. For the case with no jumps this problem is known as the H_2-control problem. From the point of view of getting a solution, it can be considered similar to the linear quadratic regulator (Subsection 4.3.2) or long run average cost (Subsection 4.3.3) control problems, although the framework of these problems are somewhat different. The section begins by presenting the concept of H_2-norm, tracing a parallel with the case with no jumps. Next we show that this definition is in fact equivalent to the one presented in (4.33) in Subsection 4.3.1. With the aid of a few auxiliary results, the section is concluded by establishing a connection between the existence of the mean square stabilizing solution for the CARE, the existence of the solution of a certain convex programming problem presented in [65] and the existence of the solution for the H_2-control problem for the discrete-time MJLS.

To introduce the H_2-norm for the discrete-time MJLS, we consider again model (4.25) with assumptions (4.26) and (4.27). We assume also that the ergodic assumption 3.31 holds. Recall from Subsection 4.3.1 that for $F = (F_1, \ldots, F_N) \in \mathbb{F}$, $\mathcal{G}_F$ represents system $\mathcal{G}$ in (4.25) when the control law is given by $u(k) = F_{\theta(k)} x(k)$. Following [65], we define:

Definition 4.7. *The H_2-norm of system $\mathcal{G}_F$ is given by*

$$\|\mathcal{G}_F\|_2^2 \triangleq \sum_{s=1}^{r} \|z_s\|_2^2$$

where z_s represents the output sequence $(z(0), z(1), \ldots)$ given by (4.25) when

1. *$x(0) = 0$ and the input sequence is given by $w = (w(0), w(1), \ldots)$ with $w(0) = e_s$, $w(k) = 0$ for $k > 0$, where $\{e_1, \ldots, e_r\}$ forms a basis for $\mathbb{C}^r$, and*
2. *$\theta(0) = i$ with probability v_i where $\{v_i; i \in \mathbb{N}\}$ is such that $v_i > 0$ for $i \in \mathbb{N}$.*

It is easy to see that the definition above is equivalent to consider system $\mathcal{G}_F$ without inputs, and with initial conditions given by $x(0) = G_j e_s$ and $\theta(0) = j$ with probability v_j for $j \in \mathbb{N}$ and $s = 1, \ldots, r$. Notice that the equivalence between Definition 4.7 and the one presented in (4.33) will be established in Subsection 4.4.3. When we restrict ourselves to the so-called deterministic case ($N = 1$ and $p_{11} = 1$), both definitions reduce to the usual H_2-norm.

4.4.2 The H_2-norm and the Grammians

There is a direct relation between the H_2-norm and the solutions of the observability and controllability Grammians. This relation is particularly important, for it will allow us to obtain the main result of this section.

For any $F \in \mathbb{F}$, let $\mathcal{O}(F) = (\mathcal{O}_1(F), \ldots, \mathcal{O}_N(F)) \in \mathbb{H}^{n+}$ be defined as

$$\mathcal{O}_i(F) \triangleq C_i^* C_i + F_i^* D_i^* D_i F_i.$$

Also for any $F \in \mathbb{F}$, let $\mathcal{S}o(F) = (\mathcal{S}o_1(F), \ldots, \mathcal{S}o_N(F)) \in \mathbb{H}^{n+}$ and $\mathcal{S}c(F) = (\mathcal{S}c_1(F), \ldots, \mathcal{S}c_N(F)) \in \mathbb{H}^{n+}$ be the unique solutions of the observability and controllability Grammians respectively (see Proposition 3.20), given by

$$\mathcal{S}o_i(F) = (A_i + B_i F_i)^* \mathcal{E}_i(\mathcal{S}o(F))(A_i + B_i F_i) + \mathcal{O}_i(F)$$
$$= \mathcal{L}_i(\mathcal{S}o(F)) + \mathcal{O}_i(F) \tag{4.40}$$

for $i \in \mathbb{N}$ and

$$\mathcal{S}c_j(F) = \sum_{i=1}^{N} p_{ij} \left[(A_i + B_i F_i)\mathcal{S}c_i(F)(A_i + B_i F_i)^* + v_i G_i G_i^* \right]$$

$$= \mathcal{T}_j(\mathcal{S}c(F)) + \sum_{i=1}^{N} p_{ij} \left[v_i G_i G_i^* \right] \tag{4.41}$$

for $j \in \mathbb{N}$, where $\mathcal{L}$ and $\mathcal{T}$ are defined as in (3.8) and (3.7) respectively with $\Gamma_i = A_i + B_i F_i$. Since $\mathcal{O}(F) \geq 0$ and $G_j G_j^* \geq 0$ for $j \in \mathbb{N}$, we have that $\mathcal{S}o(F) \geq 0$ and $\mathcal{S}c(F) \geq 0$.

With (4.40) and (4.41) we are ready to present the next result, adapted from [65] (see Definition 4.7), which establishes a connection between the H_2-norm and the Grammians.

Proposition 4.8. *For any $F \in \mathbb{F}$,*

$$\|\mathcal{G}_F\|_2^2 = \sum_{i=1}^{N} v_i \operatorname{tr}(G_i^* \mathcal{E}_i(\mathcal{S}o(F))G_i)$$

$$= \sum_{j=1}^{N} \operatorname{tr}\left(\begin{bmatrix} C_j^* C_j & 0 \\ 0 & D_j^* D_j \end{bmatrix} \begin{bmatrix} \mathcal{S}c_j(F) & \mathcal{S}c_j(F)F_j^* \\ F_j \mathcal{S}c_j(F) & F_j \mathcal{S}c_j(F)F_j^* \end{bmatrix} \right).$$

Proof. Notice that for $w(0) = e_s$, $\theta(0) = i$ with probability v_i, and $w(k) = 0$ for $k \geq 1$ we have from the observability Grammian that

$$E(z_s(k)^* z_s(k))$$
$$= E(x(k)^* \mathcal{O}_{\theta(k)}(F)x(k))$$
$$= E(x(k)^* (\mathcal{S}o_{\theta(k)}(F)$$
$$\quad - (A_{\theta(k)} + B_{\theta(k)} F_{\theta(k)})^* \mathcal{E}_{\theta(k+1)}(\mathcal{S}o(F))(A_{\theta(k)} + B_{\theta(k)} F_{\theta(k)}))x(k))$$
$$= E(E(x(k)^* (\mathcal{S}o_{\theta(k)}(F)$$
$$\quad - (A_{\theta(k)} + B_{\theta(k)} F_{\theta(k)})^* \mathcal{S}o_{\theta(k+1)}(F)(A_{\theta(k)} + B_{\theta(k)} F_{\theta(k)}))x(k) \mid \mathfrak{F}_k))$$
$$= E(x(k)^* (\mathcal{S}o_{\theta(k)(F)}x(k) - x(k+1)^* \mathcal{S}o_{\theta(k+1)}(F)x(k+1))).$$

Since from MSS $E(x(k)^* x(k))$ goes to zero as k goes to infinity and $x(1) = G_{\theta(0)}e_s$, we have that

$$\|z_s\|_2^2 = \sum_{k=1}^{\infty} E(\|z_s(k)\|^2) = E(e_s^* G_{\theta(0)}^* \mathcal{S}o_{\theta(1)}(F)G_{\theta(0)}e_s)$$

$$= \sum_{i=1}^{N} v_i e_s^* G_i^* \mathcal{E}_i(\mathcal{S}o(F))G_i e_s$$

and

$$\|\mathcal{G}_F\|_2^2 = \sum_{i=1}^{N} v_i \operatorname{tr}(G_i^* \mathcal{E}_i(\mathcal{S}o(F))G_i). \tag{4.42}$$

From the observability Grammian,

$$(A_i + B_i F_i)^* \left(\sum_{j=1}^{N} p_{ij} \mathcal{S}o_j(F) \right) (A_i + B_i F_i) + \mathcal{O}_i(F) - \mathcal{S}o_i(F) = 0$$

and from (4.42),

$$\|\mathcal{G}_F\|_2^2 = \sum_{i=1}^{N} v_i \operatorname{tr} \left[G_i^* \mathcal{E}_i(\mathcal{S}o(F)) G_i \right]$$

$$= \sum_{i=1}^{N} \operatorname{tr} \left[v_i G_i G_i^* \sum_{j=1}^{N} p_{ij} \mathcal{S}o_j(F) \right.$$

$$+ \mathcal{S}c_i(F) \left((A_i + B_i F_i)^* \left(\sum_{j=1}^{N} p_{ij} \mathcal{S}o_j(F) \right) (A_i + B_i F_i) \right.$$

$$\left. + \mathcal{O}_i(F) - \mathcal{S}o_i(F) \right) \Big].$$

Thus, from the controllability Grammian,

$$\|\mathcal{G}_F\|_2^2 = \sum_{j=1}^{N} \operatorname{tr} \left[\sum_{i=1}^{N} v_i p_{ij} G_i G_i^* \mathcal{S}o_j(F) \right.$$

$$+ \left(\begin{bmatrix} C_j^* C_j & 0 \\ 0 & D_j^* D_j \end{bmatrix} \begin{bmatrix} \mathcal{S}c_j(F) & \mathcal{S}c_j(F) F_j^* \\ F_j \mathcal{S}c_j(F) & F_j \mathcal{S}c_j(F) F_j^* \end{bmatrix} \right) \Big]$$

$$+ \sum_{j=1}^{N} \left(\sum_{i=1}^{N} \operatorname{tr} \left[p_{ij} (A_i + B_i F_i) \mathcal{S}c_i(F) (A_i + B_i F_i)^* \mathcal{S}o_j(F) \right] \right)$$

$$- \sum_{j=1}^{N} \operatorname{tr} \left[\mathcal{S}c_j(F) \mathcal{S}o_j(F) \right]$$

$$= \sum_{j=1}^{N} \operatorname{tr} \left[\sum_{i=1}^{N} p_{ij} \left((A_i + B_i F_i) \mathcal{S}c_i(F) (A_i + B_i F_i)^* \right. \right.$$

$$\left. + v_i G_i G_i^* \right) - \mathcal{S}c_j(F) \Big] \mathcal{S}o_j(F) \Big]$$

$$+ \sum_{j=1}^{N} \operatorname{tr} \left[\begin{bmatrix} C_j^* C_j & 0 \\ 0 & D_j^* D_j \end{bmatrix} \begin{bmatrix} \mathcal{S}c_j(F) & \mathcal{S}c_j(F) F_j^* \\ F_j \mathcal{S}c_j(F) & F_j \mathcal{S}c_j(F) F_j^* \end{bmatrix} \right]$$

$$= \sum_{j=1}^{N} \operatorname{tr} \left[\begin{bmatrix} C_j^* C_j & 0 \\ 0 & D_j^* D_j \end{bmatrix} \begin{bmatrix} \mathcal{S}c_j(F) & \mathcal{S}c_j(F) F_j^* \\ F_j \mathcal{S}c_j(F) & F_j \mathcal{S}c_j(F) F_j^* \end{bmatrix} \right]$$

completing the proof of the proposition. $\qquad \square$

4.4.3 An Alternative Definition for the H_2-control Problem

Let us show now the equivalence between Definition 4.7 and the one presented in (4.33). Suppose that $v_i = \pi_i$ and $w = (w(0), w(1), \ldots)$ is a wide sense white noise sequence in model (4.25). Let $F \in \mathbb{F}$ be an admissible controller, $x(k)$ as in (4.25) with $u(k) = F_{\theta(k)}x(k)$, and

$$Q_i(k) = E(x(k)x(k)^* 1_{\{\theta(k)=i\}}), \quad i \in \mathbb{N}.$$

From Chapter 3, Proposition 3.35 (recalling that $E(w(k)) = 0, E(w(k)w(k)^*) = I$),

$$Q_j(k+1) = \sum_{i=1}^{N} p_{ij}[(A_i + B_i F_i)Q_i(k)(A_i + B_i F_i)^* + \pi_i(k)G_i G_i^*].$$

Moreover, since the closed loop system is MSS, we have that $Q(k) \overset{k\uparrow\infty}{\to} Q$ (see Chapter 3, Proposition 3.36), where $Q = (Q_1, \ldots, Q_N)$ is the unique solution of the controllability Grammian (4.41), that is, $Q = \mathcal{S}c(F)$. Notice that

$$E(\|z(k)\|^2) = E(\mathrm{tr}(z(k)z(k)^*))$$

$$= \sum_{i=1}^{N} \mathrm{tr}\left(\begin{bmatrix} C_i^* C_i & 0 \\ 0 & D_i^* D_i \end{bmatrix} \begin{bmatrix} Q_i(k) & Q_i(k)F_i^* \\ F_i Q_i(k) & F_i Q_i(k)F_i^* \end{bmatrix} \right)$$

$$\overset{k\uparrow\infty}{\to} \sum_{i=1}^{N} \mathrm{tr}\left(\begin{bmatrix} C_i^* C_i & 0 \\ 0 & D_i^* D_i \end{bmatrix} \begin{bmatrix} Q_i & Q_i F_i^* \\ F_i Q_i & F_i Q_i F_i^* \end{bmatrix} \right) = \|\mathcal{G}_F\|_2^2 \quad (4.43)$$

and thus from Proposition 4.8 we have the equivalence between Definition 4.7 and the one presented in (4.33).

4.4.4 Connection Between the CARE and the H_2-control Problem

In this subsection we will establish the equivalence of:

(i) finding the stabilizing solution for the quadratic CARE (4.35);
(ii) obtaining a stabilizing controller F that minimizes the H_2-norm of system $\mathcal{G}_F$, as given by Definition 4.7 or Proposition 4.8, and
(iii) obtaining the optimal solution of a certain convex programming problem, which will be defined in the following.

We assume in this subsection that $G_i G_i^* > 0$ for each $i \in \mathbb{N}$, so that from (4.41) we have that $\mathcal{S}c(F) > 0$, and that all matrices are real. We define the two following sets:

$$\Psi \triangleq \{ W = (W_1, \ldots, W_N) \in \mathbb{H}^{n+m}; \text{for } j \in \mathbb{N},$$
$$W_j = \begin{bmatrix} W_{j1} & W_{j2} \\ W_{j2}^* & W_{j3} \end{bmatrix} \geq 0, W_{j1} > 0,$$
$$\sum_{i=1}^{N} p_{ij}(A_i W_{i1} A_i^* + B_i W_{i2}^* A_i^* + A_i W_{i2} B_i^* + B_i W_{i3} B_i^* + v_i G_i G_i^*)$$
$$- W_{j1} \leq 0\}$$

and

$$\widehat{\Psi} \triangleq \{ W = (W_1, \ldots, W_N) \subseteq \Psi; \text{for } j \in \mathbb{N},$$
$$W_{j3} = W_{j2}^* W_{j1}^{-1} W_{j2},$$
$$\sum_{i=1}^{N} p_{ij}(A_i W_{i1} A_i^* + B_i W_{i2}^* A_i^* + A_i W_{i2} B_i^* + B_i W_{i3} B_i^* + v_i G_i G_i^*)$$
$$- W_{j1} = 0 \}.$$

Note that, unlike $\widehat{\Psi}$, Ψ is a convex set. The cost function associated with the optimization problem mentioned in (iii) above is defined as

$$\lambda(W) \triangleq \sum_{j=1}^{N} \text{tr}\left(\begin{bmatrix} C_i^* C_i & 0 \\ 0 & D_i D_i^* \end{bmatrix} \begin{bmatrix} W_{i1} & W_{i2} \\ W_{i2}^* & W_{i3} \end{bmatrix} \right).$$

We introduce now two mappings, which will be used later on to connect the solutions of (i), (ii) and (iii). For any $F = (F_1, \ldots, F_N) \in \mathbb{F}$, let $\mathcal{S}c(F) = (\mathcal{S}c_1(F), \ldots, \mathcal{S}c_N(F)) > 0$ be as in (4.41). Define

$$\mathcal{K}(F) \triangleq \left(\begin{bmatrix} \mathcal{S}c_1(F) & \mathcal{S}c_1(F)F_1^* \\ F_1\mathcal{S}c_1(F) & F_1\mathcal{S}c_1(F)F_1^* \end{bmatrix}, \ldots, \begin{bmatrix} \mathcal{S}c_N(F) & \mathcal{S}c_N(F)F_N^* \\ F_N\mathcal{S}c_N(F) & F_N\mathcal{S}c_N(F)F_N^* \end{bmatrix} \right).$$

From the Schur complement (see Lemma 2.23) and (4.41) it is immediate that $\mathcal{K}(F) \in \widehat{\Psi}$. Define also for $W \in \Psi$

$$\mathcal{Y}(W) \triangleq (W_{12}^* W_{11}^{-1}, \ldots, W_{N2}^* W_{N1}^{-1}).$$

Since $W_{i3} \geq W_{i2}^* W_{i1}^{-1} W_{i2}$ for $i \in \mathbb{N}$ (from the Schur complement), we have

$$\sum_{i=1}^{N} p_{ij}((A_i + B_i W_{i2}^* W_{i1}^{-1})W_{i1}(A_i + B_i W_{i2}^* W_{i1}^{-1})^* + v_i G_i G_i^*) - W_{j1}$$

$$\leq \sum_{i=1}^{N} p_{ij}(A_i W_{i1} A_i^* + B_i W_{i2}^* A_i^* + A_i W_{i2} B_i^* + B_i W_{i3} B_i^* + v_i G_i G_i^*) - W_{j1}$$

$$\leq 0 \tag{4.44}$$

and therefore from Theorem 3.9, $\mathcal{Y}(W) \in \mathbb{F}$. Then we have defined the mappings $\mathcal{K} : \mathbb{F} \longrightarrow \widehat{\Psi}$ and $\mathcal{Y} : \Psi \longrightarrow \mathbb{F}$. The following proposition relates them.

Proposition 4.9. *The following assertions hold:*

1. $\mathcal{Y}\mathcal{K} = \mathcal{I}$,
2. $\mathcal{K}\mathcal{Y} = \mathcal{I}$ on $\widehat{\Psi}$,
3. $\mathcal{K}\mathcal{Y}(W) \leq W$ for any $W \in \Psi$,

where we recall that $\mathcal{I}$ is the identity operator.

Proof. It is immediate to check that $\mathcal{Y}(\mathcal{K}(F)) = F$, showing 1). From the uniqueness of the solution $\mathcal{S}c(F)$ of (4.41), we get that for any $W \in \widehat{\Psi}$, $\mathcal{K}(\mathcal{Y}(W)) = (W_{11}, \ldots, W_{N1})$, that is, $\mathcal{K}(\mathcal{Y}(W)) = W$, showing 2).

Let us now prove 3). We have that

$$\sum_{i=1}^{N} p_{ij}((A_i + B_i W_{i2}^* W_{i1}^{-1})\mathcal{K}_i(\mathcal{Y}(W))(A_i + B_i W_{i2}^* W_{i1}^{-1})^* + \upsilon_i G_i G_i^*)$$

$$- \mathcal{S}c_i(\mathcal{Y}(W)) = 0$$

for $j \in \mathbb{N}$. Thus, from Theorem 3.9 and (4.44), $\mathcal{S}c_j(\mathcal{Y}(W)) \leq W_{j1}$ for $j \in \mathbb{N}$. If we prove that

$$W_i - (\mathcal{K}\mathcal{Y})_i(W)$$
$$= \begin{bmatrix} W_{i1} - \mathcal{S}c_i(\mathcal{Y}(W)) & W_{i2} - \mathcal{S}c_i(\mathcal{Y}(W))W_{i1}^{-1}W_{i2} \\ W_{i2}^* - W_{i2}^* W_{i1}^{-1}\mathcal{S}c_i(\mathcal{Y}(W)) & W_{i3} - W_{i2}^* W_{i1}^{-1}\mathcal{S}c_i(\mathcal{Y}(W))W_{i1}^{-1}W_{i2} \end{bmatrix} \geq 0$$

for $i \in \mathbb{N}$ then we will have the desired result. As $\mathcal{S}c_i(\mathcal{Y}(W)) \leq W_{i1}$ we have (from the Schur complement) that

$$(W_{i1} - \mathcal{S}c_i(\mathcal{Y}(W)))(W_{i1} - \mathcal{S}c_i(\mathcal{Y}(W)))^{\dagger}(W_{i1} - \mathcal{S}c_i(\mathcal{Y}(W)))W_{i1}^{-1}W_{i2}$$
$$= W_{i2} - \mathcal{S}c_i(\mathcal{Y}(W))W_{i1}^{-1}W_{i2}$$

and

$$W_{i3} - W_{i2}^* W_{i1}^{-1}\mathcal{S}c_i(\mathcal{Y}(W))W_{i1}^{-1}W_{i2} - W_{i2}^* W_{i1}^{-1}(W_{i1} - \mathcal{S}c_i(\mathcal{Y}(W)))$$
$$(W_{i1} - \mathcal{S}c_i(\mathcal{Y}(W)))^{\dagger}(W_{i1} - \mathcal{S}c_i(\mathcal{Y}(W)))W_{i1}^{-1}W_{i2}$$
$$= W_{i3} - W_{i2}^* W_{i1}^{-1}W_{i2} \geq 0.$$

Using again the Schur complement we complete the proof.   □

The next theorem presents the main result of this section, which establishes the equivalence of problems (i), (ii) and (iii) and the connection between their solutions.

Theorem 4.10. *The following assertions are equivalent:*

1. *There exists the mean square stabilizing solution $X = (X_1, \ldots, X_N)$ for the CARE given by (4.35).*
2. *There exists $F = (F_1, \ldots, F_N) \in \mathbb{F}$ such that*

$$\|\mathcal{G}_F\|_2 = \min\{\|\mathcal{G}_K\|_2 ; K \in \mathbb{F}\}.$$

3. *There exists $W = (W_1, \ldots, W_N) \in \Psi$ such that*

$$\lambda(W) = \min\{\lambda(V); V \in \Psi\}.$$

Moreover

4. If X satisfies 1), then $\mathcal{F}(X)$ satisfies 2) and $\mathcal{K}(\mathcal{F}(X))$ satisfies 3), where $\mathcal{F}(X)$ is as in (4.24).
5. If F satisfies 2), then $\mathcal{K}(F)$ satisfies 3) and $\mathcal{S}o(F)$ satisfies 1).
6. If W satisfies 3), then $\mathcal{S}o(\mathcal{Y}(W))$ satisfies 1) where $\mathcal{Y}(W)$ satisfies 2).

Proof. The second part of the proof (assertions 4), 5) and 6)) follows immediately from the first part (equivalence of 1), 2) and 3)).

2)$\Leftrightarrow$3): From Propositions 4.8 and 4.9 it is immediate that

$$\min\{\|\,\mathcal{G}_K\,\|_2; K \in \mathbb{F}\} = \min\{\lambda(V); V \in \widehat{\varPsi}\}$$

and since $\widehat{\varPsi} \subseteq \varPsi$,

$$\min\{\lambda(V); V \in \widehat{\varPsi}\} \geq \min\{\lambda(V); V \in \varPsi\}.$$

On the other hand, from Proposition 4.9, $\mathcal{K}\mathcal{Y}(V) \leq V$, with $\mathcal{K}\mathcal{Y}(V) \in \widehat{\varPsi}$. Therefore, since $\lambda(\mathcal{K}\mathcal{Y}(V)) \leq \lambda(V)$, we have

$$\min\{\lambda(V); V \in \widehat{\varPsi}\} \leq \min\{\lambda(V); V \in \varPsi\},$$

showing that

$$\min\{\|\,\mathcal{G}_K\,\|_2; K \in \mathbb{F}\} = \min\{\lambda(V); V \in \varPsi\},$$

which proves the equivalence between 2) and 3).

1)$\Leftrightarrow$2): Suppose that X is the mean square stabilizing solution of the CARE (4.35). It is easy to show that X satisfies

$$X_i - (A_i + B_i\mathcal{F}_i(X))^*\mathcal{E}_i(X)(A_i + B_i\mathcal{F}_i(X)) = \mathcal{O}_i(\mathcal{F}(X))$$

and from the uniqueness of the solution of this equation (see Proposition 3.20), $\mathcal{S}o(\mathcal{F}(X)) = X$. For any $K \in \mathbb{F}$ and $\mathcal{S}o(K)$ as in (4.40) we have, according to Lemma A.7, that

$$(\mathcal{S}o_i(K) - X_i) - (A_i + B_iK_i)^*\mathcal{E}_i(\mathcal{S}o(K) - X)(A_i + B_iK_i)$$
$$= (K_i - \mathcal{F}_i(X))^*\mathcal{R}_i(X)(K_i - \mathcal{F}_i(X)) \quad (4.45)$$

for $i \in \mathbb{N}$. We also have, from Proposition 3.20, that $\mathcal{S}o(K) \geq X$. From Proposition 4.8,

$$\|\mathcal{G}_K\|_2^2 = \sum_{i=1}^N v_i \operatorname{tr}(G_i^*\mathcal{E}_i(\mathcal{S}o(F))G_i) \geq \sum_{i=1}^N v_i \operatorname{tr}(G_i^*\mathcal{E}_i(X)G_i) - \|\mathcal{G}_{\mathcal{F}(X)}\|_2^2,$$

showing that $\mathcal{F}(X)$ satisfies 2).

On the other hand, suppose that there exists $F = (F_1, \ldots, F_N) \in \mathbb{F}$ satisfying 2). Then clearly (A, B) is mean square stabilizable and there exists a maximal solution X to the CARE (4.35) (see Theorem A.10). Moreover, according to Theorem A.10, it is possible to find a sequence $X^l \in \mathbb{H}^{n+}$ and $K^l = \mathcal{F}(X^{l-1}) \in \mathbb{F}$, $l = 1, 2, \ldots$ such that (A.16) are satisfied and $X^l \longrightarrow X$ as $l \longrightarrow \infty$. Start with $K^0 = F$, so that $X^0 = \mathcal{S}o(F) \geq X^l \geq X$ for all l. From optimality of F, we have

$$\|\mathcal{G}_{K^1}\|_2^2 = \sum_{i=1}^N v_i \operatorname{tr}(G_i^* \mathcal{E}_i(X^1) G_i) \geq \sum_{i=1}^N v_i \operatorname{tr}(G_i^* \mathcal{E}_i(\mathcal{S}o(F)) G_i) = \|\mathcal{G}_F\|_2^2,$$

and since $\mathcal{S}o(F) \geq X^1$,

$$\sum_{i=1}^N v_i \operatorname{tr}(G_i^*(\mathcal{E}_i(\mathcal{S}o(F)) - \mathcal{E}_i(X^1)) G_i) = 0,$$

or, in other words,

$$v_i \operatorname{tr}(G_i^*(\mathcal{E}_i(\mathcal{S}o(F)) - \mathcal{E}_i(X^1)) G_i) = 0$$

for each $i \in \mathbb{N}$. Since $v_i G_i G_i^* > 0$, we can conclude that $\mathcal{E}(X^0) = \mathcal{E}(\mathcal{S}o(F)) = \mathcal{E}(X^1)$. Thus, from (4.24), $\mathcal{F}(X^0) = \mathcal{F}(X^1)$, and from the definition of X^1 in Theorem A.10 it follows that

$$\begin{aligned}
X_i^1 &= (A_i + B_i \mathcal{F}_i(X^0))^* \mathcal{E}_i(X^1)(A_i + B_i \mathcal{F}_i(X^0)) + \mathcal{O}_i(\mathcal{F}(X^0)) \\
&= (A_i + B_i \mathcal{F}_i(X^1))^* \mathcal{E}_i(X^1)(A_i + B_i \mathcal{F}_i(X^1)) + \mathcal{O}_i(\mathcal{F}(X^1)),
\end{aligned}$$

and thus X^1 satisfies the CARE (4.35). Moreover, from Theorem A.10, $\mathcal{F}(X^0) = \mathcal{F}(X^1) \in \mathbb{F}$, showing that X^1 is the mean square stabilizing solution of (4.35). Since the unique mean square stabilizing solution of (4.35) coincides with the maximal solution (Lemma A.14), $X^1 = X$ and $\mathcal{F}(X) = \mathcal{F}(X^1) \in \mathbb{F}$, proving that X is indeed the mean square stabilizing solution. □

4.5 Quadratic Control with Stochastic ℓ_2-input

4.5.1 Preliminaries

In this section we shall consider the quadratic optimal control problem of (4.25) for the case in which $w = (w(0), w(1), \ldots) \in \mathcal{C}^r$. This result will be used in the min-max H_∞-control problem to be considered in Chapter 7. First we define more precisely the set of admissible controls. Set

$$\mathcal{U} \triangleq \{ u = (u(0), u(1), \ldots) \in \mathcal{C}^m; x = (x(0), x(1), \ldots) \in \mathcal{C}^n$$
$$\text{for every } w = (w(0), w(1), \ldots) \in \mathcal{C}^r \text{ where } x \text{ is given by } (4.25) \}.$$

For any $(x_0, w, u) \in \mathcal{C}_0^n \oplus \mathcal{C}^r \oplus \mathcal{U}$, $\theta_0 \in \Theta_0$, $w = (w(0), w(1), \ldots)$, $u = (u(0), u(1), \ldots)$, set the cost function $\mathfrak{J}(\theta_0, x_0, w, u)$ as in (4.29), which is rewritten below with the dependence on the input w:

$$\mathfrak{J}(\theta_0, x_0, w, u) \triangleq \sum_{k=0}^{\infty} E\left(\|C_{\theta(k)}x(k)\|^2 + \|D_{\theta(k)}u(k)\|^2\right). \tag{4.46}$$

We want to solve the following problem:

$$\bar{\mathfrak{J}}(\theta_0, x_0, w) \triangleq \inf_{u \in \mathcal{U}} \mathfrak{J}(\theta_0, x_0, w, u). \tag{4.47}$$

Remark 4.11. The assumption that $\|w\|_2 < \infty$ can be relaxed if we consider the discounted case, which is a particular case of (4.25), (4.46) and (4.47). Indeed in this case, for $\rho \in (0, 1)$,

$$\mathfrak{J}^\rho(\theta_0, x_0, w, u) = \sum_{k=0}^{\infty} E\left(\rho^k \left(\|C_{\theta(k)}x(k)\|^2 + \|D_{\theta(k)}u(k)\|^2\right)\right)$$

and defining $x^\rho(k) = \rho^{k/2}x(k), u^\rho(k) = \rho^{k/2}u(k), w^\rho(k) = \rho^{k/2}w(k)$, we get that

$$\mathfrak{J}^\rho(\theta_0, x_0, w, u) = \sum_{k=0}^{\infty} E\left(\|C_{\theta(k)}x^\rho(k)\|^2 + \|D_{\theta(k)}u^\rho(k)\|^2\right)$$

with

$$x^\rho(k+1) = A^\rho_{\theta(k)}x^\rho(k) + B^\rho_{\theta(k)}u^\rho(k) + G^\rho_{\theta(k)}w^\rho(k),$$
$$x^\rho(0) = \rho^{1/2}x_0 \in \mathcal{C}_0^n, \theta(0) = \theta_0 \in \Theta_0$$

and $A^\rho_i = \rho^{1/2}A_i, B^\rho_i = \rho^{1/2}B_i, G^\rho_i = \rho^{1/2}G_i$. Then the problem is in the framework posed above, with the requirement that $\|w^\rho\|_2 < \infty$ (and not $\|w\|_2 < \infty$).

4.5.2 Auxiliary Result

The purpose of this section is to characterize the solution of problem (4.47) for arbitrary $w \in \mathcal{C}^r$. This will be done in Theorem 4.14 below. The proof of this theorem will follow from the next lemmas.

Suppose that there exists the mean square stabilizing solution $X = (X_1, \ldots, X_N) \in \mathbb{H}^{n+}$ as in Definition 4.4 for the CARE (4.35). For $\Gamma = (\Gamma_1, \ldots, \Gamma_N) \in \mathbb{H}^n$, we recall the definition of $\Pi\Gamma(k, l)$ presented in (C.4), as follows

$$\Pi\Gamma(k, l) = \Gamma_{\theta(k)} \ldots \Gamma_{\theta(l+1)}, \ k = l + 1, l + 2, \ldots, \quad \Pi\Gamma(l, l) = I.$$

Set

$$\Gamma_i = A_i + B_i \mathcal{F}_i(X)$$

for each $i \in \mathbb{N}$, and $r = (r(0), r(1), \ldots)$ as

$$r(k) \triangleq \sum_{\iota=0}^{\infty} E\left(\Pi\Gamma(k + \iota, k)^* X_{\theta(k+\iota+1)} G_{\theta(k+\iota)} w(k + \iota) \mid \mathfrak{F}_k\right). \tag{4.48}$$

Lemma 4.12. *r belongs to $\mathcal{C}^n$.*

Proof. Consider the system

$$\mathcal{G}_r = \begin{cases} x(k+1) = \Gamma_{\theta(k)}x(k) + \varrho(k) \\ \qquad z(k) = X_{\theta(k)}x(k) \end{cases}$$

with the input sequence $\varrho(k)$ of dimension n and the output sequence $z(k)$ of dimension n. According to (C.6), the adjoint operator $\mathcal{G}_r^* \in \mathbb{B}(\mathcal{C}^n)$ is as follows; for any $v \in \mathcal{C}^n$:

$$\mathcal{G}_r^*(v)(k) = \sum_{\iota=0}^{\infty} E\Big(\Pi\Gamma(k+\iota,k)^* X_{\theta(k+\iota+1)} v(k+\iota+1)|\mathfrak{F}_k \Big). \qquad (4.49)$$

For $w \in \mathcal{C}^r$, set $v(k) = G_{\theta(k-1)}w(k-1)$, so that $v \in \mathcal{C}^n$. Therefore we have that $\mathcal{G}_r^*(v) \in \mathcal{C}^n$, and from (4.48) and (4.49),

$$\mathcal{G}_r^*(v) = r.$$

$\square$

Define $\bar{r} \triangleq (\bar{r}(0), \dots)$ as

$$\bar{r}(k) \triangleq -(D_{\theta(k)}^* D_{\theta(k)} + B_{\theta(k)}^* \mathcal{E}_{\theta(k)}(X)B_{\theta(k)})^{-1} B_{\theta(k)}^* r(k)$$

and set $\bar{u} \triangleq (\bar{u}(0), \dots)$ and $\bar{x} \triangleq (x_0, \bar{x}(1), \dots)$ as:

$$\bar{u}(k) \triangleq \mathcal{F}_{\theta(k)}(X)\bar{x}(k) + \bar{r}(k) \qquad (4.50)$$

$$\bar{x}(k+1) \triangleq A_{\theta(k)}\bar{x}(k) + B_{\theta(k)}\bar{u}(k) + G_{\theta(k)}w(k)$$

$$= (A_{\theta(k)} + B_{\theta(k)}\mathcal{F}_{\theta(k)}(X))\bar{x}(k) + B_{\theta(k)}\bar{r}(k) + G_{\theta(k)}w(k)$$

$$\bar{x}(0) \triangleq x_0.$$

Clearly from Lemma 4.12, $\bar{r} = (\bar{r}(0), \dots) \in \mathcal{C}^m$ and from the fact that $r_\sigma(\mathcal{T}) < 1$, it follows from Theorem 3.34 that $\bar{x} \in \mathcal{C}^n$ and $\bar{u} \in \mathcal{C}^m$. Define now $\eta \triangleq (\eta(0), \eta(1), \dots)$ as

$$\eta(0) \triangleq A_{\theta(0)}^* \eta(1) + C_{\theta(0)}^* C_{\theta(0)}\bar{x}(0)$$

$$\eta(k+1) \triangleq \Big(I + \mathcal{E}_{\theta(k)}(X)B_{\theta(k)}(D_{\theta(k)}^* D_{\theta(k)})^{-1} B_{\theta(k)}^* \Big)^{-1}$$

$$\times (\mathcal{E}_{\theta(k)}(X)A_{\theta(k)}\bar{x}(k) + r(k)).$$

Since (see [6], p. 349)

$$(D_{\theta(k)}^* D_{\theta(k)})^{-1} B_{\theta(k)}^* \Big(I + \mathcal{E}_{\theta(k)}(X)B_{\theta(k)}(D_{\theta(k)}^* D_{\theta(k)})^{-1} B_{\theta(k)}^* \Big)^{-1}$$

$$= \Big(D_{\theta(k)}^* D_{\theta(k)} + B_{\theta(k)}^* \mathcal{E}_{\theta(k)}(X)B_{\theta(k)} \Big)^{-1} B_{\theta(k)}^*$$

we have that

$$\bar{u}(k) = -(D^*_{\theta(k)} D_{\theta(k)})^{-1} B^*_{\theta(k)} \eta(k+1). \qquad (4.51)$$

In what follows, set $C^*C \triangleq (C^*_1 C_1, \ldots, C^*_N C_N)$, and recall the definition of $\mathcal{R}(X)$ in (4.37).

Lemma 4.13. *For each* $k = 1, 2, \ldots,$

$$\eta(k) = E\left(A^*_{\theta(k)} \eta(k+1) \mid \mathfrak{F}_{k-1}\right) + \mathcal{E}_{\theta(k-1)}(C^*C)\bar{x}(k).$$

Proof. First of all note, from the matrix inversion lemma (see [6], p. 348), that

$$\eta(k+1) =$$

$$\left(I - \mathcal{E}_{\theta(k)}(X)B_{\theta(k)}\left(D^*_{\theta(k)} D_{\theta(k)} + B^*_{\theta(k)} \mathcal{E}_{\theta(k)}(X)B_{\theta(k)}\right)^{-1} B^*_{\theta(k)}\right)$$
$$\times \left(\mathcal{E}_{\theta(k)}(X)A_{\theta(k)}\bar{x}(k) + r(k)\right)$$

and therefore,

$$A^*_{\theta(k)} \eta(k+1) = \left(A^*_{\theta(k)} \mathcal{E}_{\theta(k)}(X)A_{\theta(k)}\right.$$

$$\left. - A^*_{\theta(k)} \mathcal{E}_{\theta(k)}(X)B_{\theta(k)} \mathcal{R}_{\theta(k)}(X)^{-1} B^*_{\theta(k)} \mathcal{E}_{\theta(k)}(X)A_{\theta(k)}\right)\bar{x}(k)$$

$$+ (A_{\theta(k)} + B_{\theta(k)} \mathcal{F}_{\theta(k)}(X))^* r(k)$$
$$= (X_{\theta(k)} - C^*_{\theta(k)} C_{\theta(k)})\bar{x}(k) + (A_{\theta(k)} + B_{\theta(k)} \mathcal{F}_{\theta(k)}(X))^* r(k). \qquad (4.52)$$

Moreover from (4.48), and the fact that $w(k)$ is $\mathfrak{F}_k$-measurable,

$$r(k) = \mathcal{E}_{\theta(k)}(X)G_{\theta(k)}w(k)$$

$$+ \sum_{j=1}^{\infty} E\left(\left(\prod_{\ell=k+j}^{k+1}(A_{\theta(\ell)} + B_{\theta(\ell)} \mathcal{F}_{\theta(\ell)}(X))^*\right) X_{\theta(k+j+1)} G_{\theta(k+j)} w(k+j) \mid \mathfrak{F}_k\right)$$

$$= \mathcal{E}_{\theta(k)}(X)G_{\theta(k)}w(k) + E\left((A_{\theta(k+1)} + B_{\theta(k+1)} \mathcal{F}_{\theta(k+1)}(X))^* r(k+1) \mid \mathfrak{F}_k\right).$$
$$(4.53)$$

From (4.52), (4.53) and the fact that $\bar{x}(k)$ is $\mathfrak{F}_{k-1}$-measurable, it follows that

$$E\left(A_{\theta(k)}^{*}\eta(k+1)\mid \mathfrak{F}_{k-1}\right)$$
$$= E\left((X_{\theta(k)} - C_{\theta(k)}^{*}C_{\theta(k)})\bar{x}(k)\mid \mathfrak{F}_{k-1}\right)$$
$$\quad + E\left((A_{\theta(k)} + B_{\theta(k)}\mathcal{F}_{\theta(k)}(X))^{*}r(k)\mid \mathfrak{F}_{k-1}\right)$$
$$= \mathcal{E}_{\theta(k-1)}(X)\bar{x}(k) - \mathcal{E}_{\theta(k-1)}(C^{*}C)\bar{x}(k) + r(k-1)$$
$$\quad - \mathcal{E}_{\theta(k-1)}(X)G_{\theta(k-1)}w(k-1)$$
$$= -\mathcal{E}_{\theta(k-1)}(C^{*}C)\bar{x}(k) + (\mathcal{E}_{\theta(k-1)}(X)A_{\theta(k-1)}\bar{x}(k-1) + r(k-1))$$
$$\quad - \mathcal{E}_{\theta(k-1)}(X)B_{\theta(k-1)}\mathcal{R}_{\theta(k-1)}(X)^{-1}B_{\theta(k-1)}^{*}$$
$$\quad \times \left(\mathcal{E}_{\theta(k-1)}(X)A_{\theta(k-1)}\bar{x}(k-1) + r(k-1)\right)$$
$$= (D_{\theta(k-1)}^{*}D_{\theta(k-1)} - \mathcal{E}_{\theta(k-1)}(X)B_{\theta(k-1)}\mathcal{R}_{\theta(k-1)}(X)^{-1}B_{\theta(k-1)}^{*}$$
$$\quad \times \left(\mathcal{E}_{\theta(k-1)}(X)A_{\theta(k-1)}\bar{x}(k-1) + r(k-1))\right) - \mathcal{E}_{\theta(k-1)}(C^{*}C)\bar{x}(k)$$
$$= \eta(k) - \mathcal{E}_{\theta(k-1)}(C^{*}C)\bar{x}(k)$$

completing the proof. ☐

4.5.3 The Optimal Control Law

We can now proceed to the main theorem of this section, which gives the solution to (4.47) and mirrors its deterministic counterpart (see, for instance, [151] and [204]).

Theorem 4.14. *Suppose that there exists the mean square stabilizing solution* $X = (X_1, \ldots, X_N) \in \mathbb{H}^{n+}$ *to the CARE (4.35). Then the optimal cost of (4.46) and (4.47) is achieved for* $u = \bar{u}$, $\bar{u} = (\bar{u}(0), \bar{u}(1), \ldots) \in \mathcal{C}^m$ *as in (4.50), and*

$$\bar{\mathfrak{J}}(\theta_0, x_0, w) = \mathfrak{J}(\theta_0, x_0, w, \bar{u})$$
$$= \sum_{k=0}^{\infty} \operatorname{Re}\left(E(\eta(k+1)^{*}G_{\theta(k)}w(k))\right) + \operatorname{Re}\left(E(x_0^{*}\eta(0))\right).$$

Proof. For any $u = (u(0), \ldots) \in \mathcal{U}$, $x_0 \in \mathcal{C}_0^n$, $\theta_0 \in \Theta_0$, $x = (x_0, x(1), \ldots) \in \mathcal{C}^n$ given by (4.25), and recalling that $x(k), \bar{x}(k)$ are $\mathfrak{F}_{k-1}$-measurable, we have from Lemma 4.13 and the definition of $\eta(0)$, that (we set below $\mathcal{E}_{\theta(-1)}(V) = V_{\theta(0)}$ for any $V = (V_1, \ldots, V_N) \in \mathbb{H}^n$):

$$\sum_{k=0}^{\infty} E(x(k)^* C_{\theta(k)}^* C_{\theta(k)} \bar{x}(k) - w(k)^* G_{\theta(k)}^* \eta(k+1) - u(k)^* B_{\theta(k)}^* \eta(k+1))$$

$$= \sum_{k=0}^{\infty} E(E(x(k)^* C_{\theta(k)}^* C_{\theta(k)} \bar{x}(k) \mid \mathfrak{F}_{k-1})$$

$$- w(k)^* G_{\theta(k)}^* \eta(k+1) - u(k)^* B_{\theta(k)}^* \eta(k+1))$$

$$= \sum_{k=0}^{\infty} E(x(k)^* \mathcal{E}_{\theta(k-1)}(C^* C) \bar{x}(k) - (w(k)^* G_{\theta(k)}^* - u(k)^* B_{\theta(k)}^*) \eta(k+1))$$

$$= \sum_{k=0}^{\infty} E(x(k)^* (\eta(k) - E(A_{\theta(k)}^* \eta(k+1) \mid \mathfrak{F}_{k-1}))$$

$$- (x(k+1)^* - x(k)^* A_{\theta(k)}^*) \eta(k+1))$$

$$= \sum_{k=0}^{\infty} E(x(k)^* \eta(k) - x(k+1)^* \eta(k+1)$$

$$- x(k)^* A_{\theta(k)}^* \eta(k+1) + x(k)^* A_{\theta(k)}^* \eta(k+1))$$

$$= \sum_{k=0}^{\infty} E(x(k)^* \eta(k) - x(k+1)^* \eta(k+1))$$

$$= E(x^*(0) \eta(0))$$

since that $|E(x(k)^* \eta(k))| \le \|x(k)\|_2 \|\eta(k)\|_2 \to 0$ as $k \to \infty$. Thus, recalling from (4.51) that $u(k)^* B_{\theta(k)}^* \eta(k+1) = -u(k)^* D_{\theta(k)}^* D_{\theta(k)} \bar{u}(k)$, we get that

$$\mathfrak{J}(\theta_0, x_0, w, u) - 2\operatorname{Re}(E(x_0^* \eta(0)))$$

$$= \sum_{k=0}^{\infty} E\Big(\left\| C_{\theta(k)} x(k) \right\|^2 + \left\| D_{\theta(k)} u(k) \right\|^2 - 2\operatorname{Re}(x(k)^* C_{\theta(k)}^* C_{\theta(k)} \bar{x}(k)$$

$$- w(k)^* G_{\theta(k)}^* \eta(k+1) + u(k)^* D_{\theta(k)}^* D_{\theta(k)} \bar{u}(k)) + \left\| C_{\theta(k)} \bar{x}(k) \right\|^2$$

$$+ \left\| D_{\theta(k)} \bar{u}(k) \right\|^2 - \left\| C_{\theta(k)} \bar{x}(k) \right\|^2 - \left\| D_{\theta(k)} \bar{u}(k) \right\|^2 \Big)$$

$$= \sum_{k=0}^{\infty} E\Big(\left\| C_{\theta(k)} (x(k) - \bar{x}(k)) \right\|^2 + \left\| D_{\theta(k)} (u(k) - \bar{u}(k)) \right\|^2 - \left\| C_{\theta(k)} \bar{x}(k) \right\|^2$$

$$- \left\| D_{\theta(k)} \bar{u}(k) \right\|^2 + 2\operatorname{Re}(w(k)^* G_{\theta(k)}^* \eta(k+1)) \Big)$$

$$= \sum_{k=0}^{\infty} E\Big(\left\| C_{\theta(k)} (x(k) - \bar{x}(k)) \right\|^2 + \left\| D_{\theta(k)} (u(k) - \bar{u}(k)) \right\|^2 \Big) - \mathfrak{J}(\theta_0, x_0, w, \bar{u})$$

$$+ \sum_{k=0}^{\infty} 2\operatorname{Re}(E(\eta(k+1)^* G_{\theta(k)} w(k))).$$

From the expression above it is clear that the minimum of (4.47) is achieved at $u = \bar{u}$. □

4.5.4 An Application to a Failure Prone Manufacturing System

Let us consider a manufacturing system producing a single commodity. The demand will be represented by a sequence of independent and identically distributed positive second order random variables $\{w(k); k = 0, 1, \ldots\}$, with mean $E(w(k)) = \gamma > 0$. It is desired to make the production meet the demand. The manufacturing system is subject to occasional breakdowns, and therefore can be at two possible states: *working* and *out of order*. The transition between these two states satisfies a two-state discrete-time Markov chain. State 1 means that the manufacturing system is out of order, whereas state 2 means that it is working. Let $x(k)$ denote the inventory of the commodity at time k with initial value x_0, $u(k)$ the total production at time k, and $w(k)$ the demand at time k. We have that the inventory at time $k + 1$ will satisfy

$$x(k + 1) = x(k) + b_{\theta(k)} u(k) - w(k), \quad x(0) = x_0,$$

where $b_1 = 0$ and $b_2 = 1$. Here the transition between the states will be defined by $p_{11} \in (0, 1)$ and $p_{22} \in (0, 1)$. Our goal is to control the production $u(k)$ at time k so that to minimize the expected discounted cost

$$\mathfrak{J}(\theta_0, x_0, w, u) = \sum_{k=0}^{\infty} \rho^k E(m\, x(k)^2 + u(k)^2)$$

where $m > 0$ and $\rho \in (0, 1)$. Following Remark 4.11, we have that the above problem can be rewritten as

$$x^\rho(k + 1) = \rho^{1/2} x^\rho(k) + \rho^{1/2} b_{\theta(k)} u^\rho(k) - \rho^{1/2} w^\rho(k) \qquad (4.54)$$

where

$$x^\rho(k) = \rho^{k/2} x(k), \quad u^\rho(k) = \rho^{k/2} u(k), \quad w^\rho(k) = \rho^{k/2} w(k).$$

Clearly $w^\rho = (w^\rho(0), w^\rho(1), \ldots)$, satisfies $\|w^\rho\|_2 < \infty$. In this case the cost to be minimized is given by

$$\mathfrak{J}^\rho(\theta_0, x_0, w, u) = \sum_{k=0}^{\infty} E(m\, x^\rho(k)^2 + u^\rho(k)^2). \qquad (4.55)$$

It is easy to verify that the system given by (4.54) and (4.55) satisfies the requirements of mean square stabilizability and detectability, so that Theorem 4.14 can be applied. Writing $X = (X_1^+, X_2^+)$ the solution of the set of two-coupled Riccati equations, we have that it is the unique positive solution of

$$X_1^+ = m + \rho(p_{11}X_1^+ + p_{12}X_2^+) \Rightarrow X_1^+ = \frac{m + \rho p_{12}X_2^+}{1 - \rho p_{11}} \tag{4.56a}$$

$$X_2^+ = m + \rho(p_{21}X_1^+ + p_{22}X_2^+) - \frac{\rho^2(p_{21}X_1^+ + p_{22}X_2^+)^2}{1 + \rho(p_{21}X_1^+ + p_{22}X_2^+)}$$

$$= \frac{m + (m+1)\rho(p_{21}X_1^+ + p_{22}X_2^+)}{1 + \rho(p_{21}X_1^+ + p_{22}X_2^+)} \tag{4.56b}$$

and

$$\mathcal{F}_1(X^+) = 0, \quad -1 < \mathcal{F}_2(X^+) = \frac{-\rho\mathcal{E}_2(X^+)}{1 + \rho\mathcal{E}_2(X^+)} < 0 \tag{4.57}$$

where

$$\mathcal{E}_i(X^+) = p_{i1}X_1^+ + p_{i2}X_2^+, \ i = 1, 2.$$

From expression (4.53) we have

$$r(k) = -\rho^{(k+1)/2}(\mathcal{E}_{\theta(k)}(X^+)w(k) + \tilde{r}_{\theta(k)}\gamma) \tag{4.58}$$

where

$$\tilde{r}_i = \sum_{j=1}^{\infty} \rho^{j/2} E\left(\prod_{\ell=j}^{1} \rho^{1/2}(1 + b_{\theta(\ell)}\mathcal{F}_{\theta(\ell)}(X)^+)X_{\theta(j+1)}^+ \mid \theta(0) = i \right), \ i = 1, 2.$$

The above expression leads to

$$\tilde{r}_1 = p_{11}\rho\mathcal{E}_1(X^+) + p_{12}\rho(1 + \mathcal{F}_2(X^+))\mathcal{E}_2(X^+) + p_{11}\rho\tilde{r}_1 + p_{12}\rho(1 + \mathcal{F}_2(X^+))\tilde{r}_2$$
$$\tilde{r}_2 = p_{21}\rho\mathcal{E}_1(X^+) + p_{22}\rho(1 + \mathcal{F}_2(X^+))\mathcal{E}_2(X^+) + p_{21}\rho\tilde{r}_1 + p_{22}\rho(1 + \mathcal{F}_2(X^+))\tilde{r}_2$$

that is,

$$\begin{bmatrix} \tilde{r}_1 \\ \tilde{r}_2 \end{bmatrix} = T^{-1}V \tag{4.59}$$

where

$$T = \begin{bmatrix} (1 - p_{11}\rho) & -p_{12}\rho(1 + \mathcal{F}_2(X^+)) \\ -p_{21}\rho & (1 - p_{22}\rho(1 + \mathcal{F}_2(X^+))) \end{bmatrix}$$

$$V = \begin{bmatrix} p_{11}\rho\mathcal{E}_1(X^+) + p_{12}\rho(1 + \mathcal{F}_2(X^+))\mathcal{E}_2(X^+) \\ p_{21}\rho\mathcal{E}_1(X^+) + p_{22}\rho(1 + \mathcal{F}_2(X^+))\mathcal{E}_2(X^+) \end{bmatrix}.$$

From (4.56), (4.57), and (4.59) we obtain X_1^+, X_2^+, $\mathcal{F}_2(X^+)$, $\tilde{r}_1$ and $\tilde{r}_2$. From (4.58) we get

$$\bar{r}(k) = \frac{\rho^{1/2}}{1 + \rho\mathcal{E}_2(X^+)} \left(\rho^{(k+1)/2}(\mathcal{E}_2(X^+)w(k) + \tilde{r}_2\gamma) \right) \mathbf{1}_{\{\theta(k)=2\}}$$

which yields the optimal control law

$$\bar{u}^{\rho}(k) = \bar{u}(k)\rho^{k/2}$$

$$= \left(\mathcal{F}_2(X^+)\bar{x}(k)\rho^{k/2} \right.$$

$$\left. + \frac{\rho^{1/2}}{1 + \rho\mathcal{E}_2(X^+)}\rho^{(k+1)/2} \left(\mathcal{E}_2(X^+)w(k) + \tilde{r}_2\gamma \right) \right) \mathbf{1}_{\{\theta(k)=2\}}$$

that is,

$$\bar{u}(k) = \left(-\mathcal{F}_2(X^+)(w(k) - \bar{x}(k)) + c\gamma \right) \mathbf{1}_{\{\theta(k)=2\}} \qquad (4.60)$$

where

$$c = \frac{\rho}{1 + \rho\mathcal{E}_2(X^+)}\tilde{r}_2$$

and

$$\bar{x}(k+1) = \bar{x}(k) + b_{\theta(k)}\bar{u}(k) - w(k). \qquad (4.61)$$

Notice that $\bar{u}(k)$ will always be positive if $x_0 \le w(0) + \frac{1}{-\mathcal{F}_2(X^+)}c\gamma$. Indeed, we have the following result.

Proposition 4.15. *If $x_0 \le w(0) + \frac{1}{-\mathcal{F}_2(X^+)}c\gamma$ then*

$$\bar{x}(k) \le w(k) + \frac{1}{-\mathcal{F}_2(X^+)}c\gamma$$

for all $k = 0, 1, \ldots$.

Proof. Applying induction on k, we have that the result holds for $k = 0$ by hypothesis. Suppose now that it holds for k. Then, by the induction hypothesis,

$$0 \le b_{\theta(k)}\bar{u}(k) \le -\mathcal{F}_2(X^+)(w(k) - \hat{x}(k)) + c\gamma$$

so that

$$\bar{x}(k+1) = \bar{x}(k) + b_{\theta(k)}\bar{u}(k) - w(k)$$
$$\le \bar{x}(k) - \mathcal{F}_2(X^+)(w(k) - \bar{x}(k)) + c\gamma - w(k)$$
$$= (1 + \mathcal{F}_2(X^+))(\bar{x}(k) - w(k)) + c\gamma$$
$$\le (1 + \mathcal{F}_2(X^+))\frac{1}{-\mathcal{F}_2(X^+)}c\gamma + c\gamma = \frac{1}{-\mathcal{F}_2(X^+)}c\gamma$$
$$\le w(k+1) + \frac{1}{-\mathcal{F}_2(X^+)}c\gamma.$$

$\square$

Thus (4.60) and (4.61) give the optimal control and optimal inventory even under the restriction that $u(k) \ge 0$ for all $k = 0, 1, \ldots$, provided that

$$x_0 \le w(0) + \frac{1}{-\mathcal{F}_2(X^+)}c\gamma.$$

A simple case is presented in the following example.

Example 4.16. Consider a manufacturing system producing a single commodity as described above, with inventory $x(k)$ satisfying $x(k+1) = x(k) + b_{\theta(k)}u(k) - w(k)$. The initial amount of the commodity is assumed to be $x(0) = x_0 = 0$. The demand is represented by the sequence of i.i.d. positive second order random variables $\{w(k); k = 0, 1, \dots, \}$, with mean $E(w(k)) = \gamma = 1$. $\{\theta(k); k = 0, 1, \dots\}$ represents a Markov chain taking values in $\{1, 2\}$, where state 1 means that the manufacturing system is out of order, with $b_1 = 0$, and state 2 means that the manufacturing system is working, with $b_2 = 1$. The transition probability matrix P of the Markov chain $\{\theta(k); k = 0, 1, \dots\}$ is given by

$$P = \begin{bmatrix} 0.1 & 0.9 \\ 0.05 & 0.95 \end{bmatrix}$$

and it is desired to find the optimal production $\{\bar{u}(k) \geq 0; k = 0, 1, \dots\}$ which minimizes the following discounted expected cost:

$$\sum_{k=0}^{\infty} 0.95^k E(x(k)^2 + u(k)^2).$$

Solving (4.56), (4.57), (4.58) and (4.59), we obtain

$$X_1^+ = 2.6283, \; X_2^+ = 1.6124, \; \mathcal{F}_2(X^+) = -0.6124, \; \tilde{r}_1 = 1.1930, \; \tilde{r}_2 = 1.1034$$

and from (4.60) the optimal control law is given by

$$\bar{u}(k) = (0.6124(w(k) - \bar{x}(k)) + 0.4063)\, \mathbf{1}_{\{\theta(k)=2\}}$$

where $\bar{x}(k)$ is given by (4.61). Note that from Proposition 4.15, $\bar{u}(k) \geq 0$ for all $k = 0, 1, \dots$.

4.6 Historical Remarks

Optimal control theory has certainly played an important role in the building of modern technological society. There are by now fairly well developed mathematical theories of optimal control (stochastic and deterministic). Although the first major theoretical works on optimal control date back to the late 1950s, a continued impetus has been given to this area either, for instance, by some relevant unresolved theoretical issues, or motivated by applications (see, e.g., [111]). For those interested in a nice introduction to the linear theory we refer, for instance, to [79] and [80] (see also the classical [112]). For recent advances we refer to [214], [224], and references therein.

A cursory examination of the literature reveals that there is by now a satisfactory collection of results dealing with quadratic (and H_2) optimal control for MJLS, for the case of complete observations of the augmented state (x, θ). An initial trickle of papers ([206] and [219]), which has lain fallow for more than a decade, grew to a fairly substantial amount of papers in the last two decades. Without any intention of being exhaustive, we mention, for instance, [2], [28], [52], [53], [56], [62], [64], [65], [78], [115], [137], [144], [147], [171], [176], [206] and [219], as a representative sample (see also [67] and [117] for the infinite countable case and [22] for the robust continuous-time case). Some issues regarding the associated coupled Riccati equation, including the continuous-time case, can be found in [1], [59], [60], [84], [90], [91], [122] and [192]. Quadratic optimal control problems of MJLS subject to constraints on the state and control variables are considered in [75]. An iterative Monte Carlo technique for deriving the optimal control of the infinite horizon linear regulator problem of MJLS for the case in which the transition probability matrix of the Markov chain is not known is analyzed in [76]. Several results on control problems for jump linear systems are presented in the recent book [77]. Other topics that deserve attention in this scenario are, for instance, those related to a risk-sensitive (see [217]) approach for the optimal control of MJLS.

5

Filtering

Filtering problems are of interest not only because of their wide number of applications but also for being the main steps in studying control problems with partial observations on the state variable. There is nowadays a huge body of theory on this subject, having the celebrated Kalman filter as one of the great achievements. This chapter is devoted to the study of filtering problems for MJLS. Two situations are considered: the first one deals with the class of Markov jump filters, which will be essential in devising a separation principle in Chapter 6. In this case the jump variable is assumed accessible. In the second situation the jump variable is not accessible, and we derive the minimum linear mean square error filter and analyze the associated stationary filter. The solutions will be obtained in terms of Riccati difference equations and algebraic Riccati equations.

5.1 Outline of the Chapter

In this chapter we deal with the filtering problem for the class of dynamical MJLS described in the previous chapters by $\mathcal{G}$. First, following the pattern on the previous chapters, we consider the case in which the Markov chain $\{\theta(k)\}$ is assumed to be known. We consider in Section 5.2 the finite horizon case, and in Section 5.3 the infinite horizon case. The results developed in this chapter will be used in a separation principle, to be discussed in Chapter 6.

Next, we discuss the case in which the Markov chain is not observed. It is well known (see, for instance, [17]) that for this case the number of filters for the optimal nonlinear state estimation increases exponentially with time, which makes this approach impractical. In Section 5.4 we provide the optimal linear minimum mean square filter (LMMSE), which can be easily implemented on line. In addition, we obtain sufficient conditions for the convergence of the error covariance matrix to a stationary value for the LMMSE. Under the assumption of mean square stability of the MJLS and ergodicity of the associated Markov chain it is shown that there exists a unique solution for the stationary Riccati filter equation and, moreover, this solution is the limit of the error covariance matrix of the LMMSE. This result is suitable for designing a time-invariant stable suboptimal filter of LMMSE for MJLS. As shown in Section 5.5, this stationary filter can also be derived from an LMI optimization problem. The advantage of this formulation is that through the LMI approach we can consider uncertainties in the parameters of the system. In Subsections 8.4.1 and 8.4.2 we present some numerical examples of the LMMSE filter with the IMM filter (see [32]), which is another suboptimal filter used to alleviate the numerical difficulties mentioned above.

5.2 Finite Horizon Filtering with $\theta(k)$ Known

We consider in this section the finite horizon minimum mean square linear Markov jump filter for a MJLS when an output $y(k)$ and jump variable $\theta(k)$ are available. It is important to stress here that we are restricting our attention to the family of linear Markov jump filters (that is, a filter that depends just on the present value of the Markov parameter), since otherwise the optimal linear mean square filter would be obtained from a sample path Kalman filter, as explained in Remark 5.2 below. The case in which the jump variable is not available will be presented in Section 5.4. On the stochastic basis $(\Omega, \mathfrak{F}, \{\mathfrak{F}_k\}, \mathcal{P})$, consider the following MJLS:

$$\mathcal{G} = \begin{cases} x(k+1) = A_{\theta(k)}(k)x(k) + B_{\theta(k)}(k)u(k) + G_{\theta(k)}(k)w(k) \\ y(k) = L_{\theta(k)}(k)x(k) + H_{\theta(k)}(k)w(k) \\ x(0) = x_0 \in \mathcal{C}_0^n, \theta(0) = \theta_0 \in \Theta_0 \end{cases} \tag{5.1}$$

where as in Subsection 4.2

$$A(k) = (A_1(k), \dots, A_N(k)) \in \mathbb{H}^n,$$
$$B(k) = (B_1(k), \dots, B_N(k)) \in \mathbb{H}^{m,n},$$
$$G(k) = (G_1(k), \dots, G_N(k)) \in \mathbb{H}^{r,n},$$

and

$$L(k) = (L_1(k), \dots, L_N(k)) \in \mathbb{H}^{n,p},$$
$$H(k) = (D_1(k), \dots, D_N(k)) \in \mathbb{H}^{r,p}.$$

The input sequence $\{w(k); k \in \mathbb{T}\}$ is a r-dimensional wide sense white noise sequence, (that is, $E(w(k)) = 0$, $E(w(k)w(k)^*) = I$, $E(w(k)w(l)^*) = 0$ for $k \neq l$), and $\{y(k); k \in \mathbb{T}\}$ is the p-dimensional sequence of measurable variables. The output and operation modes $(y(k), \theta(k)$ respectively) are known at each time k. The noise $\{w(k); k \in \mathbb{T}\}$ and the Markov chain $\{\theta(k); k \in \mathbb{T}\}$ are independent sequences, and the initial condition (x_0, θ_0) are such that x_0 and θ_0 are independent random variables with $E(x_0) = \mu_0$ and $E(x_0 x_0^*) = \mathcal{Q}_0$. The Markov chain $\{\theta(k)\}$ has time-varying transition probabilities $p_{ij}(k)$.

We assume that

$$G_i(k)H_i(k)^* = 0 \tag{5.2}$$
$$H_i(k)H_i(k)^* > 0. \tag{5.3}$$

Notice that condition (5.3) makes sure that all components of the output and their linear combinations are noisy. As we are going to see in Remark 5.1, there is no loss of generality in assuming (5.2).

We consider dynamic Markov jump filters $\mathcal{G}_K$ given by

$$\mathcal{G}_K = \begin{cases} \widehat{x}(k+1) = \widehat{A}_{\theta(k)}(k)\widehat{x}(k) + \widehat{B}_{\theta(k)}(k)y(k) \\ u(k) = \widehat{C}_{\theta(k)}(k)\widehat{x}(k) \\ \widehat{x}(0) = \widehat{x}_0. \end{cases} \tag{5.4}$$

The reason for choosing this kind of filter is that they depend just on $\theta(k)$ (rather than on the entire past history of modes $\theta(0), \ldots, \theta(k)$), so that the closed loop system is again a MJLS. In particular, as will be seen in Remark 5.6, time-invariant parameters can be considered as an approximation for the optimal solution.

Therefore in the optimal filtering problem we want to find $\widehat{A}_i(k)$, $\widehat{B}_i(k)$, $\widehat{C}_i(k)$ in (5.4) with $\widehat{x}_0$ deterministic, such as to minimize the cost

$$\sum_{k=1}^{T} E(\|v(k)\|^2)$$

where

$$v(k) = x(k) - \widehat{x}(k). \tag{5.5}$$

It will be shown that the solution to this problem is associated to a set of filtering coupled Riccati difference equations given by (5.13) below.

Remark 5.1. There is no loss of generality in considering $G_i(k)H_i(k)^* = 0$, $i \in \mathbb{N}$ as in (5.2). Indeed, notice that (5.1) can be rewritten as

$$\mathcal{G} = \begin{cases} x(k+1) = \bar{A}_{\theta(k)}(k)x(k) + \bar{G}_{\theta(k)}(k)w(k) + \bar{B}_{\theta(k)}(k)y(k) + B_{\theta(k)}(k)u(k) \\ y(k) = L_{\theta(k)}(k)x(k) + H_{\theta(k)}(k)w(k) \\ x(0) = x_0 \in \mathcal{C}_0^n, \theta(0) = \theta_0 \in \Theta_0 \end{cases} \tag{5.6}$$

where

$$\bar{A}_i(k) = A_i(k) - G_i(k)H_i(k)^*(H_i(k)H_i(k)^*)^{-1}L_i(k), \tag{5.7}$$
$$\bar{G}_i(k) = G_i(k)(I - H_i(k)^*(H_i(k)H_i(k)^*)^{-1}H_i(k)) \tag{5.8}$$

and

$$\bar{B}_i(k) = G_i(k)H_i(k)^*(H_i(k)H_i(k)^*)^{-1}. \tag{5.9}$$

Therefore the optimal filtering problem in this case is to find $\widehat{A}_i(k), \widehat{B}_i(k), \widehat{C}_i(k)$ in

$$
\mathcal{G}_K = \begin{cases}
\widehat{x}(k+1) = \widehat{A}_{\theta(k)}(k)\widehat{x}(k) + (\widehat{B}_{\theta(k)} + \bar{B}_{\theta(k)}(k))y(k) \\
u(k) = \widehat{C}_{\theta(k)}(k)\widehat{x}(k) \\
\widehat{x}(0) = \widehat{x}_0
\end{cases}
\tag{5.10}
$$

with $\widehat{x}_0$ deterministic, such that it minimizes $\sum_{k=1}^{T} E(\|v(k)\|^2)$ with $v(k)$ as in (5.5), replacing $A_i(k)$ and $G_i(k)$ by $\bar{A}_i(k)$ and $\bar{G}_i(k)$ as in (5.7) and (5.8) respectively. When we take the difference between $x(k)$ in (5.1) and $\widehat{x}(k)$ in (5.10) the term $\bar{B}_{\theta k}(k)y(k)$ vanishes, and we return to the original problem. Notice that in this case condition (5.2) would be satisfied.

Remark 5.2. It is well known that for the case in which $(y(k), \theta(k))$ are available, the best linear estimator of $x(k)$ is derived from the Kalman filter for time varying systems (see [52], [80]), since all the values of the mode of operation are known at time k. Indeed, the recursive equation for the covariance error matrix $Z(k)$ and the gain of the filter $K(k)$ would be as follows:

$$
\begin{aligned}
Z(k+1) =& A_{\theta(k)}(k)Z(k)A_{\theta(k)}(k)^* + G_{\theta(k)}(k)G_{\theta(k)}(k)^* \\
& - A_{\theta(k)}(k)Z(k)L_{\theta(k)}(k)^*(H_{\theta(k)}(k)H_{\theta(k)}(k)^* \\
& + L_{\theta(k)}(k)Z(k)L_{\theta(k)}(k)^*)^{-1}L_{\theta(k)}(k)Z(k)A_{\theta(k)}(k)^* \\
K(k) =& - A_{\theta(k)}(k)Z(k)L_{\theta(k)}(k)^*(H_{\theta(k)}(k)H_{\theta(k)}(k)^* \\
& + L_{\theta(k)}(k)Z(k)L_{\theta(k)}(k)^*)^{-1}.
\end{aligned}
\tag{5.11}
$$

As pointed out in [146], off-line computation of the filter (5.11) is inadvisable since $Z(k)$ and $K(k)$ are sample path dependent, and the number of sample paths grows exponentially in time. Indeed on the time interval $[0, T]$ it would be necessary to pre-compute $N + N^2 + \ldots + N^T = N\frac{N^T-1}{N-1}$ gains. On the other hand, the optimal filter in the form of (5.4) requires much less pre-computed gains (NT instead of $N\frac{N^T-1}{N-1}$) and depends on just $\theta(k)$ at time k, which allows us to consider, as pointed out in Remark 5.6 below, an approximation by a Markov time-invariant filter.

Define $\mathbb{J}(k) \triangleq \{i \in \mathbb{N}; \pi_i(k) > 0\}$ (recall the definition of $\pi_i(k)$ in (3.28)) and

$$
\begin{cases}
\widehat{x}_e(k+1) \triangleq A_{\theta(k)}(k)\widehat{x}_e(k) + B_{\theta(k)}(k)u(k) - M_{\theta(k)}(k)(y(k) - L_{\theta(k)}(k)\widehat{x}_e(k)) \\
\widehat{x}_e(0) \triangleq E(x_0) = \mu_0
\end{cases}
\tag{5.12}
$$

where $M_i(k)$ is defined from the following filtering coupled Riccati difference equations:

$$Y_j(k+1) \triangleq \sum_{i \in \mathbb{J}(k)} p_{ij}(k) \Big[A_i(k) Y_i(k) A_i(k)^* - A_i(k) Y_i(k) L_i(k)^*$$
$$\times (H_i(k) H_i(k)^* \pi_i(k) + L_i(k) Y_i(k) L_i(k)^*)^{-1} L_i(k) Y_i(k) A_i(k)^*$$
$$+ \pi_i(k) G_i(k) G_i(k)^* \Big], \tag{5.13}$$
$$Y_i(0) \triangleq \pi_i(0)(\mathbb{Q}_0 - \mu_0 \mu_0^*)$$

and

$$M_i(k) \triangleq \begin{cases} -A_i(k) Y_i(k) L_i(k)^* (H_i(k) H_i(k)^* \pi_i(k) + L_i(k) Y_i(k) L_i(k)^*)^{-1}, \\ \quad \text{for } i \in \mathbb{J}(k) \\ 0, \text{ for } i \notin \mathbb{J}(k). \end{cases}$$
$$\tag{5.14}$$

The associated error related with the estimator given in (5.12) is defined by

$$\widetilde{x}_e(k) \triangleq x(k) - \widehat{x}_e(k) \tag{5.15}$$

and from (5.1) and (5.12) we have that

$$\widetilde{x}_e(k+1) = [A_{\theta(k)}(k) + M_{\theta(k)}(k) L_{\theta(k)}(k)] \widetilde{x}_e(k)$$
$$+ [G_{\theta(k)}(k) + M_{\theta(k)}(k) H_{\theta(k)}(k)] w(k) \tag{5.16}$$
$$\widetilde{x}_e(0) = x_0 - E(x_0) = x(0) - \mu_0.$$

Using the identity (A.2), it follows that $Y_i(k)$ in (5.13) can also be written as

$$Y_j(k+1) = \sum_{i \in \mathbb{J}(k)} p_{ij}(k)[(A_i(k) + M_i(k) L_i(k)) Y_i(k)(A_i(k) + M_i(k) L_i(k))^*$$
$$+ \pi_i(k)(G_i(k) + M_i(k) L_i(k))(G_i(k) + M_i(k) L_i(k))^*]. \tag{5.17}$$

Set, for $i \in \mathbb{N}$,
$$\bar{Y}_i(k) \triangleq E(\widetilde{x}_e(k) \widetilde{x}_e(k)^* \mathbf{1}_{\{\theta(k)=i\}}).$$

From (5.16) and Proposition 3.35 2), we have that

$$\bar{Y}_i(k+1) = \sum_{i=1}^{N} p_{ij}(k)[(A_i(k) + M_i(k) L_i(k)) \bar{Y}_i(k)(A_i(k) + M_i(k) L_i(k))^*$$
$$+ \pi_i(k)(G_i(k) + M_i(k) L_i(k))(G_i(k) + M_i(k) L_i(k))^*] \tag{5.18}$$
$$\bar{Y}_i(0) = \pi_i(0)(\mathbb{Q}_0 - \mu_0 \mu_0^*).$$

Noticing that whenever $\pi_i(k) = 0$, we have

$$E(\widetilde{x}_e(k) \widetilde{x}_e(k)^* \mathbf{1}_{\{\theta(k)=i\}}) = 0,$$

we can rewrite (5.18) as

$$\bar{Y}_i(k+1) = \sum_{i \in \mathbb{J}(k)} p_{ij}(k)[(A_i(k) + M_i(k)L_i(k))\bar{Y}_i(k)(A_i(k) + M_i(k)L_i(k))^*$$

$$+ \pi_i(k)(G_i(k) + M_i(k)L_i(k))(G_i(k) + M_i(k)L_i(k))^*]. \qquad (5.19)$$

Comparing (5.17) and (5.19) we can check by induction on k that

$$Y_i(k) = \bar{Y}_i(k) = E(\tilde{x}_e(k)\tilde{x}_e(k)^* \mathbf{1}_{\{\theta(k)=i\}}). \qquad (5.20)$$

We have the following theorems.

Theorem 5.3. *For $\hat{x}(k), \hat{x}_e(k), \tilde{x}_e(k)$ given by (5.4), (5.12) and (5.16) respectively and $i \in \mathbb{N}$, $k = 0, 1, \ldots$, we have that*

$$E(\tilde{x}_e(k)\hat{x}_e(k)^* \mathbf{1}_{\{\theta(k)=i\}}) = 0,$$
$$E(\tilde{x}_e(k)\hat{x}(k)^* \mathbf{1}_{\{\theta(k)=i\}}) = 0.$$

Proof. Let us prove it by induction on k. For $k = 0$ it is clearly true since from the independence of $x(0)$ and $\theta(0)$, the fact that $\hat{x}(0)$ is deterministic, and $E(\tilde{x}_e(0)) = 0$, we have that $E(\tilde{x}_e(0)\hat{x}(0)^* \mathbf{1}_{\{\theta(0)=i\}}) = \pi_i(0)E(\tilde{x}_e(0))\hat{x}(0)^* = 0$, and similarly $E(\tilde{x}_e(0)\hat{x}_e(0)^* \mathbf{1}_{\{\theta(0)=i\}}) = 0$. Suppose it holds for k. Then from (5.4) and (5.16), the induction hypothesis for k, and since $w(k)$ has null mean and is not correlated with $\hat{x}(k)$ and independent from $\theta(k)$, we have,

$$E(\tilde{x}_e(k+1)\hat{x}(k+1)^* \mathbf{1}_{\{\theta(k+1)=j\}})$$

$$= \sum_{i \in \mathbb{J}(k)} p_{ij}(k)\Big[(A_i(k) + M_i(k)L_i(k))E(\tilde{x}_e(k)\hat{x}(k)^* \mathbf{1}_{\{\theta(k)=i\}})\hat{A}_i(k)^*$$

$$+ (A_i(k) + M_i(k)L_i(k))E(\tilde{x}_e(k)y(k)^* \mathbf{1}_{\{\theta(k)=i\}})\hat{B}_i(k)^*$$

$$+ (G_i(k) + M_i(k)H_i(k))E(w(k)\hat{x}(k)^* \mathbf{1}_{\{\theta(k)=i\}})\hat{A}_i(k)^*$$

$$+ (G_i(k) + M_i(k)H_i(k))E(w(k)y(k)^* \mathbf{1}_{\{\theta(k)=i\}})\hat{B}_i(k)^*\Big]$$

$$= \sum_{i \in \mathbb{J}(k)} p_{ij}(k)\Big[(A_i(k) + M_i(k)L_i(k))E(\tilde{x}_e(k)y(k)^* \mathbf{1}_{\{\theta(k)=i\}})$$

$$+ (G_i(k) + M_i(k)H_i(k))E(w(k)y(k)^* \mathbf{1}_{\{\theta(k)=i\}})\Big]\hat{B}_i(k)^*.$$

Notice that

$$y(k) = L_{\theta(k)}(k)x(k) + H_{\theta(k)}(k)w(k)$$
$$= L_{\theta(k)}(k)(\tilde{x}_e(k) + \hat{x}_e(k)) + H_{\theta(k)}(k)w(k)$$

and from the induction hypothesis on k, we have that for $i \in \mathbb{J}(k)$,

$$E(\tilde{x}_e(k)y(k)^* \mathbf{1}_{\{\theta(k)=i\}})$$
$$= E(\tilde{x}_e(k)\tilde{x}_e(k)^* \mathbf{1}_{\{\theta(k)=i\}})L_i(k)^* + E(\tilde{x}_e(k)\hat{x}_e(k)^* \mathbf{1}_{\{\theta(k)=i\}})L_i(k)^*$$
$$+ E(\tilde{x}_e(k)w(k)^* \mathbf{1}_{\{\theta(k)=i\}})H_i(k)^*$$
$$= Y_i(k)L_i(k)^*.$$

We also have that

$$E(w(k)y(k)^*\mathbf{1}_{\{\theta(k)=i\}})$$
$$= E(w(k)\widetilde{x}_e(k)^*\mathbf{1}_{\{\theta(k)=i\}})L_i(k)^* + E(w(k)\widehat{x}_e(k)^*\mathbf{1}_{\{\theta(k)=i\}})L_i(k)^*$$
$$+ E(w(k)w(k)^*\mathbf{1}_{\{\theta(k)=i\}})H_i(k)^*$$
$$= E(w(k)w(k)^*)\mathcal{P}(\theta(k) = i)H_i(k)^* = \pi_i(k)H_i(k)^*.$$

Thus, recalling that by hypothesis $G_i(k)H_i(k)^* = 0$, and from (5.14),

$$M_i(k)\left(L_i(k)Y_i(k)L_i(k)^* + \pi_i(k)H_i(k)H_i(k)^*\right) = -A_i(k)Y_i(k)L_i(k)^*,$$

we have that

$$E(\widetilde{x}_e(k + 1)\widehat{x}(k + 1)^*\mathbf{1}_{\{\theta(k+1)=j\}})$$
$$= \sum_{i\in\mathbb{J}(k)} p_{ij}(k)\Big[(A_i(k) + M_i(k)L_i(k))Y_i(k)L_i(k)^*$$
$$+ \big(G_i(k) + M_i(k)H_i(k)\big)H_i(k)^*\pi_i(k)\Big]\widehat{B}_i(k)^*$$
$$= \sum_{i\in\mathbb{J}(k)} p_{ij}(k)\Big[A_i(k)Y_i(k)L_i(k)^*$$
$$+ M_i(k)\big(L_i(k)Y_i(k)L_i(k)^* + \pi_i(k)H_i(k)H_i(k)^*\big)\Big]\widehat{B}_i(k)^*$$
$$= \sum_{i=1}^{N} p_{ij}(k)\Big[A_i(k)Y_i(k)L_i(k)^* - A_i(k)Y_i(k)L_i(k)^*\Big]\widehat{B}_i(k)^*$$
$$= 0.$$

Similarly,

$$E(\widetilde{x}_e(k + 1)\widehat{x}_e(k + 1)^*\mathbf{1}_{\{\theta(k+1)=j\}}) = 0,$$

proving the result.   □

Theorem 5.4. *Let $v(k)$ and $(Y_1(k),\ldots,Y_N(k))$ be as in (5.5) and (5.13) respectively. Then for every $k = 0, 1, \ldots$, $E(\|v(k)\|^2) \geq \sum_{i=1}^{N}\mathrm{tr}(Y_i(k))$.*

Proof. Recalling from Theorem 5.3 that

$$E(\widetilde{x}_e(k)\widehat{x}_e(k)^*\mathbf{1}_{\{\theta(k)=i\}}) = 0,$$
$$E(\widetilde{x}_e(k)\widehat{x}(k)^*\mathbf{1}_{\{\theta(k)=i\}}) = 0,$$

and from (5.20) that

$$E(\widetilde{x}_e(k)\widetilde{x}_e(k)^*\mathbf{1}_{\{\theta(k)=i\}}) = Y_i(k),$$

we have

$$E(\|v(k)\|^2) = E(\|x(k) - \widehat{x}_e(k) + \widehat{x}_e(k) - \widehat{x}(k)\|^2)$$

$$= \sum_{i=1}^{N} E((\|\widetilde{x}_e(k) + (\widehat{x}_e(k) - \widehat{x}(k)))\mathbf{1}_{\{\theta(k)=i\}}\|^2)$$

$$= \sum_{i=1}^{N} \operatorname{tr}\left(E\Big((\widetilde{x}_e(k) + (\widehat{x}_e(k) - \widehat{x}(k)))\right.$$

$$\left. \times (\widetilde{x}_e(k) + (\widehat{x}_e(k) - \widehat{x}(k)))^* \mathbf{1}_{\{\theta(k)=i\}}\Big)\right)$$

$$= \sum_{i=1}^{N} \operatorname{tr}(Y_i(k)) + E(\|\widehat{x}_e(k) - \widehat{x}(k)\|^2)$$

$$+ \sum_{i=1}^{N} \operatorname{tr}(E(\widetilde{x}_e(k)\widehat{x}_e(k)^* \mathbf{1}_{\{\theta(k)=i\}}) - E(\widetilde{x}_e(k)\widehat{x}(k)^* \mathbf{1}_{\{\theta(k)=i\}}))$$

$$= \sum_{i=1}^{N} \operatorname{tr}(Y_i(k)) + E(\|\widehat{x}_e(k) - \widehat{x}(k)\|^2) \geq \sum_{i=1}^{N} \operatorname{tr}(Y_i(k))$$

completing the proof of the theorem.   □

The next theorem is straightforward from Theorem 5.4, and shows that the solution for the optimal filtering problem can be obtained from the filtering Riccati recursive equations $Y(k) = (Y_1(k), \ldots, Y_N(k))$ as in (5.13) and gains $M(k) = (M_1(k), \ldots, M_N(k))$ as in (5.14):

Theorem 5.5. *An optimal solution for the filtering problem posed above is:*

$$\widehat{A}_i^{op}(k) = A_i(k) + M_i(k)L_i(k) + B_i(k)\widehat{C}_i(k); \quad \widehat{B}_i^{op}(k) = -M_i(k); \quad (5.21)$$

with $\widehat{C}_i(k)$ arbitrary, and the optimal cost is $\sum_{k=1}^{T} \sum_{i=1}^{N} \operatorname{tr}(Y_i(k))$.

Remark 5.6. For the case in which A_i, G_i, L_i, H_i and p_{ij} in (5.1) are time-invariant and $\{\theta(k)\}$ satisfies the ergodic assumption 3.31, so that $\pi_i(k)$ converges to $\pi_i > 0$ as k goes to infinity, the filtering coupled Riccati difference equations (5.13) lead to the following filtering coupled algebraic Riccati equations and respective gains:

$$Y_j = \sum_{i=1}^{N} p_{ij}[A_iY_iA_i^* + \pi_i G_iG_i^* - A_iY_iL_i^*(H_iH_i^*\pi_i + L_iY_iL_i^*)^{-1}L_iY_iA_i^*]$$

$$(5.22)$$

$$M_i = -A_iY_iL_i^*(H_iH_i^*\pi_i + L_iY_iL_i^*)^{-1}. \quad (5.23)$$

In Appendix A is presented a sufficient condition for the existence of a unique solution $Y = (Y_1, \ldots, Y_N) \in \mathbb{H}^{n+}$ for (5.22), and convergence of $Y(k)$ to Y.

Convergence to the stationary state is often rapid, so that the optimal filter (5.21) could be approximated by the time-invariant Markov jump filter

$$\widehat{x}(k+1) = (A_{\theta(k)} + M_{\theta(k)}L_{\theta(k)})\widehat{x}(k) - M_{\theta(k)}y(k),$$

which just requires us to keep in memory the gains $M = (M_1, \ldots, M_N)$.

5.3 Infinite Horizon Filtering with $\theta(k)$ Known

We consider in this section the infinite horizon minimum mean square Markov jump filter for a MJLS when an output $y(k)$ and jump variable $\theta(k)$ are available. As in the previous section, we are restricting our attention to the family of linear Markov jump filters. On the stochastic basis $(\Omega, \mathfrak{F}, \{\mathfrak{F}_k\}, \mathcal{P})$, consider the following time-invariant version of the MJLS $\mathcal{G}$, denoted in this subsection by $\mathcal{G}_v$, seen in (5.1):

$$\mathcal{G}_v = \begin{cases} x(k+1) = A_{\theta(k)}x(k) + B_{\theta(k)}u(k) + G_{\theta(k)}w(k) \\ y(k) = L_{\theta(k)}x(k) + H_{\theta(k)}w(k) \\ v(k) = R_{\theta(k)}^{1/2}[-F_{\theta(k)}x(k) + u(k)] \\ x(0) = x_0 \in \mathcal{C}_0^n, \theta(0) = \theta_0 \in \Theta_0. \end{cases} \tag{5.24}$$

The input sequence $\{w(k); k \in \mathbb{T}\}$ is again a r-dimensional wide sense white noise sequence. We assume that $F = (F_1, \ldots, F_N)$ stabilizes (A, B) in the mean square sense (see Chapter 3), so that for each $i \in \mathbb{N}$,

$$R_i \geq 0,$$

and set

$$\bar{F}_i = -R_i^{1/2}F_i.$$

The output and operation modes ($y(k), \theta(k)$ respectively) are known at each time k. The noise $\{w(k); k \in \mathbb{T}\}$ and the Markov chain $\{\theta(k); k \in \mathbb{T}\}$ are independent sequences, and the initial condition (x_0, θ_0) is such that x_0 and θ_0 are independent random variables with $E(x_0) = \mu_0$ and $E(x_0x_0^*) = Q_0$. The Markov chain $\{\theta(k)\}$ has time-invariant transition probabilities p_{ij}, and satisfies the ergodic assumption 3.31, so that $\pi_i(k)$ converges to $\pi_i > 0$ for each $i \in \mathbb{N}$ as k goes to infinity.

We assume that

$$H_i H_i^* > 0 \tag{5.25}$$

and, without loss of generality (see Remark 5.1), that

$$G_i H_i^* = 0. \tag{5.26}$$

We consider dynamic Markov jump filters $\mathcal{G}_K$ given by

$$\mathcal{G}_K = \begin{cases} \widehat{x}(k+1) = \widehat{A}_{\theta(k)}\widehat{x}(k) + \widehat{B}_{\theta(k)}y(k) \\ u(k) = \widehat{C}_{\theta(k)}\widehat{x}(k) \\ \widehat{x}(0) = \widehat{x}_0. \end{cases} \quad (5.27)$$

As in the previous section, the reason for choosing this kind of filter is that they depend just on $\theta(k)$ (rather than on the entire past history of modes $\theta(0), \ldots, \theta(k)$), so that the closed loop system is again a (time-invariant) MJLS.

From (5.24) and (5.27) we have that the closed loop system is

$$\begin{bmatrix} x(k+1) \\ \widehat{x}(k+1) \end{bmatrix} = \begin{bmatrix} A_{\theta(k)} & B_{\theta(k)}\widehat{C}_{\theta(k)} \\ \widehat{B}_{\theta(k)}L_{\theta(k)} & \widehat{A}_{\theta(k)} \end{bmatrix} \begin{bmatrix} x(k) \\ \widehat{x}(k) \end{bmatrix} + \begin{bmatrix} G_{\theta(k)} \\ \widehat{B}_{\theta(k)}H_{\theta(k)} \end{bmatrix} w(k)$$

$$v(k) = R_{\theta(k)}^{1/2}[-F_{\theta(k)} \quad \widehat{C}_{\theta(k)}] \begin{bmatrix} x(k) \\ \widehat{x}(k) \end{bmatrix}. \quad (5.28)$$

Writing

$$\Gamma_i \triangleq \begin{bmatrix} A_i & B_i\widehat{C}_i \\ \widehat{B}_iL_i & \widehat{A}_i \end{bmatrix}; \; \Psi_i \triangleq \begin{bmatrix} G_i \\ \widehat{B}_iH_i \end{bmatrix}; \; \Phi_i \triangleq R_i^{1/2}[-F_i \quad \widehat{C}_i]; \; \mathbf{v}(k) \triangleq \begin{bmatrix} x(k) \\ \widehat{x}(k) \end{bmatrix} \quad (5.29)$$

we have from (5.28) that the Markov jump closed loop system $\mathcal{G}_{cl}$ is given by

$$\mathcal{G}_{cl} = \begin{cases} \mathbf{v}(k+1) = \Gamma_{\theta(k)}\mathbf{v}(k) + \Psi_{\theta(k)}w(k) \\ v(k) = \Phi_{\theta(k)}\mathbf{v}(k) \end{cases} \quad (5.30)$$

with $\mathbf{v}(k)$ of dimension n_{cl}. We say that the controller $\mathcal{G}_K$ given by (5.27) is admissible if the closed loop MJLS $\mathcal{G}_{cl}$ (5.30) is MSS.

Therefore in the infinite horizon optimal filtering problem we want to find $\widehat{A}_i, \widehat{B}_i, \widehat{C}_i$ in (5.27) with $\widehat{x}_0$ deterministic, such that the closed loop system (5.28) is MSS and minimizes

$$\lim_{k\to\infty} E(\|v(k)\|^2).$$

The solution of the infinite horizon filtering problem posed above is closely related to the mean square stabilizing solution for the filtering coupled algebraic Riccati equations, defined next.

Definition 5.7 (Filtering CARE). *We say that $Y = (Y_1, \ldots, Y_N) \in \mathbb{H}^{n+}$ is the mean square stabilizing solution for the filtering CARE if it satisfies for each $j \in \mathbb{N}$*

$$Y_j = \sum_{i=1}^N p_{ij}[A_iY_iA_i^* + \pi_iG_iG_i^* - A_iY_iL_i^*(H_iH_i^*\pi_i + L_iY_iL_i^*)^{-1}L_iY_iA_i^*] \quad (5.31)$$

with $r_\sigma(\mathcal{T}) < 1$ where $\mathcal{T}(.) = (\mathcal{T}_1(.), \ldots, \mathcal{T}_N(.))$ is defined as in (3.7) with $\Gamma_i = A_i + M_i(Y)L_i$ and $M_i(Y)$ as

$$M_i(Y) \triangleq -A_i Y_i L_i^* (H_i H_i^* \pi_i + L_i Y_i L_i^*)^{-1} \tag{5.32}$$

for each $i \in \mathbb{N}$.

As mentioned before, in Appendix A conditions for the existence of stabilizing solutions for the filtering (and control) CARE are presented in terms of the concepts of mean square stabilizability and mean square detectability. Let $M = (M_1, \ldots, M_N)$ be as in (5.32) (for simplicity we drop from now on the dependence on Y). We have the following theorem which shows that the solution for the optimum filtering problem can be obtained from the mean square stabilizing solution of the filtering CARE (5.31) and the gains $M = (M_1, \ldots, M_N)$ (5.32). Recall the definition of the H_2-optimal control problem in Section 4.4 and (4.34).

Theorem 5.8. *An optimal solution for the filtering problem posed above is:*

$$\widehat{A}_i^{op} = A_i + M_i L_i + B_i F_i; \quad \widehat{B}_i^{op} = -M_i; \quad and \quad \widehat{C}_i^{op} = F_i \tag{5.33}$$

and the optimal cost is

$$\min_{G_K} \|\mathcal{G}_v\|_2^2 = \|\bar{\mathcal{G}}_v\|_2^2 = \sum_{i=1}^N \operatorname{tr}(\bar{F}_i Y_i \bar{F}_i^*). \tag{5.34}$$

Proof. Let us denote by $\widehat{x}^{op}(k), u^{op}(k)$ the sequence generated by (5.27) when $(\widehat{A}^{op}, \widehat{B}^{op}, \widehat{C}^{op})$ is as in (5.33), by $x^{op}(k)$ the sequence generated by (5.24) when we apply the control sequence $u^{op}(k)$, and $e^{op}(k) \triangleq x^{op}(k) - \widehat{x}^{op}(k)$. This leads to the following equations:

$$
\begin{aligned}
x^{op}(k+1) =& A_{\theta(k)} x^{op}(k) + B_{\theta(k)} F_{\theta(k)} \widehat{x}^{op}(k) + G_{\theta(k)} w(k) \\
\widehat{x}^{op}(k+1) =& (A_{\theta(k)} + M_{\theta(k)} L_{\theta(k)}) \widehat{x}^{op}(k) - M_{\theta(k)} (L_{\theta(k)} x^{op}(k) + H_{\theta(k)} w(k)) \\
& + B_{\theta(k)} F_{\theta(k)} \widehat{x}^{op}(k) \\
=& A_{\theta(k)} \widehat{x}^{op}(k) - M_{\theta(k)} L_{\theta(k)} e^{op}(k) - M_{\theta(k)} H_{\theta(k)} w(k) \\
& + B_{\theta(k)} F_{\theta(k)} \widehat{x}^{op}(k)
\end{aligned}
$$

and thus,

$$
\begin{aligned}
x^{op}(k+1) &= [A_{\theta(k)} + B_{\theta(k)} F_{\theta(k)}] x^{op}(k) - B_{\theta(k)} F_{\theta(k)} e^{op}(k) + G_{\theta(k)} w(k) \\
e^{op}(k+1) &= [A_{\theta(k)} + M_{\theta(k)} L_{\theta(k)}] e^{op}(k) + [G_{\theta(k)} + M_{\theta(k)} H_{\theta(k)}] w(k)
\end{aligned}
$$

that is,

$$
\begin{bmatrix} x^{op}(k+1) \\ e^{op}(k+1) \end{bmatrix} = \begin{bmatrix} A_{\theta(k)} + B_{\theta(k)} F_{\theta(k)} & -B_{\theta(k)} F_{\theta(k)} \\ 0 & A_{\theta(k)} + M_{\theta(k)} L_{\theta(k)} \end{bmatrix} \begin{bmatrix} x^{op}(k) \\ e^{op}(k) \end{bmatrix}
$$
$$
+ \begin{bmatrix} G_{\theta(k)} \\ G_{\theta(k)} + M_{\theta(k)} H_{\theta(k)} \end{bmatrix} w(k).
$$

Since Y is the mean square stabilizing solution of (5.31), and by hypothesis F stabilizes (A, B) in the mean square sense we have that the above system is MSS. Thus the closed loop system for $(\widehat{A}^{op}, \widehat{B}^{op}, \widehat{C}^{op})$ is MSS. We also have that

$$v^{op}(k) = \bar{F}_{\theta(k)}(x^{op}(k) - \widehat{x}^{op}(k)) = \bar{F}_{\theta(k)}e^{op}(k).$$

Writing

$$P_i^{op}(k) = E(e^{op}(k)e^{op}(k)^* \mathbf{1}_{\{\theta(k)=i\}})$$

it follows that (recall that $G_i H_i^* = 0$)

$$P_j^{op}(k+1) = \sum_{i=1}^{N} p_{ij}[(A_i + M_i L_i)P_i^{op}(k)(A_i + M_i L_i)^*$$
$$+ \pi_i(k)(G_i G_i^* + M_i H_i H_i^* M_i^*)]$$

and $P^{op}(k) \overset{k\uparrow\infty}{\to} P^{op}$ (see Chapter 3, Proposition 3.36), where

$$P_j^{op} = \sum_{i=1}^{N} p_{ij}[(A_i + M_i L_i)P_i^{op}(A_i + M_i L_i)^* + \pi_i(G_i G_i^* + M_i H_i H_i^* M_i^*)].$$

But notice that from the CARE (5.31) and (5.32),

$$Y_j = \sum_{i=1}^{N} p_{ij}[(A_i + M_i L_i)Y_i(A_i + M_i L_i)^* + \pi_i(G_i G_i^* + M_i H_i H_i^* M_i^*)]$$

and from uniqueness of the controllability Grammian (4.41), $Y = P^{op}$. Thus,

$$\|\mathcal{G}_v^{op}\|_2^2 = \sum_{i=1}^{N} \mathrm{tr}(\bar{F}_i Y_i \bar{F}_i^*).$$

Consider now any $(\widehat{A}, \widehat{B}, \widehat{C})$ such that the closed loop system (5.30) is MSS. From (4.43) and Theorem 5.4,

$$E(\|v(k)\|^2) = \sum_{i=1}^{N} \mathrm{tr}(\Phi_i P_i(k)\Phi_i^*) \geq \sum_{i=1}^{N} \mathrm{tr}(\bar{F}_i Y_i(k)\bar{F}_i^*),$$

where $P(k) = (P_1(k), \ldots, P_N(k))$ is

$$P_i(k) = E(\mathbf{v}(k)\mathbf{v}(k)^* \mathbf{1}_{\{\theta(k)=i\}}), \quad i \in \mathbb{N}$$

and $Y(k) = (Y_1(k), \ldots, Y_N(k))$ as in (5.13). From Proposition 3.35 in Chapter 3,

$$P_j(k+1) = \sum_{i=1}^{N} p_{ij}[\Gamma_i P_i(k)\Gamma_i^* + \pi_i(k)\Psi_i\Psi_i^*].$$

Moreover from Proposition 3.36 in Chapter 3, $P(k) \overset{k\uparrow\infty}{\to} P$ with $P = (P_1, \ldots, P_N)$ the unique solution of the controllability Grammian (4.41). It is shown in Appendix A, Proposition A.23, that $Y(k) \overset{k\uparrow\infty}{\to} Y$, and thus we have that

$$\lim_{k\to\infty} E(\|v(k)\|^2) = \sum_{i=1}^{N} \operatorname{tr}(\Phi_i P_i \Phi_i^*)$$

$$\geq \lim_{k\to\infty} \sum_{i=1}^{N} \operatorname{tr}(\bar{F}_i Y_i(k) \bar{F}_i^*)$$

$$= \sum_{i=1}^{N} \operatorname{tr}(\bar{F}_i Y_i \bar{F}_i^*)$$

proving the desired result.   □

5.4 Optimal Linear Filter with $\theta(k)$ Unknown

5.4.1 Preliminaries

In this section we consider again model (5.1) with all matrices and variables real, and $u(k) = 0$ for all $k = 0, 1, \ldots$, and it is desired to estimate the state vector $x(k)$ from the observations $y(k)$. The main difference here with respect to the previous sections is that we shall assume that the Markov chain $\theta(k)$ is not known. In this case it is well known that the optimal nonlinear filter for this case is obtained from a bank of Kalman filters, which requires exponentially increasing memory and computation with time ([17]). To limit the computational requirements, sub-optimal filters have been proposed in the literature such as, for instance, in [3], [17], [32], [51], [101], [210] and [211], among other authors, under the hypothesis of Gaussian distribution for the disturbances, and by [226] and [227] for the non-Gaussian case. In some of the papers mentioned before the authors considered non-linear sub-optimal estimators, which require on-line calculations. In this section we shall present the optimal linear minimum mean square filter (LMMSE) for MJLS, derived in [58]. This filter has dimension Nn (recall that n is the dimension of the state variable and N is the number of states of the Markov chain). The advantage of this formulation is that it leads to a time-varying linear filter easy to implement and in which all calculations (the gain matrices) can be performed off-line. Moreover it can be applied to a broader class of systems than the linear models with output uncertainty studied in [138] and [180]. Examples and comparisons of this filter are presented in Subsection 8.4.1.

As we shall see, the error covariance matrix obtained from the LMMSE can be written in terms of a recursive Riccati difference equation of dimension Nn, added with an additional term that depends on the second moment matrix of the state variable. This extra term would be zero for the case with no jumps. Conditions to guarantee the convergence of the error covariance matrix to the stationary solution of an Nn dimensional algebraic Riccati equation are also derived, as well as the stability of the stationary filter. These results allow us to design a time-invariant (a fixed gain matrix) stable sub-optimal filter of LMMSE for MJLS.

In this section it will be convenient to introduce the following notation. For any sequence of second order random vectors $\{r(k)\}$ we define the "centered" random vector $r^c(k)$ as $r(k) - E(r(k))$, $\widehat{r}(k|t)$ the best affine estimator of $r(k)$ given $\{y(0), \ldots, y(t)\}$, and $\widetilde{r}(k|t) = r(k) - \widehat{r}(k|t)$. Similarly $\widehat{r}^c(k|t)$ is the best linear estimator of $r^c(k)$ given $\{y^c(0), \ldots, y^c(t)\}$ and $\widetilde{r}^c(k|t) = r^c(k) - \widehat{r}^c(k|t)$. It is well known (cf. [80], p. 109) that

$$\widehat{r}(k|t) = \widehat{r}^c(k|t) + E(r(k)) \tag{5.35}$$

and, in particular, $\widetilde{r}^c(k|t) = \widetilde{r}(k|t)$. We shall denote by $\mathfrak{L}(y^k)$ the linear subspace spanned by $y^k \triangleq (y(k)^* \cdots y(0)^*)^*$ (see [80]), that is, a random variable $r \in \mathfrak{L}(y^k)$ if $r = \sum_{i=0}^{k} \alpha(i)^* y(i)$ for some $\alpha(i) \in \mathbb{R}^m, i = 0, \ldots, k$.

We recall that for the second order random vectors r and s taking values in $\mathbb{R}^n$, the inner product $< .; . >$ is defined as

$$< r; s > = E(s^* r)$$

and therefore we say that r and s are orthogonal if $< r; s > = E(s^* r) = 0$. For $t \leq k$, the best linear estimator $\widehat{r}^c(k|t) = (\widehat{r}_1^c(k|t), \ldots, \widehat{r}_n^c(k|t))^*$ of the random vector $r^c(k) = (r_1^c(k), \ldots, r_n^c(k))^*$ is the projection of $r^c(k)$ onto the subspace $\mathfrak{L}((y^c)^t)$ and satisfies the following properties (cf. [80], p. 108 and 113):

1. $\widehat{r}_j^c(k|t) \in \mathfrak{L}((y^c)^t), j = 1, \ldots, n$
2. $\widetilde{r}_j(k|t)$ is orthogonal to $\mathfrak{L}((y^c)^t), j = 1, \ldots, n$
3. if $\mathrm{cov}((y^c)^t)$ is non-singular then

$$\widehat{r}^c(k|t) = E(r^c(k)(y^c)^{t*}) \, \mathrm{cov}((y^c)^t)^{-1}(y^c)^t \tag{5.36}$$

$$\widehat{r}^c(k|k) = \widehat{r}^c(k|k-1) + E(\widehat{r}^c(k)\widetilde{y}(k|k-1)^*)$$
$$\times E(\widetilde{y}(k|k-1)\widetilde{y}(k|k-1)^*)^{-1}(y^c(k) - \widehat{y}^c(k|k-1)) \tag{5.37}$$

5.4.2 Optimal Linear Filter

On the stochastic basis $(\Omega, \mathfrak{F}, \{\mathfrak{F}_k\}, \mathcal{P})$, consider the following MJLS:

$$\mathcal{G} = \begin{cases} x(k+1) = A_{\theta(k)}(k)x(k) + G_{\theta(k)}(k)w(k) \\ y(k) = L_{\theta(k)}(k)x(k) + H_{\theta(k)}(k)w(k) \\ x(0) = x_0 \in \mathcal{C}_0^n, \theta(0) = \theta_0 \in \Theta_0 \end{cases} \tag{5.38}$$

with the same assumptions as in Section 5.2, except that the $\theta(k)$ is not known at each time k. In particular we recall the notation for the initial condition as in (4.4). For $k \geq 0$ and $j \in \mathbb{N}$ define

$$z_j(k) \triangleq x(k)\mathbf{1}_{\{\theta(k)=j\}} \in \mathbb{R}^n$$

$$z(k) \triangleq \begin{pmatrix} z_1(k) \\ \vdots \\ z_N(k) \end{pmatrix} \in \mathbb{R}^{Nn}$$

and recall from (3.3) that $q(k) = E(z(k))$. Define also $\widehat{z}(k|k-1)$ as the projection of $z(k)$ onto $\mathcal{L}(y^{k-1})$ and

$$\widetilde{z}(k|k-1) \triangleq z(k) - \widehat{z}(k|k-1).$$

The second-moment matrices associated to the above variables are

$$Q_i(k) \triangleq E(z_i(k)z_i(k)^*) \in \mathbb{B}(\mathbb{R}^n), i \in \mathbb{N},$$

$$Z(k) \triangleq E(z(k)z(k)^*) = \text{diag}[Q_i(k)] \in \mathbb{B}(\mathbb{R}^{Nn}),$$

$$\widehat{Z}(k|l) \triangleq E(\widehat{z}(k|l)\widehat{z}(k|l)^*) \in \mathbb{B}(\mathbb{R}^{Nn}), 0 \le l \le k,$$

$$\widetilde{Z}(k|l) \triangleq E(\widetilde{z}(k|l)\widetilde{z}(k|l)^*) \in \mathbb{B}(\mathbb{R}^{Nn}), 0 \le l \le k.$$

We consider the following augmented matrices

$$\mathsf{A}(k) \triangleq \begin{bmatrix} p_{11}(k)A_1(k) & \cdots & p_{N1}(k)A_N(k) \\ \vdots & \ddots & \vdots \\ p_{1N}(k)A_1(k) & \cdots & p_{NN}(k)A_N(k) \end{bmatrix} \in \mathbb{B}(\mathbb{R}^{Nn}) \tag{5.39}$$

$$\mathsf{H}(k) \triangleq [H_1(k)\pi_1(k)^{1/2} \cdots H_N(k)\pi_N(k)^{1/2}] \in \mathbb{B}(\mathbb{R}^{Nr}, \mathbb{R}^p), \tag{5.40}$$

$$\mathsf{L}(k) \triangleq [L_1(k) \cdots L_N(k)] \in \mathbb{B}(\mathbb{R}^{Nn}, \mathbb{R}^p), \tag{5.41}$$

$$\mathsf{G}(k) \triangleq \text{diag}[[(p_{1j}(k)\pi_1(k))^{1/2}G_1(k) \cdots (p_{Nj}(k)\pi_N(k))^{1/2}G_N(k)]]$$

$$\in \mathbb{B}(\mathbb{R}^{N^2 r}, \mathbb{R}^{Nn}). \tag{5.42}$$

We recall that we still assume that conditions (5.2) and (5.3) hold. We present now the main result of [58], derived from geometric arguments as in [80].

Theorem 5.9. *Consider the system represented by (5.38). Then the LMMSE* $\widehat{x}(k|k)$ *is given by*

$$\widehat{x}(k|k) = \sum_{i=1}^{N} \widehat{z}_i(k|k) \tag{5.43}$$

where $\widehat{z}(k|k)$ satisfies the recursive equation

$$\widehat{z}(k|k) = \widehat{z}(k|k-1) + \widetilde{Z}(k|k-1)\mathsf{L}(k)^* \left(\mathsf{L}(k)\widetilde{Z}(k|k-1)\mathsf{L}(k)^*\right.$$

$$\left. + \mathsf{H}(k)\mathsf{H}(k)^*\right)^{-1}(y(k) - \mathsf{L}(k)\widehat{z}(k|k-1)) \tag{5.44}$$

$$\widehat{z}(k|k-1) = \mathsf{A}(k-1)\widehat{z}(k-1|k-1), \; k \ge 1 \tag{5.45}$$

$$\widehat{z}(0|-1) = q(0) = \begin{pmatrix} \mu_0\pi_1(0) \\ \vdots \\ \mu_0\pi_N(0) \end{pmatrix}.$$

The positive semi-definite matrices $\widetilde{Z}(k|k-1) \in \mathbb{B}(\mathbb{R}^{Nn})$ *are obtained from*

$$\widetilde{Z}(k|k-1) = Z(k) - \widehat{Z}(k|k-1) \tag{5.46}$$

where $Z(k) = \mathrm{diag}[Q_j(k)]$ *are given by the recursive equation*

$$Q_j(k+1) = \sum_{i=1}^{N} p_{ij}(k)A_i(k)Q_i(k)A_i(k)^* + \sum_{i=1}^{N} p_{ij}(k)\pi_i(k)G_i(k)G_i(k)^*,$$
$$Q_j(0) = \mathbb{Q}_0\pi_j(0), \quad j \in \mathbb{N} \tag{5.47}$$

and $\widehat{Z}(k|k-1)$ *are given by the recursive equation*

$$\widehat{Z}(k|k) = \widehat{Z}(k|k-1) + \widehat{Z}(k|k-1)\mathsf{L}(k)^* \big(\mathsf{L}(k)\widetilde{Z}(k|k-1)\mathsf{L}(k)^*$$
$$+ \mathsf{H}(k)\mathsf{H}(k)^*\big)^{-1}\mathsf{L}(k)\widehat{Z}(k|k-1) \tag{5.48}$$
$$\widehat{Z}(k|k-1) = \mathsf{A}(k-1)\widehat{Z}(k-1|k-1)\mathsf{A}(k-1)^*, \tag{5.49}$$
$$\widehat{Z}(0|-1) = q(0)q(0)^*.$$

Proof. See Appendix B.   □

Remark 5.10. Notice that in Theorem 5.9 the inverse of

$$\mathsf{L}(k)\widetilde{Z}(k|k-1)\mathsf{L}(k)^* + \mathsf{H}(k)\mathsf{H}(k)^*$$

is well defined since for each $k = 0,1,\ldots$ there exists $\iota(k) \in \mathbb{N}$ such that $\pi_{\iota(k)}(k) > 0$ and from condition (5.3)

$$\mathsf{L}(k)\widetilde{Z}(k|k-1)\mathsf{L}(k)^* + \mathsf{H}(k)\mathsf{H}(k)^* \geq \mathsf{H}(k)\mathsf{H}(k)^*$$
$$= \sum_{i=1}^{N} \pi_i(k)H_i(k)H_i(k)^*$$
$$\geq \pi_{\iota(k)}(k)H_{\iota(k)}(k)H_{\iota(k)}(k)^* > 0.$$

In Theorem 5.9 the term $\widetilde{Z}(k|k-1)$ is expressed in (5.46) as the difference between $Z(k)$ and $\widehat{Z}(k|k-1)$, which are obtained from the recursive equations (5.47) and (5.48). In the next lemma we shall write $\widetilde{Z}(k|k-1)$ directly as a recursive Riccati equation, with an additional term that depends on the second moment matrices $Q_i(k)$. Notice that this extra term would be zero for the case in which there are no jumps ($N = 1$). We define the linear operator $\mathfrak{V}(\cdot, k) : \mathbb{H}^n \to \mathbb{B}(\mathbb{R}^{Nn})$ as follows: for $\Upsilon = (\Upsilon_1, \ldots, \Upsilon_N) \in \mathbb{H}^n$,

$$\mathfrak{V}(\Upsilon, k) \triangleq \mathrm{diag}\left[\sum_{i=1}^{N} p_{ij}(k)A_i(k)\Upsilon_i A_i(k)^*\right] - \mathsf{A}(k)(\mathrm{diag}[\Upsilon_i])\mathsf{A}(k)^*. \tag{5.50}$$

Notice that if $\Upsilon = (\Upsilon_1, \ldots, \Upsilon_N) \in \mathbb{H}^{n+}$ then $\mathfrak{V}(\Upsilon, k) \geq 0$. Indeed, consider $v = (v_1^* \cdots v_N^*)^* \in \mathbb{R}^{Nn}$. Then it is easy to check that

$$v^* \mathfrak{V}(\Upsilon, k)v = \sum_{i=1}^{N} E\Bigg(\big(v_{\theta(k+1)} - E(v_{\theta(k+1)}|\theta(k) = i)\big)^* A_i \Upsilon_i$$

$$\times A_i^* \big(v_{\theta(k+1)} - E(v_{\theta(k+1)}|\theta(k) = i)\big)|\theta(k) = i \Bigg) \geq 0.$$

We have the following lemma:

Lemma 5.11. $\widetilde{Z}(k|k-1)$ *satisfies the following recursive Riccati equation*

$$\widetilde{Z}(k+1|k) = A(k)\widetilde{Z}(k|k-1)A(k)^* + \mathfrak{V}(Q(k),k) + G(k)G(k)^*$$
$$- A(k)\widetilde{Z}(k|k-1)L(k)^* (L(k)\widetilde{Z}(k|k-1)L(k)^* + H(k)H(k)^*)^{-1}$$
$$\times L(k)\widetilde{Z}(k|k-1)A(k)^*. \tag{5.51}$$

where $Q(k) = (Q_1(k), \ldots, Q_N(k)) \in \mathbb{H}^{n+}$ *are given by the recursive equation* *(5.47).*

Proof. See Appendix B. □

As mentioned before, for the case with no jumps ($N = 1$) we would have $\mathfrak{V}(Q(k), k) = 0$ and therefore (5.51) would reduce to the standard recursive Ricatti equation for the Kalman filter.

5.4.3 Stationary Linear Filter

Equations (5.47) and (5.51) describe a recursive Riccati equation for $\widetilde{Z}(k|k-1)$. We shall establish now its convergence when $k \to \infty$. We assume that all matrices in System (5.38) and the transition probabilities p_{ij} are time-invariant, System (5.38) is mean square stable (MSS) according to Definition 3.8 in Chapter 3, and that the Markov chain $\{\theta(k)\}$ is ergodic (see assumption 3.31). We recall that from this hypothesis, it follows that $\lim_{k \to \infty} \mathcal{P}(\theta(k) = i)$ exists, it is independent from $\theta(0)$, and

$$0 < \pi_i = \lim_{k \to \infty} \mathcal{P}(\theta(k) = i) = \lim_{k \to \infty} \pi_i(k).$$

We redefine the matrices $A, H(k), L, G(k)$ defined in (5.39), (5.40), (5.41) and (5.42) as

$$A = \begin{bmatrix} p_{11}A_1 & \cdots & p_{N1}A_N \\ \vdots & \cdots & \vdots \\ p_{1N}A_1 & \cdots & p_{NN}A_N \end{bmatrix} \in \mathbb{B}(\mathbb{R}^{Nn}), \tag{5.52}$$

$$H(k) = [H_1 \pi_1(k)^{1/2} \cdots H_N \pi_N(k)^{1/2}] \in \mathbb{B}(\mathbb{R}^{Nr}, \mathbb{R}^p), \tag{5.53}$$

$$L = [L_1 \cdots L_N] \in \mathbb{B}(\mathbb{R}^{Nn}, \mathbb{R}^p), \tag{5.54}$$

$$G(k) \triangleq \text{diag}[[(p_{1j}\pi_1(k))^{1/2}G_1 \cdots (p_{Nj}\pi_N(k))^{1/2}G_N]] \in \mathbb{B}(\mathbb{R}^{N^2 r}, \mathbb{R}^{Nn}), \tag{5.55}$$

and define

$$H \triangleq [H_1 \pi_1^{1/2} \cdots H_N \pi_N^{1/2}] \in \mathbb{B}(\mathbb{R}^{Nr}, \mathbb{R}^p), \tag{5.56}$$

$$G \triangleq \mathrm{diag}[[(p_{1j}\pi_1)^{1/2}G_1 \cdots (p_{Nj}\pi_N)^{1/2}G_N]] \in \mathbb{B}(\mathbb{R}^{N^2 r}, \mathbb{R}^{Nn}). \tag{5.57}$$

Since we are considering the time-invariant case, we have that the operator $\mathfrak{V}$ defined in (5.50) is also time-invariant, and therefore we can drop the time dependence. From MSS and ergodicity of the Markov chain, and from Proposition 3.36 in Chapter 3, $Q(k) \to Q$ as $k \to \infty$, where $Q = (Q_1, \ldots, Q_N) \in \mathbb{H}^{n+}$ is the unique solution that satisfies:

$$Z_j = \sum_{i=1}^{N} p_{ij}(A_i Z_i A_i^* + \pi_i G_i G_i^*), j \in \mathbb{N}. \tag{5.58}$$

In what follows we define for any matrix $Z \in \mathbb{B}(\mathbb{R}^{Nn})$, $Z \geq 0$, $\mathsf{T}(Z) \in \mathbb{B}(\mathbb{R}^p, \mathbb{R}^{Nn})$ as:

$$\mathsf{T}(Z) \triangleq -\mathsf{A}Z\mathsf{L}^*(\mathsf{L}Z\mathsf{L}^* + \mathsf{H}\mathsf{H}^*)^{-1}. \tag{5.59}$$

As in Remark 5.10, we have that $\mathsf{L}Z\mathsf{L}^* + \mathsf{H}\mathsf{H}^* > 0$ and thus the above inverse is well defined.

The following theorem establishes the asymptotic convergence of $\widetilde{Z}(k|k-1)$. The idea of the proof is first to show that there exists a unique positive semi-definite solution P for the algebraic Riccati equation, and then prove that for some positive integer $\kappa > 0$, there exists lower and upper bounds, $R(k)$ and $P(k+\kappa)$ respectively, for $\widetilde{Z}(k+\kappa|k+\kappa-1)$, such that it squeezes asymptotically $\widetilde{Z}(k+\kappa|k+\kappa-1)$ to P.

Theorem 5.12. *Suppose that the Markov chain $\{\theta(k)\}$ is ergodic and that System (5.38) is MSS. Consider the algebraic Riccati equation given by:*

$$Z = \mathsf{A}Z\mathsf{A}^* + \mathsf{G}\mathsf{G}^* - \mathsf{A}Z\mathsf{L}^*(\mathsf{L}Z\mathsf{L}^* + \mathsf{H}\mathsf{H}^*)^{-1}\mathsf{L}Z\mathsf{A}^* + \mathfrak{V}(Q) \tag{5.60}$$

where $Q = (Q_1, \ldots, Q_N) \in \mathbb{H}^{n+}$ satisfies (5.58). Then there exists a unique positive semi-definite solution $P \in \mathbb{B}(\mathbb{R}^{Nn})$ to (5.60). Moreover,

$$r_\sigma(\mathsf{A} + \mathsf{T}(P)\mathsf{L}) < 1$$

and for any $Q(0) = (Q_1(0), \ldots, Q_N(0))$ with $Q_i(0) \geq 0$, $i \in \mathbb{N}$, and $\widetilde{Z}(0|-1) = \mathrm{diag}[Q_i(0)] - q(0)q(0)^* \geq 0$ we have that $\widetilde{Z}(k+1|k)$ given by (5.47) and (5.51) satisfies

$$\widetilde{Z}(k+1|k) \overset{k \to \infty}{\to} P.$$

Proof. See Appendix B.   □

5.5 Robust Linear Filter with $\theta(k)$ Unknown

5.5.1 Preliminaries

The main advantage of the scheme presented in Section 5.4 is that it is very simple to implement, and all calculations can be performed off-line. Moreover the stationary filter presented in Subsection 5.4.3 is a discrete-time-invariant linear system. Another advantage is, as presented in this section, that we can consider uncertainties in the parameters of the system through, for instance, a LMI approach. As in the previous section, we assume that only an output of the system is available, and therefore the values of the jump parameter are not known. It is desired to design a dynamic linear filter such that the closed loop system is mean square stable and minimizes the stationary expected value of the square error. We consider uncertainties on the parameters of the possible modes of operation of the system. A LMI approach, based on [129], is proposed to solve the problem. Examples and comparisons of this filter are presented in Subsection 8.4.2.

We initially consider the filtering problem of a discrete-time MJLS with no uncertainties on the modes of operation of the system. It will be shown that the stationary mean square error for the filter obtained from the LMI formulation, to be presented in Subsection 5.5.3, and the one obtained from the ARE approach, presented in Subsection 5.4.3, lead to the same value. The case with uncertainties is considered within the LMI formulation in Subsection 5.5.4.

In what follows we denote, for $V \in \mathbb{B}(\mathbb{R}^n)$, $\mathrm{dg}[V] \in \mathbb{B}(\mathbb{R}^{Nn})$ the block diagonal matrix formed by V in the diagonal and zero elsewhere, that is,

$$\mathrm{dg}[V] \triangleq \begin{bmatrix} V & \cdots & 0 \\ \vdots & \ddots & \vdots \\ 0 & \cdots & V \end{bmatrix}.$$

5.5.2 Problem Formulation

For the case with no uncertainties, we consider the time-invariant version of the model (5.38) as follows:

$$\begin{aligned}
x(k+1) &= A_{\theta(k)}x(k) + G_{\theta(k)}w(k) \\
y(k) &= L_{\theta(k)}x(k) + H_{\theta(k)}w(k) \\
v(k) &= Jx(k)
\end{aligned} \tag{5.61}$$

where we assume the same assumptions as in Subsection 5.4.3, including that the Markov chain $\{\theta(k)\}$ is ergodic, and that System (5.61) is MSS. Here $y(k)$ is the output variable, which is the only information available from the evolution of the system, since we assume that the jump variable $\theta(k)$ is not known. We wish to design a dynamic estimator $\widehat{v}(k)$ for $v(k)$ of the following form:

$$\widehat{z}(k+1) = A_f \widehat{z}(k) + B_f y(k)$$
$$\widehat{v}(k) = J_f \widehat{z}(k) \tag{5.62}$$
$$e(k) = v(k) - \widehat{v}(k)$$

where $A_f \in \mathbb{B}(\mathbb{R}^{n_f})$, $B_f \in \mathbb{B}(\mathbb{R}^p, \mathbb{R}^{n_f})$, $J_f \in \mathbb{B}(\mathbb{R}^{n_f}, \mathbb{R}^{n_o})$, and $e(k)$ denotes the estimation error. Defining

$$x_e(k) \triangleq \begin{bmatrix} x(k) \\ \widehat{z}(k) \end{bmatrix}$$

we have from (5.61) and (5.62) that

$$x_e(k+1) = \begin{bmatrix} A_{\theta(k)} & 0 \\ B_f L_{\theta(k)} & A_f \end{bmatrix} x_e(k) + \begin{bmatrix} G_{\theta(k)} \\ B_f H_{\theta(k)} \end{bmatrix} w(k) \tag{5.63}$$
$$e(k) = [J - J_f] x_e(k)$$

which is a discrete-time MJLS. We are interested in filters such that (5.63) is MSS. From the fact that System (5.61) is MSS, it follows that System (5.63) is MSS if and only if $r_\sigma(A_f) < 1$ (see Appendix B, Proposition B.3).

We recall that MSS of (5.61) is equivalent to $r_\sigma(\mathcal{T}) < 1$ (see Chapter 3), where the operator $\mathcal{T} \in \mathbb{B}(\mathbb{H}^n)$ presented in (3.7) is $\mathcal{T}(Q) = (\mathcal{T}_1(Q), \ldots, \mathcal{T}_N(Q))$ with $\mathcal{T}_j(Q) = \sum_{i=1}^N p_{ij} A_i Q_i A_i^*$ for $Q = (Q_1, \ldots, Q_N) \in \mathbb{H}^n$. For $k \geq 0$ and $j \in \mathbb{N}$, we recall that

$$Q_i(k) = E(z_i(k) z_i(k)^*) \in \mathbb{B}(\mathbb{R}^n), i \in \mathbb{N},$$
$$Q(k) = (Q_1(k), \ldots, Q_N(k)) \in \mathbb{H}^{n+}, \tag{5.64}$$
$$Z(k) = \mathrm{diag}[Q_i(k)].$$

We consider the augmented matrices $\mathsf{A}, \mathsf{H}(k), \mathsf{L}, \mathsf{G}(k), \mathsf{H}, \mathsf{G}$ as in (5.52), (5.53), (5.54), (5.55), (5.56), (5.57) respectively, and define

$$\widehat{Z}(k) \triangleq E(\widehat{z}(k)\widehat{z}(k)^*),$$
$$U_i(k) \triangleq E(z_i(k)\widehat{z}(k)^*),$$
$$U(k) \triangleq \begin{bmatrix} U_1(k) \\ \vdots \\ U_N(k) \end{bmatrix}$$

and

$$\Delta_i \triangleq \begin{bmatrix} p_{i1}(1-p_{i1}) & -p_{i1}p_{i2} & \cdots & -p_{i1}p_{iN} \\ -p_{i1}p_{i2} & p_{i2}(1-p_{i2}) & \cdots & -p_{i2}p_{iN} \\ \vdots & & \cdots & \vdots \\ -p_{i1}p_{iN} & -p_{iN}p_{i2} & \cdots & p_{iN}(1-p_{iN}) \end{bmatrix}, \quad i \in \mathbb{N}.$$

It was proved in Chapter 3, Proposition 3.35, that

$$Q_j(k+1) = T_j(Q(k)) + \sum_{i=1}^{N} \pi_i(k)p_{ij}G_iG_i^*. \tag{5.65}$$

It follows from (5.50) and (5.65) that

$$Z(k+1) = AZ(k)A^* + \mathfrak{V}(Q(k)) + G(k)G(k)^*. \tag{5.66}$$

The next result will be useful in the following.

Proposition 5.13. $\Delta_i \geq 0$.

Proof. See Appendix B. $\qquad\square$

Note that for any $Q = (Q_1, \ldots, Q_N) \in \mathbb{H}^n$,

$$\mathfrak{V}(Q) = \begin{bmatrix} \sum_{i=1}^{N} p_{i1}A_iQ_iA_i^* & \cdots & 0 \\ \vdots & \ddots & \vdots \\ 0 & \cdots & \sum_{i=1}^{N} p_{iN}A_iQ_iA_i^* \end{bmatrix}$$

$$- \begin{bmatrix} \sum_{i=1}^{N} p_{i1}^2 A_iQ_iA_i^* & \cdots & \sum_{i=1}^{N} p_{i1}p_{iN}A_iQ_iA_i^* \\ \vdots & \ddots & \vdots \\ \sum_{i=1}^{N} p_{i1}p_{iN}A_iQ_iA_i^* & \cdots & \sum_{i=1}^{N} p_{iN}^2 A_iQ_iA_i^* \end{bmatrix}$$

$$= \sum_{i=1}^{N} \begin{bmatrix} p_{i1}(1-p_{i1})I & \cdots & -p_{i1}p_{iN}I \\ \vdots & \ddots & \vdots \\ -p_{i1}p_{iN}I & \cdots & p_{iN}(1-p_{iN})I \end{bmatrix} \begin{bmatrix} A_iQ_iA_i^* & \cdots & 0 \\ \vdots & \ddots & \vdots \\ 0 & \cdots & A_iQ_iA_i^* \end{bmatrix}$$

and given that $\Delta_i \geq 0$, we have that the square root matrix $\Delta_i^{1/2} \geq 0$ (so that $\Delta_i = \Delta_i^{1/2}\Delta_i^{1/2}$) is well defined. It follows that (recall that $\otimes$ represents the Kronecker product)

$$\mathfrak{V}(Q) = \sum_{i=1}^{N} (\Delta_i^{1/2} \otimes I)(\Delta_i^{1/2} \otimes I) \, \mathrm{dg}[A_iQ_iA_i^*]$$

$$= \sum_{i=1}^{N} (\Delta_i^{1/2} \otimes I) \, \mathrm{dg}[A_i] \, \mathrm{dg}[Q_i] \, \mathrm{dg}[A_i^*](\Delta_i^{1/2} \otimes I).$$

Therefore, writing $\mathsf{D}_i \triangleq (\Delta_i^{1/2} \otimes I) \, \mathrm{dg}[A_i]$, we have from (5.66) that:

$$Z(k+1) = AZ(k)A^* + \sum_{i=1}^{N} \mathsf{D}_i \, \mathrm{dg}[Q_i(k)]\mathsf{D}_i^* + G(k)G(k)^*. \tag{5.67}$$

Defining

$$S(k) \triangleq E\left(\begin{bmatrix} z(k) \\ \hat{z}(k) \end{bmatrix} \begin{bmatrix} z(k)^* & \hat{z}(k)^* \end{bmatrix}\right) = \begin{bmatrix} Z(k) & U(k) \\ U(k)^* & \hat{Z}(k) \end{bmatrix}, \tag{5.68}$$

we have the following proposition:

Proposition 5.14. *For $k = 0, 1, 2, \ldots$,*

$$S(k+1) = \begin{bmatrix} A & 0 \\ B_f L & A_f \end{bmatrix} S(k) \begin{bmatrix} A & 0 \\ B_f L & A_f \end{bmatrix}^* + \sum_{i=1}^{N} \begin{bmatrix} D_i \\ 0 \end{bmatrix} dg[Q_i(k)] \begin{bmatrix} D_i^* & 0 \end{bmatrix}$$

$$+ \begin{bmatrix} G(k) \\ 0 \end{bmatrix} \begin{bmatrix} G(k)^* & 0 \end{bmatrix} + \begin{bmatrix} 0 \\ B_f H(k) \end{bmatrix} \begin{bmatrix} 0 & H(k)^* B_f^* \end{bmatrix}. \tag{5.69}$$

Proof. See Appendix B. □

The next result guarantees the convergence of $S(k)$ as defined in (5.68) and (5.69) to a $S \geq 0$ when $k \to \infty$.

Proposition 5.15. *Consider $S(k)$ given by (5.68) and (5.69) and that $r_\sigma(\mathcal{T}) < 1$, $r_\sigma(A_f) < 1$. Then $S(k) \overset{k \to \infty}{\longrightarrow} S \geq 0$, with S of the following form:*

$$S = \begin{bmatrix} Z & U \\ U^* & \hat{Z} \end{bmatrix}, \qquad Z = \begin{bmatrix} Q_1 & \cdots & 0 \\ \vdots & \ddots & \vdots \\ 0 & \cdots & Q_N \end{bmatrix} \geq 0.$$

Moreover, S is the only solution of the equation in V

$$V = \begin{bmatrix} A & 0 \\ B_f L & A_f \end{bmatrix} V \begin{bmatrix} A & 0 \\ B_f L & A_f \end{bmatrix}^* + \sum_{i=1}^{N} \begin{bmatrix} D_i \\ 0 \end{bmatrix} dg[X_i] \begin{bmatrix} D_i^* & 0 \end{bmatrix}$$

$$+ \begin{bmatrix} G \\ 0 \end{bmatrix} \begin{bmatrix} G^* & 0 \end{bmatrix} + \begin{bmatrix} 0 \\ B_f H \end{bmatrix} \begin{bmatrix} 0 & H^* B_f^* \end{bmatrix} \tag{5.70}$$

with

$$V = \begin{bmatrix} X & R \\ R^* & \hat{X} \end{bmatrix}, \qquad X = \begin{bmatrix} X_1 & \cdots & 0 \\ \vdots & \ddots & \vdots \\ 0 & \cdots & X_N \end{bmatrix}. \tag{5.71}$$

Furthermore if V satisfies

$$V \geq \begin{bmatrix} A & 0 \\ B_f L & A_f \end{bmatrix} V \begin{bmatrix} A & 0 \\ B_f L & A_f \end{bmatrix}^* + \sum_{i=1}^{N} \begin{bmatrix} D_i \\ 0 \end{bmatrix} dg[X_i] \begin{bmatrix} D_i^* & 0 \end{bmatrix}$$

$$+ \begin{bmatrix} G \\ 0 \end{bmatrix} \begin{bmatrix} G^* & 0 \end{bmatrix} + \begin{bmatrix} 0 \\ B_f H \end{bmatrix} \begin{bmatrix} 0 & H^* B_f^* \end{bmatrix} \tag{5.72}$$

then $V \geq S$.

Proof. See Appendix B.   □

Thus we will be interested in finding (A_f, B_f, J_f) such that $r_\sigma(A_f) < 1$ and minimizes

$$
\lim_{k\to\infty} E(\|e(k)\|^2) = \mathrm{tr}\left(\lim_{k\to\infty} E\big(e(k)e(k)^*\big)\right)
$$

$$
= \mathrm{tr}\left(\begin{bmatrix} J & -J_f \end{bmatrix} \lim_{k\to\infty} E\big(x_e(k)x_e(k)^*\big) \begin{bmatrix} J^* \\ -J_f^* \end{bmatrix}\right)
$$

$$
= \mathrm{tr}\left(\begin{bmatrix} J & -J_f \end{bmatrix}\right.
$$

$$
\times \lim_{k\to\infty} E\left(\begin{bmatrix} \sum_{i=1}^N z_i(k) \\ \widehat{z}(k) \end{bmatrix} \begin{bmatrix} \sum_{i=1}^N z_i(k)^* & \widehat{z}(k)^* \end{bmatrix}\right) \left.\begin{bmatrix} J^* \\ -J_f^* \end{bmatrix}\right)
$$

$$
= \mathrm{tr}\left(\begin{bmatrix} J & -J_f \end{bmatrix} \begin{bmatrix} I \cdots I & 0 \\ 0 & I \end{bmatrix}\right.
$$

$$
\times \lim_{k\to\infty} E\left(\begin{bmatrix} z(k) \\ \widehat{z}(k) \end{bmatrix} \begin{bmatrix} z(k)^* & \widehat{z}(k)^* \end{bmatrix}\right) \begin{bmatrix} I \\ \vdots \\ I \\ 0 \end{bmatrix} \begin{matrix} \\ 0 \\ \\ I \end{matrix} \left.\begin{bmatrix} J^* \\ -J_f^* \end{bmatrix}\right)
$$

$$
= \mathrm{tr}\left(\begin{bmatrix} J & \cdots & J & -J_f \end{bmatrix} \lim_{k\to\infty} S(k) \begin{bmatrix} J^* \\ \vdots \\ J^* \\ -J_f^* \end{bmatrix}\right)
$$

$$
= \mathrm{tr}\left(\begin{bmatrix} J & \cdots & J & -J_f \end{bmatrix} S \begin{bmatrix} J^* \\ \vdots \\ J^* \\ -J_f^* \end{bmatrix}\right) \tag{5.73}
$$

where the last equality follows from Proposition 5.15 and

$$
S = \begin{bmatrix} Z & U \\ U^* & \widehat{Z} \end{bmatrix}, \qquad Z = \mathrm{diag}[Q_i],
$$

which satisfies, according to Proposition 5.15, the equation

$$
S = \begin{bmatrix} A & 0 \\ B_f L & A_f \end{bmatrix} S \begin{bmatrix} A & 0 \\ B_f L & A_f \end{bmatrix}^* + \sum_{i=1}^N \begin{bmatrix} D_i \\ 0 \end{bmatrix} \mathrm{dg}[Q_i] \begin{bmatrix} D_i^* & 0 \end{bmatrix}
$$

$$
+ \begin{bmatrix} G \\ 0 \end{bmatrix} \begin{bmatrix} G^* & 0 \end{bmatrix} + \begin{bmatrix} 0 \\ B_f H \end{bmatrix} \begin{bmatrix} 0 & H^* B_f^* \end{bmatrix}. \tag{5.74}
$$

In the following section we shall formulate this problem as an LMI optimization problem.

5.5.3 LMI Formulation of the Filtering Problem

From (5.73) and (5.74), the filtering problem can be written, using the Schur complement (see Lemma 2.23), with an $\epsilon > 0$ arbitrarily small precision, as follows (which is not in an LMI optimization formulation yet):

$$\min \operatorname{tr}(W)$$

subject to

$$\mathsf{S} = \begin{bmatrix} Z & U \\ U^* & \hat{Z} \end{bmatrix} > 0, \quad Z = \operatorname{diag}[Q_i], \quad i \in \mathbb{N} \tag{5.75}$$

$$\begin{bmatrix} \mathsf{S} & \mathsf{S}\begin{bmatrix} J & \cdots & J & -J_f \end{bmatrix}^* \\ \begin{bmatrix} J & \cdots & J & -J_f \end{bmatrix}\mathsf{S} & W \end{bmatrix} > 0 \tag{5.76}$$

$$\begin{bmatrix} \mathsf{S} & 0 & 0 & 0 & 0 & 0 & \mathsf{S}\begin{bmatrix} A & 0 \\ B_f \mathsf{L} & A_f \end{bmatrix}^* \\ 0 & \operatorname{dg}[Q_1] & 0 & 0 & 0 & 0 & \operatorname{dg}[Q_1]\begin{bmatrix} D_1^* & 0 \end{bmatrix} \\ 0 & 0 & \ddots & 0 & 0 & 0 & \vdots \\ 0 & 0 & 0 & \operatorname{dg}[Q_N] & 0 & 0 & \operatorname{dg}[Q_N]\begin{bmatrix} D_N^* & 0 \end{bmatrix} \\ 0 & 0 & 0 & 0 & I & 0 & \begin{bmatrix} G^* & 0 \end{bmatrix} \\ 0 & 0 & 0 & 0 & 0 & I & \begin{bmatrix} 0 & H^* B_f^* \end{bmatrix} \\ \begin{bmatrix} A & 0 \\ B_f \mathsf{L} & A_f \end{bmatrix}\mathsf{S} & \begin{bmatrix} D_1 \\ 0 \end{bmatrix}\operatorname{dg}[Q_1] & \cdots & \begin{bmatrix} D_N \\ 0 \end{bmatrix}\operatorname{dg}[Q_N] & \begin{bmatrix} G \\ 0 \end{bmatrix} & \begin{bmatrix} 0 \\ B_f H \end{bmatrix} & \mathsf{S} \end{bmatrix} > 0. \tag{5.77}$$

Indeed, suppose that $\bar{\mathsf{S}} > 0$, $\bar{W} > 0$, (A_f, B_f, J_f) satisfy (5.75), (5.76) and (5.77). Then from the Schur complement,

$$\bar{W} > \operatorname{tr}\begin{bmatrix} J & \cdots & J & -J_f \end{bmatrix}\bar{\mathsf{S}}\begin{bmatrix} J & \cdots & J & -J_f \end{bmatrix}^* \tag{5.78}$$

and

$$\bar{\mathsf{S}} > \begin{bmatrix} A & 0 \\ B_f \mathsf{L} & A_f \end{bmatrix}\bar{\mathsf{S}}\begin{bmatrix} A & 0 \\ B_f \mathsf{L} & A_f \end{bmatrix}^* + \sum_{i=1}^{N}\begin{bmatrix} D_i \\ 0 \end{bmatrix}\operatorname{dg}[\bar{Q}_i]\begin{bmatrix} D_i^* & 0 \end{bmatrix}$$

$$+ \begin{bmatrix} G \\ 0 \end{bmatrix}\begin{bmatrix} G^* & 0 \end{bmatrix} + \begin{bmatrix} 0 \\ B_f H \end{bmatrix}\begin{bmatrix} 0 & H^* B_f^* \end{bmatrix}. \tag{5.79}$$

From Proposition B.4 we have that System (5.63) is MSS. From Proposition 5.15 we know that $\mathsf{S}(k) \overset{k \to \infty}{\to} \mathsf{S}$, and (5.73) and (5.74) hold. Furthermore, from Proposition 5.15, we have that $\bar{\mathsf{S}} \geq \mathsf{S}$. Since we want to minimize $\operatorname{tr}(W)$, it is clear that for (A_f, B_f, J_f) fixed, the best solution would be, in the limit, S, W satisfying (5.78) and (5.79) with equality. However, as will be shown next, it is more convenient to work with the strict inequality restrictions (5.75)–(5.77).

Theorem 5.16. *The problem of finding* S, W, *and* (A_f, B_f, J_f) *such that minimizes* $\mathrm{tr}(\mathsf{W})$ *and satisfies (5.75)–(5.77) is equivalent to:*

$$\min \mathrm{tr}(\mathsf{W})$$

subject to

$$\mathsf{X} = \mathrm{diag}[\mathsf{X}_i],$$

$$\begin{bmatrix} & & & \begin{bmatrix} J^* \\ \vdots \\ J^* \end{bmatrix} - J_{aux}^* \\ \mathsf{X} & \mathsf{X} & & \\ & & & \begin{bmatrix} J^* \\ \vdots \\ J^* \end{bmatrix} \\ \mathsf{X} & \mathsf{Y} & & \\ \begin{bmatrix} J & \cdots & J \end{bmatrix} - J_{aux} \begin{bmatrix} J & \cdots & J \end{bmatrix} & W \end{bmatrix} > 0, \qquad (5.80)$$

$$\begin{bmatrix} \mathsf{X} & \mathsf{X} & 0 & 0 & 0 & 0 & 0 & A^*\mathsf{X} & L^*F^* + A^*\mathsf{Y} + R^* \\ \mathsf{X} & \mathsf{Y} & 0 & 0 & 0 & 0 & 0 & A^*\mathsf{X} & L^*F^* + A^*\mathsf{Y} \\ 0 & 0 & \mathrm{dg}[\mathsf{X}_1] & 0 & 0 & 0 & 0 & D_1^*\mathsf{X} & D_1\mathsf{Y} \\ 0 & 0 & 0 & \ddots & 0 & 0 & 0 & \vdots & \vdots \\ 0 & 0 & 0 & 0 & \mathrm{dg}[\mathsf{X}_N] & 0 & 0 & D_N^*\mathsf{X} & D_N^*\mathsf{Y} \\ 0 & 0 & 0 & 0 & 0 & I & 0 & G^*\mathsf{X} & G^*\mathsf{Y} \\ 0 & 0 & 0 & 0 & 0 & 0 & I & 0 & H^*F^* \\ \mathsf{X}A & \mathsf{X}A & \mathsf{X}D_1 & \cdots & \mathsf{X}D_N & \mathsf{X}G & 0 & \mathsf{X} & \mathsf{X} \\ \mathsf{Y}A + FL + R & \mathsf{Y}A + FL & \mathsf{Y}D_1 & \cdots & \mathsf{Y}D_N & \mathsf{Y}G & FH & \mathsf{X} & \mathsf{Y} \end{bmatrix} > 0,$$

$$(5.81)$$

which are now LMI since the variables are

$$\mathsf{X}_i, \quad i \in \mathbb{N}, \quad \mathsf{Y}, \mathsf{W}, R, F, J_{aux}.$$

Once we have $\mathsf{X}, \mathsf{Y}, R, F, J_{aux}$, *we recover* $\mathsf{S}, A_f, B_f, J_f$ *as follows. Choose a non-singular* $(Nn \times Nn)$ *matrix* U, *make* $Z = \mathsf{X}^{-1} = \mathrm{diag}[\mathsf{X}_i^{-1}] = \mathrm{diag}[Q_i]$, *and choose* $\widehat{Z} > U^*ZU$. *Define* $V = \mathsf{Y}(\mathsf{Y}^{-1} - Z)U^{*-1}$ *(which is non-singular since from (5.80),* $\mathsf{X} > \mathsf{X}\mathsf{Y}^{-1}\mathsf{X} \Rightarrow Z = \mathsf{X}^{-1} > \mathsf{Y}^{-1}$). *Then*

$$A_f = V^{-1}R(U^*\mathsf{X})^{-1}, \qquad B_f = V^{-1}F, \qquad J_f = J_{aux}(U^*\mathsf{X})^{-1}. \qquad (5.82)$$

Proof. See Appendix B.   □

The next results show that the stationary mean square error obtained by the filter derived from the optimal solution of the LMI formulation coincides with the one obtained from the associated filtering ARE derived in Theorem 5.12. Let us show first that any solution of the LMI problem will lead to an expected stationary error greater than the one obtained from the filter generated by the ARE approach (5.60). Recall from Chap. 3 that $r_\sigma(\mathcal{T}) < 1$ implies that there exists a unique solution $Q = (Q_1, \ldots, Q_N) \in \mathbb{H}^n$ satisfying (5.58).

Proposition 5.17. *An optimal solution for the stationary filtering problem is*

$$\widehat{A}_f = A + T(P)L, \quad \widehat{B}_f = -T(P), \quad \widehat{J}_f = \begin{bmatrix} J \cdots J \end{bmatrix}$$

where P is the unique positive semi-definite solution for the filtering ARE (5.60).

Proof. See Appendix B. □

The above proposition shows that any feasible solution of the LMI problem will lead to filters such that the stationary mean square error will be greater than the stationary mean square error obtained from the filtering ARE. Next let us show that ϵ approximations of the filtering ARE (5.60) and (5.58) lead to a feasible solution for the LMI optimization problem (5.80) and (5.81). This shows that the optimal solution of the LMI will lead to a filter with associated stationary mean square error smaller than the one obtained from any filter derived by ϵ approximations of the ARE approach. The combination of these two results show that the filter obtained from the optimal LMI solution and from the ARE approach will lead to the same stationary mean square error.

Indeed, let us show that a feasible solution to (5.80) and (5.81) can be obtained from ϵ approximation of (5.58) and (5.60). Let $Q_\epsilon = (Q_{\epsilon 1}, \dots Q_{\epsilon N})$ be the unique solution satisfying

$$Q_{\epsilon j} = T_j(Q_\epsilon) + \sum_{i=1}^{N} p_{ij} \pi_i G_i G_i^* + 2\epsilon I$$

which, according to (5.50), can be rewritten as

$$Z_\epsilon = A Z_\epsilon A^* + GG^* + \mathfrak{V}(Q_\epsilon) + 2\epsilon I \tag{5.83}$$

where $Z_\epsilon = \mathrm{diag}[Q_{\epsilon i}]$.

Proposition 5.18. *Consider P_ϵ the unique positive-definite solution of the ARE (in V_ϵ)*

$$V_\epsilon = A V_\epsilon A^* + GG^* - A V_\epsilon L^* (L V_\epsilon L^* + HH^*)^{-1} L V_\epsilon A^* + \mathfrak{V}(Q_\epsilon) + \epsilon I \tag{5.84}$$

and write $T(P_\epsilon)$ as in (5.59). Then the choice

$$R = -(YA + FL)(I - Y^{-1}X), \qquad F = P_\epsilon^{-1} T(P_\epsilon)$$
$$X_i = Q_{\epsilon i}^{-1}, \quad i \in \mathbb{N}, \quad X = \mathrm{diag}[X_i]$$
$$Y = P_\epsilon^{-1}$$

is feasible for (5.80) and (5.81). Moreover, this choice leads to

$$A_f = A + T(P_\epsilon)L$$
$$B_f = -T(P_\epsilon)$$
$$J_f = \begin{bmatrix} J \cdots J \end{bmatrix}.$$

Proof. See Appendix B. □

5.5.4 Robust Filter

We assume now that the matrices A, G, L, H, are not exactly known but instead there are known matrices A^j, G^j, L^j, H^j, $j = 1, \ldots, \rho$ such that for $0 \le \lambda_j \le 1$, $\sum_{j=1}^{\rho} \lambda_j = 1$, we have that

$$A = \sum_{j=1}^{\rho} \lambda_j A^j, \qquad G = \sum_{j=1}^{\rho} \lambda_j G^j,$$

$$L = \sum_{j=1}^{\rho} \lambda_j L^j, \qquad H = \sum_{j=1}^{\rho} \lambda_j H^j. \qquad (5.85)$$

Our final result in this chapter is as follows:

Theorem 5.19. *Suppose that the following LMI optimization problem has an $(\epsilon-)$ optimal solution $\bar{X}, \bar{Y}, \bar{W}, \bar{R}, \bar{F}, \bar{J}_{aux}$:*

$$\min \operatorname{tr}(W)$$

subject to

$$X = \operatorname{diag}[X_i], \quad i \in \mathbb{N}, \qquad (5.86)$$

$$\begin{bmatrix} X & X & \left[J^* \cdots J^* \right]^* - J^*_{aux} \\ X & Y & \left[J^* \cdots J^* \right]^* \\ \left[J \cdots J \right] - J_{aux} \left[J \cdots J \right] & & W \end{bmatrix} > 0, \qquad (5.87)$$

and for $j = 1, \ldots, \rho$,

$$\begin{bmatrix} X & X & 0 & 0 & 0 & 0 & 0 & A^{j^*}X & L^{j^*}F^* + A^{j^*}Y + R^* \\ X & Y & 0 & 0 & 0 & 0 & 0 & A^{j^*}X & L^{j^*}F^* + A^{j^*}Y \\ 0 & 0 & \operatorname{dg}[X_1] & 0 & 0 & 0 & 0 & D_1^{j^*}X & D_1^{j^*}Y \\ 0 & 0 & 0 & \ddots & 0 & 0 & 0 & \vdots & \vdots \\ 0 & 0 & 0 & 0 & \operatorname{dg}[X_N] & 0 & 0 & D_N^{j^*}X & D_N^{j^*}Y \\ 0 & 0 & 0 & 0 & 0 & I & 0 & G^{j^*}X & G^{j^*}Y \\ 0 & 0 & 0 & 0 & 0 & 0 & I & 0 & H^{j^*}F^* \\ XA^j & XA^j & XD_1^j & \cdots & XD_N^j & XG^j & 0 & X & X \\ YA^j + FL^j + R & YA^j + FL^j & YD_1^j & \cdots & YD_N^j & YG^j & FH^j & X & Y \end{bmatrix} > 0. \qquad (5.88)$$

Then for the filter given as in (5.82) we have that System (5.63) is MSS, and $\lim_{k \to \infty} E(\|e(k)\|^2) \le \operatorname{tr}(\bar{W})$.

Proof. Since (5.88) holds for each $j = 1, \ldots, \rho$, and the real matrices A, L, G, H are as in (5.85), we have that (by taking the sum of (5.88) multiplied by λ_j, over j from 1 to ρ) that (5.88) also holds for A, L, G, H, D_i. By taking the inverse transformations of similarity, we conclude that (5.76) and (5.77) (and thus (5.78) and (5.79)) hold for some $\bar{S} > 0$ and $\bar{W}$. From Proposition B.4, System (5.63) is MSS. From Proposition 5.15 and (5.73) and (5.74),

$$\lim_{k \to \infty} E(\|e(k)\|^2) \leq \mathrm{tr}\left(\begin{bmatrix} J & \cdots & J & -J_f \end{bmatrix} \bar{S} \begin{bmatrix} J^* & \cdots & J^* & -J_f^* \end{bmatrix}^* \right) \leq \mathrm{tr}(\bar{W}).$$

□

5.6 Historical Remarks

From the earlier work of N. Wiener to the famous Fujisaki–Kallianpur–Kunita equation, filtering theory has achieved a remarkable degree of development, having in the celebrated Kalman filter one of its pillars due, in particular, to its fundamental importance in applications. Although the theoretical machinery available to deal with nonlinear estimation problems is by now rather considerable, there are yet many challenging questions in this area. One of these has to do with the fact that the description of the optimal nonlinear filter can rarely be given in terms of a closed finite system of stochastic differential equations, i.e., the so-called finite filters. This, among others issues, provides an interesting research topic in filtering and makes it a highly active area. For an introduction to the linear filtering theory, the readers are referred, for instance, to [79] and [80] (see also [149] for an authoritative account of the nonlinear filtering theory and [81] for a nice introduction).

We can certainly say that one of the reasons for the amount of literature dealing with the filtering problem for MJLS has to do with the fact that the optimal nonlinear filter, for the case in which both x and θ are not accessible, is infinite dimensional (in the sense used in the continuous-time nonlinear filtering theory). For the discrete-time case, the optimal estimator requires exponentially increasing memory and computation with time, which makes this optimal approach impractical. Due to this, a number of sub-optimal filters have flourished in the literature. The filtering problem for MJLS has been addressed in [3], [5], [17], [30], [31], [32], [51], [99], [210] and [211], among others, usually under the hypothesis of Gaussian distribution for the disturbances. In [58], was obtained the LMMSE for MJLS (described in this book), based on estimating $x(k)\mathbf{1}_{\{\theta(k)=i\}}$ instead of estimating directly $x(k)$. One of the advantages of this formulation is that it can be applied to a broader class of systems than the linear models with output uncertainty studied in [138] and [180]. An associated stationary filter for the LMMSE, also described in this book, was obtained in [70]. Besides the fact that the LMMSE, and the associated stationary filter produces a filter which bears those desirable properties of the Kalman filter (a recursive scheme suitable for computer implementation which allows some offline computation that alleviates the computational burden) it can, in addition, contemplate uncertainties in the parameters through, for instance, an LMI approach. This robust version was analyzed in [69] and is also described in this book. See also [86], [87], [88], [94], [95], [100], [102], [104], and [166], which include sampling algorithms, H_∞, nonlinear, and robust filters for MJLS.

6

Quadratic Optimal Control with Partial Information

The LQG control problem is one of the most popular in the control community. In the case with partial observations, the celebrated separation principle establishes that the solution of the quadratic optimal control problem for linear systems is similar to the complete observation case, except for the fact that the state variable is substituted by its estimation derived from the Kalman filter. Thus the solution of the LQG problem can be obtained from the Riccati equations associated to the filtering and control problems. The main purpose of this chapter is to trace a parallel with the standard LQG theory and study the quadratic optimal control problem of MJLS when the jump variable is available to the controller. It is shown that a result similar to the separation principle can be derived, and the solution obtained from two sets of coupled Riccati equations, one associated to the filtering problem, and the other to the control problem.

6.1 Outline of the Chapter

In this chapter we deal with the quadratic optimal control problem with partial information on the state variable $x(k)$ for the class of dynamical MJLS described in the previous chapters by $\mathcal{G}$. We assume that an output variable $y(k)$ and the jump parameters $\theta(k)$ are available to the controller. It is desired to design a dynamic Markov jump controller such that the closed loop system minimizes the quadratic functional cost of the system over a finite and infinite horizon period of time. As in the case with no jumps, we show that for the finite horizon case an optimal controller can be obtained from two coupled Riccati difference equations, one associated to the control problem (see (4.14)), and the other one associated to the filtering problem (see (5.13)). Similarly, for the infinite horizon case, an optimal controller can be obtained from the stabilizing solutions of the control CARE (4.35) and the filtering CARE (5.31). These results could be seen as a separation principle for the finite and infinte horizon quadratic optimal control problems for discrete-time

MJLS. When there is only one mode of operation our results coincide with the traditional separation principle for the LQG control of discrete-time linear systems.

As pointed out in [146] and Chapter 5, the optimal x-state estimator for this case, when the input $\{w(k); k \in \mathbb{T}\}$ is a Gaussian white noise sequence, is a Kalman filter (see Remark 5.2 and (5.11)) for a time varying system. Therefore the gains for the LQG optimal controller will be sample path dependent (see the optimal controller presented in [52], which is based on this filter). In order to get around with the sample path dependence, the authors in [146] propose a Markov jump filter (that is, a filter that depends just on the present value of the Markov parameter), based on a posterior estimate of the jump parameters. Notice however that no proof of optimality for this class of Markov filters is presented in [146], since the authors are mainly interested in the steady-state convergence properties of the filter. In this chapter, by restricting ourselves to a class of dynamic Markov jump controllers, we present the results derived in [73] and [74] to obtain a proof of optimality, following an approach similar to the standard theory for Kalman filter and LQG control (see, for instance, [14] and [80]) to develop a separation principle for MJLS . A key point in our formulation is the introduction of the indicator function for the Markov parameter in the orthogonality between the estimation error and the state estimation, presented in Theorem 5.3. This generalizes the standard case, in which there is only one mode of operation, so that the indicator function in this case would be always equal to one. The introduction of the indicator function for the Markov parameter in Theorem 5.3 is essential for obtaining the principle of separation, presented in Sections 6.2 for the finite horizon case, and 6.3 for the infinite horizon case.

6.2 Finite Horizon Case

6.2.1 Preliminaries

We consider in this section the finite horizon optimal control problem for a MJLS when an output $y(k)$ and jump variable $\theta(k)$ are available. As in Section 5.2, we are restricting our attention to the family of linear Markov jump controllers (that is, a controller that depends just on the present value of the Markov parameter) as in (5.4), since otherwise the optimal linear mean square controller would be obtained from a sample path dependent Kalman filter, as explained in Remark 5.2, and presented in [52]. On the stochastic basis $(\Omega, \mathfrak{F}, \{\mathfrak{F}_k\}, \mathcal{P})$, consider the following MJLS:

$$\mathcal{G} = \begin{cases} x(k+1) = A_{\theta(k)}(k)x(k) + B_{\theta(k)}(k)u(k) + G_{\theta(k)}(k)w(k) \\ y(k) = L_{\theta(k)}(k)x(k) + H_{\theta(k)}(k)w(k) \\ z(k) = C_{\theta(k)}(k)x(k) + D_{\theta(k)}(k)u(k) \end{cases} \tag{6.1}$$

where, as before, $\mathbb{T} = \{0, \ldots, T\}$, $\{x(k); k \in \mathbb{T}\}$ represents the n-dimensional state vector, $\{u(k); k \in \mathbb{T}\}$ the m-dimensional control sequence, $\{w(k); k \in \mathbb{T}\}$ an r-dimensional wide sense white noise sequence (recall that $E(w(k)) = 0$, $E(w(k)w(k)^*) = I$, $E(w(k)w(l)^*) = 0$ for $k \neq l$), $\{y(k); k \in \mathbb{T}\}$ the p-dimensional sequence of measurable variables and $\{z(k); k \in \mathbb{T}\}$ the q-dimensional output sequence. We assume that conditions (4.5), (4.6), (5.2) and (5.3) hold, and that the output and operation modes $(y(k), \theta(k)$ respectively) are known at each time k. The noise $\{w(k); k \in \mathbb{T}\}$ and the Markov chain $\{\theta(k); k \in \mathbb{T}\}$ are independent sequences, and the initial conditions $(x(0), \theta(0))$ are independent random variables with $E(x(0)) = \mu_0$, $E(x(0)x(0)^*) = \mathbb{Q}_0$.

We denote in this chapter $\mathfrak{G}_k$ as the σ-field generated by the random variables $\{y(t), \theta(t); t = 0, \ldots, k\}$. Thus the filtration $\{\mathfrak{G}_k; k \in \mathbb{T}\}$ is such that $\mathfrak{G}_k \subset \mathfrak{G}_{k+1} \subset \mathfrak{F}$.

We consider dynamic Markov jump controllers $\mathcal{G}_K$ given by

$$\mathcal{G}_K = \begin{cases} \widehat{x}(k+1) = \widehat{A}_{\theta(k)}(k)\widehat{x}(k) + \widehat{B}_{\theta(k)}(k)y(k) \\ u(k) = \widehat{C}_{\theta(k)}(k)\widehat{x}(k). \end{cases} \tag{6.2}$$

As already mentioned in Chapter 5, the reason for choosing this kind of controller is that they depend just on $\theta(k)$ (rather than on the entire past history of modes $\theta(0), \ldots, \theta(k)$), so that the closed loop system is again a Markov jump system.

The quadratic cost associated to the closed loop system $\mathcal{G}_{cl}$ with control law $u = (u(0), \ldots, u(T-1))$ given by (6.2) and initial conditions (x_0, θ_0) is denoted, as in Section 4.2, by $\mathfrak{J}(\theta_0, x_0, u)$, and is given by (4.10), that is, as

$$\mathfrak{J}(\theta_0, x_0, u) = \sum_{k=0}^{T-1} E(\|z(k)\|^2) + E(x(T)^* V_{\theta(T)} x(T)) \tag{6.3}$$

with $V = (V_1, \ldots, V_N) \in \mathbb{H}^{n+}$. Therefore, the finite horizon optimal quadratic control problem we want to study is: find

$$\widehat{A}(k) = (\widehat{A}_1(k), \ldots, \widehat{A}_N(k)) \in \mathbb{H}^n,$$
$$\widehat{B}(k) = (\widehat{B}_1(k), \ldots, \widehat{B}_N(k)) \in \mathbb{H}^{p,n},$$
$$\widehat{C}(k) = (\widehat{C}_1(k), \ldots, \widehat{C}_N(k)) \in \mathbb{H}^{n,m}$$

in (6.2), such that the control law $u = (u(0), \ldots, u(T-1))$ given by (6.2) produces minimal cost $\mathfrak{J}(\theta_0, x_0, u)$. We recall that this minimal cost is denoted by $\widetilde{\mathfrak{J}}(\theta_0, x_0)$.

6.2.2 A Separation Principle

Let us solve the optimal control problem posed in Subsection 6.2.1. First of all we notice that

$$E(\|z(k)\|^2) = E(x(k)^* C_{\theta(k)}(k)^* C_{\theta(k)}(k) x(k))$$
$$+ E(\|D_{\theta(k)}(k) u(k)\|^2)$$
$$= \sum_{i=1}^{N} \operatorname{tr} \left(C_i(k)^* C_i(k) E(x(k) x(k)^* \mathbf{1}_{\{\theta(k)=i\}}) \right)$$
$$+ E(\|D_{\theta(k)}(k) u(k)\|^2) \tag{6.4}$$

and for any control law $u = (u(0), \ldots, u(T-1))$ given by (6.2), we have from (5.15) that $x(k) = \widetilde{x}_e(k) + \widehat{x}_e(k)$, and from (5.20) and Theorem 5.3,

$$E(x(k) x(k)^* \mathbf{1}_{\{\theta(k)=i\}}) = E(\widetilde{x}_e(k) \widetilde{x}_e(k)^* \mathbf{1}_{\{\theta(k)=i\}}) + E(\widehat{x}_e(k) \widehat{x}_e(k)^* \mathbf{1}_{\{\theta(k)=i\}})$$
$$= Y_i(k) + E(\widehat{x}_e(k) \widehat{x}_e(k)^* \mathbf{1}_{\{\theta(k)=i\}})$$

so that (6.4) can be rewritten as

$$E(\|z(k)\|^2) = E(\|\widehat{z}_e(k)\|^2) + \sum_{i=1}^{N} \operatorname{tr}(C_i(k) Y_i(k) C_i(k)^*)$$

where $\widehat{z}_e(k) = C_{\theta(k)}(k) \widehat{x}_e(k) + D_{\theta(k)}(k) u(k)$. Similarly, we have that

$$E(x(T)^* V_{\theta(T)} x(T)) = E(\widehat{x}_e(T)^* V_{\theta(T)} \widehat{x}_e(T)) + \sum_{i=1}^{N} \operatorname{tr}(V_i Y_i(T))$$

and thus

$$\mathfrak{J}(\theta_0, x_0, u) = \sum_{k=0}^{T-1} E(\|\widehat{z}_e(k)\|^2) + E(\widehat{x}_e(T)^* V_{\theta(T)} \widehat{x}_e(T))$$
$$+ \sum_{k=0}^{T-1} \left[\sum_{i=1}^{N} [\operatorname{tr}(C_i(k) Y_i(k) C_i(k)^*) + \operatorname{tr}(V_i Y_i(T))] \right] \tag{6.5}$$

where the terms with $Y_i(k)$ do not depend on the control variable u. Therefore minimizing (6.5) is equivalent to minimizing

$$\mathcal{J}_e(u) = \sum_{k=0}^{T-1} E(\|\widehat{z}_e(k)\|^2) + E(\widehat{x}_e(T)^* V_{\theta(T)} \widehat{x}_e(T))$$

subject to

$$\begin{cases} \widehat{x}_e(k+1) = A_{\theta(k)}(k) \widehat{x}_e(k) + B_{\theta(k)}(k) u(k) - M_{\theta(k)}(k) \nu(k) \\ \widehat{x}_e(0) = E(x_0) = \mu_0 \end{cases}$$

where

$$\nu(k) = y(k) - L_{\theta(k)}(k) \widehat{x}_e(k) = L_{\theta(k)}(k) \widetilde{x}_e(k) + H_{\theta(k)}(k) w(k)$$

and $u(k)$ is given by (6.2). Let us show now that $\{\nu(k); k \in \mathbb{T}\}$ satisfies (4.2), (4.7), (4.8) and (4.9). Set

$$\Xi_i(k) = L_i(k)Y_i(k)L_i(k)^* + H_i(k)H_i(k)^*.$$

Indeed we have from Theorem 5.3 and the fact that $\{w(k); k \in \mathbb{T}\}$ is a wide sense white noise sequence independent of the Markov chain $\{\theta(k); k \in \mathbb{T}\}$ and initial condition $x(0)$, that

$$
\begin{aligned}
E(\nu(k)\nu(k)^*\mathbf{1}_{\{\theta(k)=i\}}) =& E((L_i(k)\widetilde{x}_e(k) + H_i(k)w(k)) \\
& \times (L_i(k)\widetilde{x}_e(k) + H_i(k)w(k))^*\mathbf{1}_{\{\theta(k)=i\}}) \\
=& L_i(k)Y_i(k)L_i(k)^* + H_i(k)H_i(k)^* \\
=& \Xi_i(k), \\
E(\nu(k)\widehat{x}_e(k)^*\mathbf{1}_{\{\theta(k)=i\}}) =& L_i(k)E(\widetilde{x}_e(k)\widehat{x}_e(k)^*\mathbf{1}_{\{\theta(k)=i\}}) \\
& + H_i(k)E(w(k)\widehat{x}_e(k)^*\mathbf{1}_{\{\theta(k)=i\}}) \\
=& 0, \\
E(\nu(k)u(k)^*\mathbf{1}_{\{\theta(k)=i\}}) =& L_i(k)E(\widetilde{x}_e(k)\widehat{x}(k)^*\mathbf{1}_{\{\theta(k)=i\}})\widehat{C}_i(k)^* \\
& + H_i(k)E(w(k)\widehat{x}(k)^*\mathbf{1}_{\{\theta(k)=i\}})\widehat{C}_i(k)^* \\
=& 0,
\end{aligned}
$$

for $k \in \mathbb{T}$. Moreover for any measurable functions f and g, we have from the fact that $\nu(k)$ and $\theta(k)$ are $\mathfrak{G}_k$-measurable,

$$E(f(\nu(k))g(\theta(k+1))|\mathfrak{G}_k) = f(\nu(k)) \times \sum_{j=1}^{N} p_{\theta(k)j}(k)g(j).$$

Thus the results of Section 4.2 can be applied and we have the following theorem.

Theorem 6.1 (A Separation Principle for MJLS). *An optimal solution for the control problem posed in Subsection 6.2.1 is obtained from (4.14) and (5.13). The gains (4.15) and (5.14) lead to the following optimal solution:*

$$
\begin{aligned}
\widehat{A}_i^{op}(k) &= A_i(k) + M_i(k)L_i(k) + B_i(k)F_i(k) \\
\widehat{B}_i^{op}(k) &= -M_i(k) \\
\widehat{C}_i^{op}(k) &= F_i(k)
\end{aligned}
$$

and the optimal cost is

$$
\begin{aligned}
\bar{\mathfrak{J}}(\theta_0, x_0) = \sum_{i=1}^{N} &\Bigg[\pi_i(0)\mu_0^* X_i(0)\mu_0 + \sum_{k=0}^{T-1} \mathrm{tr}\left(C_i(k)Y_i(k)C_i(k)^*\right) \\
&+ \mathrm{tr}\left(V_i Y_i(T)\right) + \sum_{k=0}^{T-1} \pi_i(k)\,\mathrm{tr}\left(M_i(k)\Xi_i(k)M_i(k)^*\mathcal{E}_i(X(k+1),k)\right) \Bigg].
\end{aligned}
$$

Remark 6.2. Notice that the choice of the Markov jump structure for the filter as in (5.4) was crucial to obtain the orthogonality derived in Theorem 5.3, and the separation principle presented here. Other choices for the structure of the filter, with more information on the Markov chain, would lead to other notions of orthogonality and "separation principle". Therefore the separation principle presented here is a direct consequence of the choice of the Markov structure for the filter.

Remark 6.3. From Remarks 4.3 and 5.6 we have that a time-invariant Markov controller approximation for the optimal filter in Theorem 6.1 would be given by the steady-state solutions (4.23) and (5.22), with gains (4.24) and (5.23), provided that the convergence conditions presented in Appendix A were satisfied.

6.3 Infinite Horizon Case

6.3.1 Preliminaries

We consider in this section a time-invariant version of model $\mathcal{G}$ presented in (6.1) of Subsection 6.2.1:

$$\mathcal{G} = \begin{cases} x(k+1) = A_{\theta(k)}x(k) + B_{\theta(k)}u(k) + G_{\theta(k)}w(k) \\ y(k) = L_{\theta(k)}x(k) + H_{\theta(k)}w(k) \\ z(k) = C_{\theta(k)}x(k) + D_{\theta(k)}u(k) \end{cases} \tag{6.6}$$

with $\{\theta(k)\}$ an ergodic Markov chain (see Assumption 3.31) with transition probability p_{ij}, so that $\pi_i(k) = \mathcal{P}(\theta(k) = i) \overset{k\uparrow\infty}{\to} \pi_i > 0$ for all $i \in \mathbb{N}$. Here $\mathbb{T} = \{0, 1, \dots\}$. As before we assume that $C_i^* D_i = 0$, $D_i^* D_i > 0$, $G_i H_i^* = 0$, $H_i H_i^* > 0$ for $i \in \mathbb{N}$, and that the output and operation modes $(y(k), \theta(k)$ respectively) are known at each time k. As in Section 5.3, we consider dynamic Markov jump controllers $\mathcal{G}_K$ given by

$$\mathcal{G}_K = \begin{cases} \widehat{x}(k+1) = \widehat{A}_{\theta(k)}\widehat{x}(k) + \widehat{B}_{\theta(k)}y(k) \\ u(k) = \widehat{C}_{\theta(k)}\widehat{x}(k). \end{cases} \tag{6.7}$$

From (6.6) and (6.7) we have that the closed loop system is

$$\begin{bmatrix} x(k+1) \\ \widehat{x}(k+1) \end{bmatrix} = \begin{bmatrix} A_{\theta(k)} & B_{\theta(k)}\widehat{C}_{\theta(k)} \\ \widehat{B}_{\theta(k)}L_{\theta(k)} & \widehat{A}_{\theta(k)} \end{bmatrix} \begin{bmatrix} x(k) \\ \widehat{x}(k) \end{bmatrix} + \begin{bmatrix} G_{\theta(k)} \\ \widehat{B}_{\theta(k)}H_{\theta(k)} \end{bmatrix} w(k)$$

$$z(k) = [C_{\theta(k)} \quad D_{\theta(k)}\widehat{C}_{\theta(k)}] \begin{bmatrix} x(k) \\ \widehat{x}(k) \end{bmatrix}. \tag{6.8}$$

Writing, as in (5.29),

$$\Gamma_i \triangleq \begin{bmatrix} A_i & B_i\widehat{C}_i \\ \widehat{B}_iL_i & \widehat{A}_i \end{bmatrix}; \ \Psi_i \triangleq \begin{bmatrix} G_i \\ \widehat{B}_iH_i \end{bmatrix}; \ \Phi_i \triangleq [C_i \ \ D_i\widehat{C}_i]; \ \mathbf{v}(k) \triangleq \begin{bmatrix} x(k) \\ \widehat{x}(k) \end{bmatrix} \quad (6.9)$$

we have from (6.8) that the closed loop MJLS $\mathcal{G}_{cl}$ is given by

$$\mathcal{G}_{cl} = \begin{cases} \mathbf{v}(k+1) = \Gamma_{\theta(k)}\mathbf{v}(k) + \Psi_{\theta(k)}w(k) \\ z(k) = \Phi_{\theta(k)}\mathbf{v}(k) \end{cases} \quad (6.10)$$

with $\mathbf{v}(k)$ of dimension n_{cl}. We say that the controller $\mathcal{G}_K$ given by (6.7) is admissible if the closed loop MJLS $\mathcal{G}_{cl}$ (6.10) is MSS according to Definition 3.8 in Chapter 3.

6.3.2 Definition of the H_2-control Problem

According to Definition 4.7, the H_2-norm of the closed loop system $\mathcal{G}_{cl}$, denoted by $\|\mathcal{G}\|_2$, given by (6.10) with $\mathbf{v}(0) = 0$ is defined as:

$$\|\mathcal{G}\|_2^2 = \sum_{s=1}^{r} \|z_s\|_2^2 \quad (6.11)$$

where we recall that

$$\|z_s\|_2^2 = \sum_{k=1}^{\infty} E(\|z_s(k)\|^2) \quad (6.12)$$

and $z_s = (z_s(0), z_s(1), \ldots)$ represents the output sequence when the input w is given by:
a) $w = (w(0), w(1), \ldots)$, $w(0) = e_s$, $w(k) = 0$ for each $k > 0$, with $e_s \in \mathbb{R}^r$ the unitary vector formed by 1 at the s-th position, 0 elsewhere and,
b) $\theta(0) = i$ with probability π_i.

Here we have modified Definition 4.7 by considering the initial distribution for the Markov chain as $v_i = \pi_i$. Since the system (6.10) is MSS we have from Theorem 3.34 in Chapter 3 that the norms $\|\mathcal{G}\|_2^2$ and $\|z_s\|_2^2$ in (6.11) and (6.12) are finite. As mentioned before, for the deterministic case (with $N = 1$ and $p_{11} = 1$), the definition above coincides with the usual H_2-norm.

According to the results in Subsection 4.4.2, the H_2-norm can be computed from the unique solution of the discrete coupled Grammians of observability and controllability. Let $S = (S_1, \ldots, S_N) \in \mathbb{H}^n$ and $P = (P_1, \ldots, P_N) \in \mathbb{H}^n$ be the unique solution (see Proposition 3.20) of the observability and controllability Grammians

$$S_i = \Gamma_i^* \mathcal{E}_i(S)\Gamma_i + \Phi_i^* \Phi_i, \quad i \in \mathbb{N} \quad (6.13)$$

$$P_j = \sum_{i=1}^{N} p_{ij}[\Gamma_i P_i \Gamma_i^* + \pi_i \Psi_i \Psi_i^*], \quad j \in \mathbb{N}. \quad (6.14)$$

The next proposition characterizes the H_2-norm in terms of the solutions of the Grammians of observability and controllability, and the proof is as in Proposition 4.8.

Proposition 6.4. *We have that*

$$\|\mathcal{G}\|_2^2 = \sum_{i=1}^{N}\sum_{j=1}^{N}\pi_i p_{ij}\,\mathrm{tr}(\Psi_i^* S_j \Psi_i) = \sum_{i=1}^{N}\mathrm{tr}(\Phi_i P_i \Phi_i^*). \tag{6.15}$$

Proof. See Proposition 4.8.   □

Therefore, the optimal H_2-control problem we want to study is: find $(\widehat{A},\widehat{B},\widehat{C})$ in (6.7), where $\widehat{A} = (\widehat{A}_1,\ldots,\widehat{A}_N)$, $\widehat{B} = (\widehat{B}_1,\ldots,\widehat{B}_N)$, $\widehat{C} = (\widehat{C}_1,\ldots,\widehat{C}_N)$, such that the closed loop MJLS $\mathcal{G}_{cl}$ (6.10) is MSS and minimizes $\|\mathcal{G}\|_2^2$.

As seen in Subsection 4.4.3, an alternative definition for the H_2-control problem would be as follows. Suppose that in model (6.6) $w = \{w(0),w(1),\ldots\}$ is a wide sense white noise sequence (that is, $E(w(k)) = 0$, $E(w(k)w(k)^*) = I$, $E(w(k)w(l)^*) = 0$ for $k \neq l$). Let $\mathcal{G}_K$ be an admissible controller given by (6.7), $\mathbf{v}(k)$ be as in (6.10) and

$$P_i(k) = E(\mathbf{v}(k)\mathbf{v}(k)^*\mathbf{1}_{\{\theta(k)=i\}}), \quad i \in \mathbb{N}. \tag{6.16}$$

As seen in Proposition 3.35 of Chapter 3,

$$P_j(k+1) = \sum_{i=1}^{N}p_{ij}[\Gamma_i P_i(k)\Gamma_i^* + \pi_i(k)\Psi_i\Psi_i^*].$$

Moreover, since the closed loop system is MSS, we have that $P(k) \overset{k\uparrow\infty}{\to} P$ (see Proposition 3.36 in Chapter 3), where $P = (P_1,\ldots,P_N)$ is the unique solution (see Proposition 3.20) of the controllability Grammian (6.14). Notice that

$$
\begin{aligned}
E(\|z(k)\|^2) &= E(\mathrm{tr}(z(k)z(k)^*)) \\
&= \mathrm{tr}(E(\Phi_{\theta(k)}\mathbf{v}(k)\mathbf{v}(k)^*\Phi_{\theta(k)}^*)) \\
&= \sum_{i=1}^{N}\mathrm{tr}(E(\Phi_i(\mathbf{v}(k)\mathbf{v}(k)^*i_{\{\theta(k)=i\}})\Phi_i^*)) \\
&= \sum_{i=1}^{N}\mathrm{tr}(\Phi_i P_i(k)\Phi_i^*) \overset{k\uparrow\infty}{\to} \sum_{i=1}^{N}\mathrm{tr}(\Phi_i P_i \Phi_i^*) \\
&= \|\mathcal{G}\|_2^2
\end{aligned}
$$

and thus from Proposition 6.4 an alternative definition for the H_2-control problem is to find $(\widehat{A},\widehat{B},\widehat{C})$ in system (6.7), where $\widehat{A} = (\widehat{A}_1,\ldots,\widehat{A}_N)$, $\widehat{B} = (\widehat{B}_1,\ldots,\widehat{B}_N)$, $\widehat{C} = (\widehat{C}_1,\ldots,\widehat{C}_N)$, such that the closed loop MJLS $\mathcal{G}_{cl}$ (6.10) is MSS and minimizes $\lim_{k\to\infty}\|z(k)\|_2^2$.

6.3.3 A Separation Principle for the H_2-control of MJLS

In Chapter 5 we used the filtering CARE (5.31) and (5.32) for solving the linear optimal filtering problem. In this subsection we consider the control CARE (4.35) and (4.36) for solving the H_2-control problem posed in Subsection 6.3.2. Suppose that there exists $X = (X_1, \ldots, X_N) \in \mathbb{H}^{n+}$ the mean square stabilizing solution (see Definition 4.4) of the optimal control CARE (4.35), and let $F = (F_1, \ldots, F_N)$, with $F_i = \mathcal{F}_i(X)$ and $\mathcal{F}_i(X)$ be as in (4.36). We write $R_i = \mathcal{R}_i(X)$ for $i \in \mathbb{N}$, where $\mathcal{R}_i(X)$ is as in (4.37). Let us return to the MJLS $\mathcal{G}$ defined in (6.6), with $\mathbb{T} = \{\ldots, -1, 0, 1, \ldots\}$ and control law of the following form:

$$u(k) = \nu(k) + F_{\theta(k)}x(k)$$

so that

$$
\begin{cases}
x(k+1) = \widetilde{A}_{\theta(k)}x(k) + B_{\theta(k)}\nu(k) + G_{\theta(k)}w(k) \\
z(k) = \widetilde{C}_{\theta(k)}x(k) + D_{\theta(k)}\nu(k) \\
y(k) = L_{\theta(k)}x(k) + H_{\theta(k)}w(k)
\end{cases}
$$

where $\widetilde{A}_i = A_i + B_iF_i$, $\widetilde{C}_i = C_i + D_iF_i$. We can decompose the above system such that

$$
\begin{cases}
x(k) = x_1(k) + x_2(k), \\
z(k) = z_1(k) + z_2(k)
\end{cases}
$$

where

$$
\mathcal{G}_c = \begin{cases}
x_1(k+1) = \widetilde{A}_{\theta(k)}x_1(k) + G_{\theta(k)}w(k) \\
z_1(k) = \widetilde{C}_{\theta(k)}x_1(k)
\end{cases} \tag{6.17}
$$

$$
\mathcal{G}_U = \begin{cases}
x_2(k+1) = \widetilde{A}_{\theta(k)}x_2(k) + B_{\theta(k)}R_{\theta(k)}^{-1/2}v(k) \\
z_2(k) = \widetilde{C}_{\theta(k)}x_2(k) + D_{\theta(k)}R_{\theta(k)}^{-1/2}v(k)
\end{cases} \tag{6.18}
$$

and $v(k) = R_{\theta(k)}^{1/2}\nu(k), v = \{v(k); k \in \mathbb{T}\}$. Notice that system $\mathcal{G}_c$ does not depend on the control $u(k)$, and that

$$z(k) = \mathcal{G}_c(w)(k) + \mathcal{G}_U(v)(k).$$

The next theorem establishes the principle of separation for H_2-control of MJLS. In what follows we recall that $\|\mathcal{G}\|_2$ represents the H_2-norm of (6.6) under a control law of the form (6.7).

Theorem 6.5. *Consider System (6.6) and Markov jump mean square stabilizing controllers as in (6.7). Suppose that there exist the mean square stabilizing solutions $Y = (Y_1, \ldots, Y_N)$ and $X = (X_1, \ldots, X_N)$ for the filtering and control CARE as in (5.31) and (4.35) respectively, and let $M = (M_1, \ldots, M_N)$*

and $F = (F_1, \dots, F_N)$ be as in (5.32) and (4.36) respectively. Then an optimal solution for the H_2-control problem is given by $\widehat{A}^{op} = (\widehat{A}_1^{op}, \dots, \widehat{A}_N^{op}), \widehat{B}^{op} = (\widehat{B}_1^{op}, \dots, \widehat{B}_N^{op}), \widehat{C}^{op} = (\widehat{C}_1^{op}, \dots, \widehat{C}_N^{op})$ as in (5.33) of Theorem 5.8. Moreover the value of the H_2-norm for this control is

$$\min_{G_K} \|\mathcal{G}\|_2^2 = \sum_{i=1}^N \pi_i \operatorname{tr}(G_i^* \mathcal{E}_i(X) G_i) + \sum_{i=1}^N \operatorname{tr}(R_i^{1/2} F_i Y_i F_i^* R_i^{1/2}).$$

Proof. Consider the closed loop system (6.8) and (6.10), rewritten below for convenience, with the output $v(k) = R_{\theta(k)}^{1/2} \nu(k) = R_{\theta(k)}^{1/2}(u(k) - F_{\theta(k)} x(k))$,

$$\mathcal{G}_v = \begin{cases} \mathbf{v}(k+1) = \Gamma_{\theta(k)} \mathbf{v}(k) + \Psi_{\theta(k)} w(k) \\ v(k) = R_{\theta(k)}^{1/2}[-F_{\theta(k)} \ \widehat{C}_{\theta(k)}] \begin{bmatrix} x(k) \\ \widehat{x}(k) \end{bmatrix} = \Phi_{\theta(k)} \mathbf{v}(k) \end{cases} \tag{6.19}$$

where $\Gamma = (\Gamma_1, \dots, \Gamma_N)$ and $\Psi = (\Psi_1, \dots, \Psi_N)$ are as in (6.9) and $\Phi = (\Phi_1, \dots, \Phi_N), \Phi_i = R_i^{1/2}[-F_i \ \widehat{C}_i], i \in \mathbb{N}$. We have from (6.17) and (6.18) that

$$z(k) = \mathcal{G}(w)(k) = \mathcal{G}_c(w)(k) + \mathcal{G}_U(\mathcal{G}_v(w))(k).$$

The norm of the operator $\mathcal{G}$ applied to w can be written as

$$\begin{aligned} \|\mathcal{G}(w)\|_2^2 &= \langle \mathcal{G}_c(w) + \mathcal{G}_U(\mathcal{G}_v(w)); \mathcal{G}_c(w) + \mathcal{G}_U(\mathcal{G}_v(w)) \rangle \\ &= \|\mathcal{G}_c(w)\|_2^2 + \langle \mathcal{G}_U^* \mathcal{G}_c(w); \mathcal{G}_v(w) \rangle + \langle \mathcal{G}_v(w); \mathcal{G}_U^* \mathcal{G}_c(w) \rangle \\ &\quad + \langle \mathcal{G}_U^* \mathcal{G}_U \mathcal{G}_v(w); \mathcal{G}_v(w) \rangle. \end{aligned}$$

We recall from (6.11) and (6.12) that

$$\|\mathcal{G}\|_2^2 = \sum_{s=1}^r \|\mathcal{G}(w_s)\|_2^2$$

where

$$w_s(k) = \begin{cases} e_s, & k = 0 \\ 0, & k \neq 0 \end{cases}$$

and e_s is a vector with 1 at the sth position and zero elsewhere. Notice now that from (C.11) in Proposition C.3,

$$\mathcal{G}_U^* \mathcal{G}_c(w_s)(t) = \begin{cases} R_{\theta(t)}^{-1/2} B_{\theta(t)}^* E\left[\left[\widetilde{A}_{\theta(t+1)}^* \dots \widetilde{A}_{\theta(0)}^*\right] \mathcal{E}_{\theta(0)}(X) G_{\theta(0)} e_s | \mathfrak{F}_t\right], & t \leq 0 \\ 0, & t > 0 \end{cases}$$

and since

$$\mathcal{G}_v(w)(t) = \Phi_{\theta(t)}\left\{ \sum_{l=-\infty}^{t-1} \left[\Gamma_{\theta(t-1)} \dots \Gamma_{\theta(l+1)}\right] \Psi_{\theta(l)} w(l) \right\}$$

we have that

$$
\mathcal{G}_v(w_s)(t) = \begin{cases} \Phi_{\theta(t)} \Big[\Gamma_{\theta(t-1)} \dots \Gamma_{\theta(1)} \Big] \Psi_{\theta(0)} e_s \ , t > 0 \\ 0 \qquad\qquad\qquad\qquad\qquad\quad , t \leq 0. \end{cases}
$$

Thus,

$$
\langle \mathcal{G}_U^* \mathcal{G}_c(w_s); \mathcal{G}_v(w_s) \rangle = 0.
$$

Furthermore from Proposition C.3, $\mathcal{G}_U^* \mathcal{G}_U = I$, and thus,

$$
\langle \mathcal{G}_U^* \mathcal{G}_U \mathcal{G}_v(w); \mathcal{G}_v(w) \rangle = \| \mathcal{G}_v(w) \|_2^2.
$$

This leads to

$$
\| \mathcal{G} \|_2^2 = \sum_s \| \mathcal{G}(w_s) \|_2^2 = \| \mathcal{G}_c \|_2^2 + \| \mathcal{G}_v \|_2^2
$$

and since $\mathcal{G}_c$ does not depend on u,

$$
\min_{G_K} \| \mathcal{G} \|_2^2 = \| \mathcal{G}_c \|_2^2 + \min_{G_K} \| \mathcal{G}_v \|_2^2 \tag{6.20}
$$

where $\mathcal{G}_c$, $\mathcal{G}_v$, and $\mathcal{G}_K$ are as in (6.17), (6.19) (or (5.24)), and (6.7) respectively. But the solution of $\min_{G_K} \| \mathcal{G}_v \|_2^2$ is as in Theorem 5.8. Therefore, from Proposition 6.4 and Theorem 5.8

$$
\| \mathcal{G}_c \|_2^2 = \sum_{i=1}^N \pi_i \operatorname{tr}(G_i^* \mathcal{E}_i(X) G_i) \tag{6.21}
$$

and

$$
\min_{G_K} \| \mathcal{G}_v \|_2^2 = \sum_{i=1}^N \operatorname{tr}(R_i^{1/2} F_i Y_i F_i^* R_i^{1/2}) \tag{6.22}
$$

completing the proof of the theorem. □

Remark 6.6. Notice that as for the deterministic case [195] and [229], we have from (6.20) that the H_2-norm can be written as the sum of two H_2-norms: the first one does not depend on the control u and has value given by (6.21), and the second one is equivalent to the infinite horizon optimal filtering problem, and has optimal value given by (6.22).

6.4 Historical Remarks

Although more than thirty years have elapsed since the seminal papers by Sworder and Wonham, MJLS has been largely wedded to the complete observation scenario. Of course, this is not a privilege of MJLS. Except for the linear case, where a *separation principle* applies, there is enormous difficulty in treating optimal control problems with partial information. This is due, in part, to the high degree of nonlinearity that is introduced into the problem when such approaches, which try to recast the problem as one with complete observations (via the state information), are carried out. Even for relatively favorable structures, the optimal problem is almost intractable. In order to tackle these difficulties, many suboptimal approaches have been devised, including the use of the certainty equivalence principle (where it does not apply optimally). For those interested in the subject we refer to [221], where the key idea of separation was introduced, and [79] (see also [21] and [224]).

Regarding MJLS, we can mention [45], [46], [65], [83], [89], [92], [98], [101], [114], [116], [185], [188], and [223], as some works dealing with partial observations. Notice that, in our scenario, there are the following possibilities, with increasing degree of difficulty: (i) the Markov chain is observed, but not the state x; (ii) the state is observed, but not the chain; and finally (iii) none of them is observed. For most of the papers mentioned above, linear feedback controls in a restricted complexity setting have been adopted. Although the results in this book advance a theory that parallels the linear, including an attempt to discuss duality and a separation principle, the problem with partial information is certainly a topic which deserves further attention in the future.

7

H_∞-Control

In this chapter we consider an H_∞-like theory for MJLS. We follow an approach based on a worst-case design problem, tracing a parallel with the time domain formulation used for studying the standard H_∞ theory. We analyze the special case of state feedback, which we believe gives the essential tools for further studies in the MJLS scenario. A recursive algorithm for the H_∞-control CARE is included.

7.1 Outline of the Chapter

In this chapter we consider the state-feedback discrete-time H_∞-control problem for the class of MJLS. It is assumed that the controller has access to both the state variables and jump variables.

We obtain a necessary and sufficient condition for the existence of a state-feedback controller that stabilizes in the mean square sense a MJLS and ensures that, for any ℓ_2-additive disturbance sequence applied to the system, the output is smaller than some prespecified bound. In the deterministic set-up, this problem would be equivalent to the H_∞-control problem in the time-domain formulation. Using the concept of mean square stabilizability and detectability, see Definitions 3.40 and 3.41, the necessary and sufficient condition is derived in terms of a set of coupled algebraic Riccati equations. The technique of proof follows the one used in the literature for H_∞-control in a state space formulation (see [203] and [209]). The proof of necessity relies on a representation result by [222], for the minimum of quadratic forms within translations of Hilbert subspaces, see Lemma 7.10. The Hilbert space technique employed here provides some powerful tools for analyzing the H_∞-control problem of MJLS, allowing us to show that the condition derived is not only sufficient but also necessary.

It is not our intention to go into the classical theory since there is by now a large number of books on this subject (see, e.g., [126], [136], [140], [203], [228], [229]). We rather prefer to develop in detail the specific aspects required to tackle the MJLS H_∞-control problem, giving the interested readers the essential tools to go further on this subject. In the last section of this chapter we give a brief historical overview of the H_∞-theory, emphasizing the key contributions to the theory, and suggesting further references for those interested in this subject.

The chapter content is as follows. In Section 7.2 the H_∞-control problem for MJLS is precisely defined, together with the main characterization result. The proofs of sufficiency and necessity are presented in Section 7.3. Section 7.4 presents a numerical technique for obtaining the desired solution for the CARE.

7.2 The MJLS H_∞-like Control Problem

7.2.1 The General Problem

On the stochastic basis $(\Omega, \mathfrak{F}, \{\mathfrak{F}_k\}, \mathcal{P})$, consider the following MJLS

$$
\mathcal{G} = \begin{cases}
x(k+1) = A_{\theta(k)}x(k) + B_{\theta(k)}u(k) + G_{\theta(k)}w(k) \\
y(k) = L_{\theta(k)}x(k) + H_{\theta(k)}w(k) \\
z(k) = C_{\theta(k)}x(k) + D_{\theta(k)}u(k)
\end{cases}
\tag{7.1}
$$

where $\mathbb{T} = \{0, 1, \ldots\}$ and, as before, $x = \{x(k); k \in \mathbb{T}\}$ represents the n-dimensional state vector, $u = \{u(k); k \in \mathbb{T}\}$ the m-dimensional control sequence in $\mathcal{C}^m$, $w = \{w(k); k \in \mathbb{T}\}$ a r-dimensional disturbance sequence in $\mathcal{C}^r$, $y = \{y(k); k \in \mathbb{T}\}$ the p-dimensional sequence of measurable variables and $z = \{z(k); k \in \mathbb{T}\}$ the q-dimensional output sequence. Unless stated otherwise, the spaces used throughout this chapter are defined as in Chapter 2 (for instance, $\mathcal{C}^m$ is defined as in Section 2.3).

In the spirit of the deterministic case, we can view w as any unknown finite-energy stochastic disturbance signal which adversely affects the to-be-controlled output z from the desired value 0. The general idea is to attenuate the effect of the disturbances in z via a control action u based on a dynamic feedback. In addition, the feedback control has to be chosen in such a way that the closed loop system $\mathcal{G}_{cl}$ is stabilized in an appropriate notion of stability. The effect of the disturbances on the to-be-controlled output z is then described by a perturbation operator $\mathcal{G}_{cl} : w \to z$, which (for zero initial state) maps any finite energy disturbance signal w (the ℓ_2-norm of the signal is finite) into the corresponding finite energy output signal z of the closed loop system. The size of this linear operator is measured by the induced norm and the larger this norm is, the larger is the effect of the unknown disturbance w on the to-be-controlled output z, i.e., $\|\mathcal{G}_{cl}\|$ measures the influence of the disturbances in the worst case scenario. The problem is to determine whether or not, for a given $\gamma > 0$, there exists a stabilizing controller achieving $\|\mathcal{G}_{cl}\| < \gamma$. Moreover, we want to know how such controllers, if they exist, can be constructed.

The standard H_∞ design leads to controllers of the *worst case* type in the sense that emphasis is focused on minimizing the effect of the disturbances which produce the largest effect on the system output.

In this chapter we shall consider just the case in which $y(k) = x(k)$, and we shall look for state-feedback controllers, that is, controllers of the form $u(k) = F_{\theta(k)}x(k)$. For this case we shall denote the closed loop map $\mathcal{G}_{cl}$ by Z_F.

Remark 7.1. In the deterministic linear continuous-time case, the norm $\|Z_F\|$ is given by the H_∞-norm (which is a norm in the Hardy space H_∞, i.e. a space which consists of all functions that are analytic and bounded in the open right half complex plane) of the associated real rational transfer functions in the space H_∞. This is the reason for the control theory dealing with this problem be known as H_∞-control theory. In our context, we use this terminology just to refer to the the disturbance attenuation aspect of the problem.

Remark 7.2. Notice that our model allows for unknown stochastic instead of deterministic disturbance signals. This is, in general, a necessary extension of the deterministic case if the underlying systems itself is random. This has to do with the fact that, seen in the context of differential game as a min-max problem (worst case design), the disturbance that maximizes the problem is given in terms of the state (or output) of the system and therefore is stochastic (restricted to deterministic disturbance, the maximum would not be achieved).

7.2.2 H_∞ Main Result

Consider now the following version of (7.1):

$$\mathcal{G} = \begin{cases} x(k+1) = A_{\theta(k)}x(k) + B_{\theta(k)}u(k) + G_{\theta(k)}w(k) \\ z(k) = C_{\theta(k)}x(k) + D_{\theta(k)}u(k) \end{cases} \tag{7.2}$$

where we assume that both $x(k)$ and $\theta(k)$ are known at each time k, and again without loss of generality that conditions (4.26) and (4.27) hold (later on we assume, for notational simplicity, that $D_i^* D_i = I$ for each $i \in \mathbb{N}$).

Our problem can be roughly formulated as follows: design a linear state-feedback control $u(k) = F_{\theta(k)}x(k) + q(k)$ such that the closed loop system $Z_F : w \to z$ has the following properties: (i) the feedback control $F_\theta : x \to u$ has to be chosen in such a way that the closed loop system $\mathcal{G}_{cl}$ is mean square stable, and (ii) the operator Z_F is γ-dissipative for some given attenuation level $\gamma > 0$ (we can guarantee a prescribed bound for the operator Z_F), i.e., the H_∞-problem here can be viewed as a worst-case design problem in the sense that one seeks to guarantee a bound on the 2-norm of the signal z independently of which ℓ_2-disturbance signal actually occurs.

Before going into the technicalities of the main result, we need to introduce some basic elements which will be essential in the treatment that we shall be

adopting here. First notice that, from (7.2) and the definition of the probabilistic space in Section 2.3, we have that $x_k = (x(0), \ldots, x(k)) \in \mathcal{C}_k^n$ for every
$k \in \mathbb{T}$. For $(x_0, w, q) \in \mathcal{C}_0^n \oplus \mathcal{C}^r \oplus \mathcal{C}^m$, $\theta_0 \in \Theta_0$, and $F = (F_1, \ldots, F_N) \in \mathbb{H}^{m,n}$,
define the linear operator $X_F(\theta_0, .)$ from $\mathcal{C}_0^n \oplus \mathcal{C}^r \oplus \mathcal{C}^m$ to $\mathcal{C}^n$ as

$$X_F(\theta_0, x_0, w, q) \triangleq x = (x_0, x(1), \ldots) \tag{7.3}$$

where $x = (x_0, x(1), \ldots)$ is defined as in (7.2) with

$$u(k) = F_{\theta(k)} x(k) + q(k) \quad \text{and} \quad q = (q(0), q(1), \ldots) \in \mathcal{C}^m.$$

From Theorem 3.34 and Theorem 3.9, it is clear that if $F = (F_1, \ldots, F_N)$
stabilizes (A, B) in the mean square sense then $x \in \mathcal{C}^n$ and

$$X_F(\theta_0, .) \in \mathbb{B}(\mathcal{C}_0^n \oplus \mathcal{C}^r \oplus \mathcal{C}^m, \mathcal{C}^n).$$

We have that $z \in \mathcal{C}^q$ and the following bounded linear operator $Z_F(\theta_0, .) \in$
$\mathbb{B}(\mathcal{C}_0^n \oplus \mathcal{C}^r \oplus \mathcal{C}^m, \mathcal{C}^q)$ is well defined:

$$Z_F(\theta_0, x_0, w, q) \triangleq z = (z(0), z(1), \ldots). \tag{7.4}$$

We need also to define the operators $X_F^0(\theta_0, .)$ and $Z_F^0(\theta_0, .)$ in $\mathbb{B}(\mathcal{C}^r, \mathcal{C}^n)$ and
$\mathbb{B}(\mathcal{C}^r, \mathcal{C}^q)$, respectively, as:

$$X_F^0(\theta_0, w) \triangleq X_F(\theta_0, 0, w, 0) \quad \text{and} \quad Z_F^0(\theta_0, w) \triangleq Z_F(\theta_0, 0, w, 0). \tag{7.5}$$

We may now formulate the H_∞-control problem here as follows:

The H_∞-control problem: given $\delta > 0$, find F that stabilizes (A, B) in the
mean square sense and such that

$$\sup_{\theta_0 \in \Theta_0} \left\| Z_F^0(\theta_0, .) \right\| < \delta$$

that is,

$$\left\| Z_F^0(\theta_0, w) \right\|_2^2 = \sum_{k=0}^{\infty} E(\left\| C_{\theta(k)} x(k) \right\|^2 + \left\| D_{\theta(k)} u(k) \right\|^2)$$

$$< \delta^2 \sum_{k=0}^{\infty} E(\left\| w(k) \right\|^2) = \delta^2 \left\| w(k) \right\|_2^2$$

for every $w \in \mathcal{C}^r$ different from 0 and $\theta_0 \in \Theta_0$.

The following theorem, to be proved in the following sections, gives a
solution to the above problem. For notational simplicity and without loss of
generality, we shall assume in the remainder of this chapter that $D_i^* D_i = I$.

Theorem 7.3. *Suppose that (C, A) is mean square detectable and consider $\delta > 0$ fixed. Then there exists $F = (F_1, \dots, F_N)$ such that it stabilizes (A, B) in the mean square sense, and such that*

$$\sup_{\theta_0 \in \Theta_0} \left\| Z_F^0(\theta_0, .) \right\| < \delta$$

if and only if there exists $P = (P_1, \dots, P_N) \in \mathbb{H}^{n+}$ satisfying the following conditions:

1. $\delta^2 I - G_i^* \mathcal{E}_i(P) G_i > \alpha^2 I$ *for all $i \in \mathbb{N}$ and some $\alpha > 0$.*
2.

$$
\begin{aligned}
P_i =& C_i^* C_i + A_i^* \mathcal{E}_i(P) A_i \\
& - A_i^* \mathcal{E}_i(P) \left[B_i \; \tfrac{1}{\delta} G_i \right] \left(\begin{bmatrix} I & 0 \\ 0 & -I \end{bmatrix} + \begin{bmatrix} B_i^* \\ \tfrac{1}{\delta} G_i^* \end{bmatrix} \mathcal{E}_i(P) \left[B_i \; \tfrac{1}{\delta} G_i \right] \right)^{-1} \\
& \times \begin{bmatrix} B_i^* \\ \tfrac{1}{\delta} G_i^* \end{bmatrix} \mathcal{E}_i(P) A_i \\
=& C_i^* C_i + (A_i + B_i F_i + \tfrac{1}{\delta} G_i U_i)^* \mathcal{E}_i(P)(A_i + B_i F_i + \tfrac{1}{\delta} G_i U_i) \\
& + F_i^* F_i - U_i^* U_i,
\end{aligned}
$$

for all $i \in \mathbb{N}$, where

$$F_i = -(I + B_i^* \mathcal{E}_i(P) B_i)^{-1} B_i^* \mathcal{E}_i(P)(A_i + \tfrac{1}{\delta} G_i U_i)$$

$$U_i = (I - \tfrac{1}{\delta^2} G_i^* \mathcal{E}_i(P) G_i)^{-1} (\tfrac{1}{\delta} G_i^*) \mathcal{E}_i(P)(A_i + B_i F_i)$$

that is,

$$
\begin{aligned}
F_i =& -(I + B_i^* \mathcal{E}_i(P) B_i + \tfrac{1}{\delta^2} B_i^* \mathcal{E}_i(P) G_i (I - \tfrac{1}{\delta^2} G_i^* \mathcal{E}_i(P) G_i)^{-1} G_i^* \mathcal{E}_i(P) B_i)^{-1} \\
& \times (B_i^* (I + \tfrac{1}{\delta^2} \mathcal{E}_i(P) G_i (I - \tfrac{1}{\delta^2} G_i^* \mathcal{E}_i(P) G_i)^{-1} G_i^*) \mathcal{E}_i(P) A_i) \\
=& -(I + B_i^* \mathcal{E}_i(P)[I - \tfrac{1}{\delta^2} G_i G_i^* \mathcal{E}_i(P)]^{-1} B_i)^{-1} \\
& \times (B_i^* \mathcal{E}_i(P)[I - \tfrac{1}{\delta^2} G_i G_i^* \mathcal{E}_i(P)]^{-1} A_i) \\
U_i =& (I - \tfrac{1}{\delta^2} G_i^* \mathcal{E}_i(P) G_i + \tfrac{1}{\delta^2} G_i^* \mathcal{E}_i(P) B_i (I + B_i^* \mathcal{E}_i(P) B_i)^{-1} B_i^* \mathcal{E}_i(P) G_i)^{-1} \\
& \times (\tfrac{1}{\delta} G_i^* (I - \mathcal{E}_i(P) B_i (I + B_i^* \mathcal{E}_i(P) B_i)^{-1} B_i^*) \mathcal{E}_i(P) A_i) \\
=& \tfrac{1}{\delta} (I - \tfrac{1}{\delta^2} G_i^* \mathcal{E}_i(P)[I + B_i B_i^* \mathcal{E}_i(P)]^{-1} G_i)^{-1} \\
& \times (G_i^* \mathcal{E}_i(P)[I + B_i B_i^* \mathcal{E}_i(P)]^{-1} A_i).
\end{aligned}
$$

3. $r_\sigma(\mathcal{T}) < 1$ where $\mathcal{T}$ is as defined in (3.7) with $\Gamma_i = A_i + B_i F_i + \frac{1}{\delta} G_i U_i$.

Remark 7.4. It is not difficult to check that, when restricted to the deterministic case (that is, a Markov chain with a single state), the above result reduces to some known results in the current literature (cf. [203]). In addition, if we take $\delta \to \infty$ in (ii) above, we obtain the CARE presented in (4.35), which provides the characterization of the LQR control problem for MJLS. This is analogous to what is found in the deterministic H_∞-control case.

7.3 Proof of Theorem 7.3

7.3.1 Sufficient Condition

We prove in this section the sufficiency part of Theorem 7.3. Note that the hypothesis that (C, A) is mean square detectable is not required now. The proof will require the following propositions:

Proposition 7.5. *Suppose that 1), 2) and 3) of Theorem 7.3 hold. Then $F = (F_1, \ldots, F_N)$ stabilizes (A, B).*

Proof. Set

$$\Lambda_i = A_i + B_i F_i = A_i + \begin{bmatrix} B_i & \frac{1}{\delta} G_i \end{bmatrix} \begin{bmatrix} F_i \\ 0 \end{bmatrix}, \quad \Lambda = (\Lambda_1, \ldots, \Lambda_N) \in \mathbb{H}^n.$$

From 1) and 2) we have that for every $i \in \mathbb{N}$,

$$P_i = C_i^* C_i + F_i^* F_i - U_i^* U_i + (A_i + B_i F_i + \frac{1}{\delta} G_i U_i)^* \mathcal{E}_i(P)(A_i + B_i F_i + \frac{1}{\delta} G_i U_i)$$

$$= C_i^* C_i + F_i^* F_i + U_i^* (I - \frac{G_i^* \mathcal{E}_i(P) G_i}{\delta^2}) U_i + (A_i + B_i F_i)^* \mathcal{E}_i(P)(A_i + B_i F_i)$$

$$\geq \Lambda_i^* \mathcal{E}_i(P) \Lambda_i + \frac{\alpha^2}{\delta^2} \left(\begin{bmatrix} F_i \\ U_i \end{bmatrix} - \begin{bmatrix} F_i \\ 0 \end{bmatrix} \right)^* \left(\begin{bmatrix} F_i \\ U_i \end{bmatrix} - \begin{bmatrix} F_i \\ 0 \end{bmatrix} \right). \tag{7.6}$$

Recall from 3) that $r_\sigma(\mathcal{T}) = r_\sigma(\mathcal{L}) < 1$ where $\mathcal{L}$ and $\mathcal{T}$ are defined as in (3.8) and (3.7) respectively, with $\Gamma_i = A_i + B_i F_i + \frac{1}{\delta} G_i U_i$, which can also be written as

$$\Gamma_i = A_i + B_i F_i + \frac{1}{\delta} G_i U_i = A_i + \begin{bmatrix} B_i & \frac{1}{\delta} G_i \end{bmatrix} \begin{bmatrix} F_i \\ U_i \end{bmatrix}.$$

From Lemma A.8 and (7.6) we have that $r_\sigma(\bar{\mathcal{L}}) < 1$, where $\bar{\mathcal{L}} \in \mathbb{B}(\mathbb{H}^n)$ is defined as in (A.9), that is, $\bar{\mathcal{L}}(.) = (\bar{\mathcal{L}}_1(.), \ldots, \bar{\mathcal{L}}_N(.))$, with

$$\bar{\mathcal{L}}_i(.) = \Lambda_i^* \mathcal{E}(.) \Lambda_i.$$

From Definition 3.40 and Theorem 3.9, we have that $F = (F_1, \ldots, F_N)$ stabilizes (A, B). $\square$

For the remaining of this section we consider, for arbitrary $\theta_0 \in \Theta_0$ and any $w \in \mathcal{C}^r$, $x = (0, x(1), \dots) = X_F^0(\theta_0, w)$, and $z = (0, z(1), \dots) = Z_F^0(\theta_0, w)$.

Proposition 7.6. *Suppose that 1), 2) and 3) of Theorem 7.3 hold. Then*

$$\|z\|_2^2 = \delta^2(\|w\|_2^2 - \|r\|_2^2) \tag{7.7}$$

where $r = (r(0), r(1), \dots) \in \mathcal{C}^r$ is defined as

$$r(k) = (I - \frac{1}{\delta^2}G_{\theta(k)}^*\mathcal{E}_{\theta(k)}(P)G_{\theta(k)})^{\frac{1}{2}}(\frac{1}{\delta}U_{\theta(k)}x(k) - w(k))).$$

Proof. From 2) of the theorem we have that

$$P_i = C_i^*C_i + F_i^*F_i + (A_i + B_iF_i)^*\mathcal{E}_i(P)(A_i + B_iF_i) + U_i^*(I - \frac{1}{\delta^2}G_i^*\mathcal{E}_i(P)G_i)U_i. \tag{7.8}$$

Since

$$x(k+1) = (A_{\theta(k)} + B_{\theta(k)}F_{\theta(k)})x(k) + G_{\theta(k)}w(k)$$
$$x(0) = 0$$

we get that

$$\begin{aligned}
\left\|P_{\theta(k+1)}^{1/2}x(k+1)\right\|_2^2 &= E(x(k+1)^*P_{\theta(k+1)}x(k+1)) \\
&= E((E((A_{\theta(k)} + B_{\theta(k)}F_{\theta(k)})x(k) + G_{\theta(k)}w(k))^*P_{\theta(k+1)} \\
&\quad \times ((A_{\theta(k)} + B_{\theta(k)}F_{\theta(k)})x(k) + G_{\theta(k)}w(k))) \mid \mathfrak{F}_k)) \\
&= E(x(k)^*(A_{\theta(k)} + B_{\theta(k)}F_{\theta(k)})^*\mathcal{E}_{\theta(k)}(P) \\
&\quad \times (A_{\theta(k)} + B_{\theta(k)}F_{\theta(k)})x(k) \\
&\quad + w(k)^*G_{\theta(k)}^*\mathcal{E}_{\theta(k)}(P)(A_{\theta(k)} + B_{\theta(k)}F_{\theta(k)})x(k) \\
&\quad + x(k)^*(A_{\theta(k)} + B_{\theta(k)}F_{\theta(k)})^*\mathcal{E}_{\theta(k)}(P)G_{\theta(k)}w(k) \\
&\quad + w(k)^*G_{\theta(k)}^*\mathcal{E}_{\theta(k)}(P)G_{\theta(k)}w(k)).
\end{aligned} \tag{7.9}$$

Using (7.8) in (7.9) it is straightforward to show that

$$\left\|P_{\theta(k+1)}^{1/2}x(k+1)\right\|_2^2 = \left\|P_{\theta(k)}^{1/2}x(k)\right\|_2^2 - \|z(k)\|_2^2 + \delta^2\|w(k)\|_2^2 - \delta^2\|r(k)\|_2^2.$$

Recalling that $x(0) = 0$ and $\|x(\ell)\|_2 \to 0$ as $\ell \to \infty$ (since F stabilizes (A, B) by Proposition 7.5), we get that

$$\begin{aligned}
0 &\le \sum_{k=0}^{\ell-1}\left(\left\|P_{\theta(k+1)}^{1/2}x(k+1)\right\|_2^2 - \left\|P_{\theta(k)}^{1/2}x(k)\right\|_2^2\right) \\
&= \left\|P_{\theta(\ell)}^{1/2}x(\ell)\right\|_2^2 \\
&\le \|P\|_{max}\|x(\ell)\|_2^2 \to 0,
\end{aligned}$$

so that

$$0 = \sum_{k=0}^{\infty}(-\|z(k)\|_2^2 + \delta^2 \|w(k)\|_2^2 - \delta^2 \|r(k)\|_2^2)$$

showing (7.7).   □

Define the operator $\widetilde{\mathcal{W}}(\theta_0,.) \in \mathbb{B}(\mathcal{C}^r)$ as

$$\widetilde{\mathcal{W}}(\theta_0,.) \triangleq \frac{1}{\delta}UX_F^0(\theta_0,.) - I,$$

that is, for $w \in \mathcal{C}^r$,

$$\widetilde{\mathcal{W}}(\theta_0, w) = \widetilde{w} = (\widetilde{w}(0), \widetilde{w}(1), \ldots)$$

where $\widetilde{w}(k) = \frac{1}{\delta}U_{\theta(k)}x(k) - w(k)$, $k \in \mathbb{T}$.

In the next proposition and in the proof of Lemma 7.8 we shall drop, for notational simplicity, the dependence of the operators in θ_0.

Proposition 7.7. *Suppose that 1), 2) and 3) of Theorem 7.3 hold. Then* $\widetilde{\mathcal{W}}$ *is invertible.*

Proof. For $w \in \mathcal{C}^r$, define the operator $\widetilde{Y}$ as $\widetilde{Y}(w) = (\widetilde{y}(0), \widetilde{y}(1), \ldots)$ where

$$\widetilde{y}(k+1) = (A_{\theta(k)} + B_{\theta(k)}F_{\theta(k)} + \frac{1}{\delta}G_{\theta(k)}U_{\theta(k)})\widetilde{y}(k) - G_{\theta(k)}w(k), \quad \widetilde{y}(0) = 0.$$

Note that, since $r_\sigma(\mathcal{T}) < 1$ (condition 3) of the theorem), we have from Theorem 3.9 that $\widetilde{Y} \in \mathbb{B}(\mathcal{C}^r, \mathcal{C}^n)$ and $\left\|\widetilde{Y}\right\| \le a$ for some $a \ge 0$ and all $\theta_0 \in \Theta_0$. Define now $\widetilde{W}_{inv} \in \mathbb{B}(\mathcal{C}^r)$ as

$$\widetilde{W}_{inv} \triangleq \frac{1}{\delta}U\widetilde{Y} - I,$$

that is, for $w \in \mathcal{C}^r$, $\widetilde{W}_{inv}(w) = \widetilde{s} = (\widetilde{s}(0), \widetilde{s}(1), \ldots)$ where $\widetilde{s}(k) = \frac{1}{\delta}U_{\theta(k)}\widetilde{y}(k) - w(k)$, $k \in \mathbb{T}$. From these definitions it is easy to verify that $\widetilde{Y}(w) = X_F^0(\widetilde{s})$ and $\widetilde{Y}(\widetilde{w}) = X_F^0(w)$. Let us show that

$$\widetilde{W}\widetilde{W}_{inv} = \widetilde{W}_{inv}\widetilde{W} = I.$$

Indeed,

$$\widetilde{W}\widetilde{W}_{inv}(w) = \widetilde{W}(\widetilde{s}) = \frac{1}{\delta}UX_F^0(\widetilde{s}) - \widetilde{s} = \frac{1}{\delta}U\widetilde{Y}(w) - (\frac{1}{\delta}U\widetilde{Y}(w) - w) = w,$$

and

$$\widetilde{W}_{inv}\widetilde{W}(w) = \widetilde{W}_{inv}(\widetilde{w}) = \frac{1}{\delta}U\widetilde{Y}(\widetilde{w}) - \widetilde{w} = \frac{1}{\delta}UX_F^0(w) - (\frac{1}{\delta}UX_F^0(w) - w) = w,$$

showing that $\widetilde{W}^{-1} = \widetilde{W}_{inv} \in \mathbb{B}(\mathcal{C}^r)$.   □

We can now prove the sufficiency of the theorem.

Lemma 7.8. *Suppose that 1), 2) and 3) of Theorem 7.3 hold. Then F stabilizes (A, B) and*

$$\sup_{\theta_0 \in \Theta_0} \left\| Z_F^0(\theta_0, .) \right\| < \delta.$$

Proof. Consider $\alpha > 0$ as in 1), and $\alpha_1 > (\alpha/\delta)$ such that $\left\| \widetilde{W}^{-1} \right\| \leq \alpha_1$ for every $\theta_0 \in \Theta_0$. Since

$$\frac{1}{\alpha_1} \|w\|_2 \leq \left\| \widetilde{W}^{-1} \right\|^{-1} \|w\|_2 \leq \left\| \widetilde{W}(w) \right\|_2$$

we conclude that

$$
\begin{aligned}
\|r\|_2^2 &= \sum_{k=0}^{\infty} E\left(\left(\frac{1}{\delta} U_{\theta(k)} x(x) - w(k) \right)^* \left(I - \frac{1}{\delta^2} G_{\theta(k)}^* \mathcal{E}_{\theta(k)}(P) G_{\theta(k)} \right) \right. \\
&\qquad \left. \times \left(\frac{1}{\delta} U_{\theta(k)} x(k) - w(k) \right) \right) \\
&\geq \left(\frac{\alpha}{\delta} \right)^2 \sum_{k=0}^{\infty} E\left(\left\| \frac{1}{\delta} U_{\theta(k)} x(k) - w(k) \right\|^2 \right) \\
&= \left(\frac{\alpha}{\delta} \right)^2 \sum_{k=0}^{\infty} E\left(\left\| \widetilde{W}(w)(k) \right\|^2 \right) \\
&= \left(\frac{\alpha}{\delta} \right)^2 \left\| \widetilde{W}(w) \right\|_2^2 \geq \left(\frac{\alpha}{\delta \alpha_1} \right)^2 \|w\|_2^2
\end{aligned}
$$

and therefore, from (7.7) with $w \neq 0$ in $\mathcal{C}^r$,

$$
\begin{aligned}
\|z\|_2^2 &= \delta^2 \left(\|w\|_2^2 - \|r\|_2^2 \right) \\
&\leq \delta^2 \left(\|w\|_2^2 - \left(\frac{\alpha}{\delta \alpha_1} \right)^2 \|w\|_2^2 \right) \\
&= \delta^2 \left(1 - \left(\frac{\alpha}{\delta \alpha_1} \right)^2 \right) \|w\|_2 \\
&< \delta^2 \|w\|_2^2,
\end{aligned}
$$

proving the desired result. $\qquad\square$

7.3.2 Necessary Condition

We show in this section the necessity part of Theorem 7.3, as stated in the next lemma.

Lemma 7.9. *Suppose that* (C, A) *is mean square detectable, there exists* $\bar{K} = (\bar{K}_1, \ldots, \bar{K}_N) \in \mathbb{H}^{n,m}$ *that stabilizes* (A, B), *and*

$$\sup_{\theta_0 \in \Theta_0} \left\| Z^0_{\bar{K}}(\theta_0, .) \right\| < \delta$$

for some $\delta > 0$. *Then there exists* $P = (P_1, \ldots, P_N) \in \mathbb{H}^{n+}$ *satisfying conditions 1), 2) and 3) of Theorem 7.3.*

From Corollary A.16, the fact that (C, A) is mean square detectable, and (A, B) is mean square stabilizable, there exists a unique $X = (X_1, \ldots, X_N) \in \mathbb{H}^{n+}$ such that

$$X_i = C_i^* C_i + A_i^* \mathcal{E}_i(X) A_i - A_i^* \mathcal{E}_i(X) B_i (I + B_i^* \mathcal{E}_i(X) B_i)^{-1} B_i^* \mathcal{E}_i(X) A_i$$
$$= C_i^* C_i + (A_i + B_i \mathcal{F}_i(X))^* \mathcal{E}_i(X)(A_i + B_i \mathcal{F}_i(X)) + \mathcal{F}_i(X)^* \mathcal{F}_i(X)$$

where $\mathcal{F}(X) = (\mathcal{F}_1(X), \ldots, \mathcal{F}_N(X))$ (defined as in (4.36) with $D_i^* D_i = I$) stabilizes (A, B). For any $(x_0, w, q) \in \mathcal{C}_0^n \oplus \mathcal{C}^r \oplus \mathcal{C}^m$, $\theta_0 \in \Theta_0$, $w = (w(0), \ldots)$, $q = (q(0), \ldots)$, we set (as in (7.3) and (7.4))

$$x = (x(0), \ldots) = X_{\mathcal{F}(X)}(\theta_0, x_0, w, q),$$
$$z = (z(0), \ldots) = Z_{\mathcal{F}(X)}(\theta_0, x_0, w, q),$$

and

$$\mathcal{J}_\delta(\theta_0, x_0, w, q) = \sum_{k=0}^{\infty} E\left(\left\| C_{\theta(k)} x(k) \right\|^2 + \left\| u(k) \right\|^2 - \delta^2 \left\| w(k) \right\|^2 \right)$$

$$= \sum_{k=0}^{\infty} E\left(\left\| Q_{\theta(k)}^{1/2} x(k) \right\|^2 + \left\| \mathcal{F}_{\theta(k)}(X) x(k) + q(k) \right\|^2 \right.$$

$$\left. - \delta^2 \left\| w(k) \right\|^2 \right)$$

$$= \left\| Z_{\mathcal{F}(X)}(\theta_0, x_0, w, q) \right\|_2^2 - \delta^2 \left\| w \right\|_2^2$$

and we want to solve the following minimax problem:

$$\hat{\mathcal{J}}_\delta(\theta_0, x_0) = \sup_{w \in \mathcal{C}^r} \inf_{q \in \mathcal{C}^m} \mathcal{J}_\delta(\theta_0, x_0, w, q). \tag{7.10}$$

We shall first solve the minimization problem,

$$\tilde{\mathcal{J}}_\delta(\theta_0, x_0, w) = \inf_{q \in \mathcal{C}^m} \mathcal{J}_\delta(\theta_0, x_0, w, q).$$

In order to solve the above problem, we define $\mathcal{R}(\theta_0, .) : \mathcal{C}^r \to \mathcal{C}^n$ and $\mathcal{Q}(\theta_0, .) : \mathcal{C}^r \to \mathcal{C}^m$ in the following way: for $w = (w(0), w(1), \ldots) \in \mathcal{C}^r$,

$$\mathcal{R}(\theta_0, w) \triangleq r = (r(0), r(1), \ldots)$$

where (see (4.48))

$$r(k) \triangleq$$

$$\sum_{j=0}^{\infty} E\left(\left(\prod_{l=k+j}^{k+1} (A_{\theta(l)} + B_{\theta(l)} \mathcal{F}_{\theta(l)}(X))^* \right) X_{\theta(k+j+1)} G_{\theta(k+j)} w(k+j) \mid \mathfrak{F}_k \right)$$

for $k \geq 0$, and

$$\mathcal{Q}(\theta_0, w) \triangleq \widetilde{q} = (\widetilde{q}(0), \widetilde{q}(1), \ldots)$$

where

$$\widetilde{q}(k) \triangleq - \left(I + B_{\theta(k)}^* \mathcal{E}_{\theta(k)}(X) B_{\theta(k)} \right)^{-1} B_{\theta(k)}^* r(k), \ \ k \geq 0.$$

As seen in Lemma 4.12, Theorem 4.14 and (4.50), we have that $\mathcal{R}(\theta_0, .) \in \mathbb{B}(\mathcal{C}^r, \mathcal{C}^n)$, $\mathcal{Q}(\theta_0, .) \in \mathbb{B}(\mathcal{C}^r, \mathcal{C}^m)$, and

$$\widetilde{\mathcal{J}}_\delta(\theta_0, x_0, w) = \mathcal{J}_\delta(\theta_0, x_0, w, \widetilde{q}) = \inf_{q \in \mathcal{C}^m} \mathcal{J}(\theta_0, x_0, w, q) \qquad (7.11)$$

where $\widetilde{q} = \mathcal{Q}(\theta_0, w)$.

We shall now move on to the maximization problem, that is,

$$\widehat{\mathcal{J}}_\delta(\theta_0, x_0) = \widetilde{\mathcal{J}}_\delta(\theta_0, x_0, \widehat{w}) = \sup_{w \in \mathcal{C}^r} \widetilde{\mathcal{J}}_\delta(\theta_0, x_0, w). \qquad (7.12)$$

We need the following result, due to Yakubovich [222], also presented in [151].

Lemma 7.10. *Consider $\mathcal{H}$ a Hilbert space and a quadratic form $\mathcal{J}(\zeta) = \langle \mathcal{S}\zeta; \zeta \rangle$, $\zeta \in \mathcal{H}$ and $\mathcal{S} \in \mathbb{B}(\mathcal{H})$ self adjoint. Let $\mathcal{M}_0$ be a closed subspace of $\mathcal{H}$ and $\mathcal{M}$ a translation of $\mathcal{M}_0$ by an element $m \in \mathcal{H}$ (i.e., $\mathcal{M} = \mathcal{M}_0 + m$). If*

$$\inf_{\zeta \in \mathcal{M}_0} \frac{\langle \mathcal{S}\zeta; \zeta \rangle}{\langle \zeta; \zeta \rangle} > 0$$

then there exits a unique element $\widehat{\zeta} \in \mathcal{M}$ such that

$$\mathcal{J}(\widehat{\zeta}) = \inf_{\zeta \in \mathcal{M}} \mathcal{J}(\zeta)$$

where $\widehat{\zeta} = p + m$, with $p = \mathcal{G}m \in \mathcal{M}_0$ for some $\mathcal{G} \in \mathbb{B}(\mathcal{H})$.

For $\theta_0 \in \Theta_0$ fixed, define the following operators $\widetilde{X}(\theta_0, .)$, $\bar{X}(\theta_0, .)$ in $\mathbb{B}(\mathcal{C}_0^n \oplus \mathcal{C}^r, \mathcal{C}^n)$, and $\widetilde{Z}(\theta_0, .)$, $\bar{Z}(\theta_0, .)$ in $\mathbb{B}(\mathcal{C}_0^n \oplus \mathcal{C}^r, \mathcal{C}^q)$; for $x_0 \in \mathcal{C}_0^n$, $w \in \mathcal{C}^r$, and $\bar{K}$ as in Lemma 7.9,

$$\widetilde{X}(\theta_0, x_0, w) \triangleq X_{\mathcal{F}(X)}(\theta_0, x_0, w, \mathcal{Q}(\theta_0, w)),$$
$$\widetilde{Z}(\theta_0, x_0, w) \triangleq Z_{\mathcal{F}(X)}(\theta_0, x_0, w, \mathcal{Q}(\theta_0, w)),$$
$$\bar{X}(\theta_0, x_0, w) \triangleq X_{\bar{K}}(\theta_0, x_0, w, 0),$$
$$\bar{Z}(\theta_0, x_0, w) \triangleq Z_{\bar{K}}(\theta_0, x_0, w, 0).$$

Note from (7.12) that, for any $q \in \mathcal{C}^m$ and every $(x_0, w) \in \mathcal{C}_0^n \oplus \mathcal{C}^r$,

$$\left\| \widetilde{Z}(\theta_0, x_0, w) \right\|_2^2 - \delta^2 \|w\|_2^2 \leq \left\| Z_{\mathcal{F}(X)}(\theta_0, x_0, w, q) \right\|_2^2 - \delta^2 \|w\|_2^2$$

and choosing $q = (q(0), \ldots)$ as $q(k) = -\mathcal{F}_{\theta(k)}(X)x(k) + \bar{K}_{\theta(k)}x(k)$, we get that $Z_{\mathcal{F}(X)}(\theta_0, x_0, w, q) = \bar{Z}(\theta_0, x_0, w)$ (note that indeed $q \in \mathcal{C}^m$) and thus

$$\left\| \widetilde{Z}(\theta_0, x_0, w) \right\|_2^2 - \delta^2 \|w\|_2^2 \leq \left\| \bar{Z}(\theta_0, x_0, w) \right\|_2^2 - \delta^2 \|w\|_2^2. \tag{7.13}$$

From Lemma 7.9, there exists $\alpha > 0$ such that for all $w \in \mathcal{C}^r$,

$$\sup_{\theta_0 \in \Theta_0} \left\{ \frac{\left\| Z_K^0(\theta_0, w) \right\|_2^2}{\|w\|_2^2} \right\} < \delta^2 - \alpha^2 \tag{7.14}$$

and thus, from (7.13) and (7.14), for every $w \in \mathcal{C}^r$ and arbitrary $\theta_0 \in \Theta_0$,

$$\alpha^2 \|w\|_2^2 \leq \delta^2 \|w\|_2^2 - \left\| \bar{Z}(\theta_0, 0, w) \right\|_2^2 \leq \delta^2 \|w\|_2^2 - \left\| \widetilde{Z}(\theta_0, 0, w) \right\|_2^2 \tag{7.15}$$

that is,

$$\inf_{w \in \mathcal{C}^r} \left\{ \frac{\delta^2 \|w\|_2^2 - \left\| \widetilde{Z}(\theta_0, 0, w) \right\|_2^2}{\|w\|_2^2} \right\} \geq \alpha^2. \tag{7.16}$$

Proposition 7.11. *Consider $\theta_0 \in \Theta_0$ fixed. Under the hypothesis of Lemma 7.9, for each $x_0 \in \mathcal{C}_0^n$, there exists a unique element $\widehat{w} \in \mathcal{C}^r$ such that*

$$\widehat{\mathcal{J}}_\delta(\theta_0, x_0) = \widetilde{\mathcal{J}}_\delta(\theta_0, x_0, \widehat{w}) = \sup_{w \in \mathcal{C}^r} \widetilde{\mathcal{J}}(\theta_0, x_0, w).$$

Moreover, for some $\mathcal{W}(\theta_0, .) \in \mathbb{B}(\mathcal{C}_0^n, \mathcal{C}^r)$, $\widehat{w} = \mathcal{W}(\theta_0, x_0)$.

Proof. We have that

$$\widetilde{\mathcal{J}}_\delta(\theta_0, x_0, w) = \left\| \widetilde{Z}(\theta_0, x_0, w) \right\|_2^2 - \delta^2 \|w\|_2^2$$

$$= \left\langle \widetilde{Z}^* \widetilde{Z}(\theta_0, x_0, w); (x_0, w) \right\rangle - \left\langle (0, \delta^2 w); (x_0, w) \right\rangle$$

$$= \left\langle \mathcal{S}(\theta_0, x_0, w); (x_0, w) \right\rangle$$

where $\mathcal{S}(\theta_0, x_0, w) = \widetilde{Z}^* \widetilde{Z}(\theta_0, x_0, w) - (0, \delta^2 w, 0)$. We have that $\mathcal{C}_0^n \oplus \mathcal{C}^r$ is a Hilbert space and $\mathcal{S} \in \mathbb{B}(\mathcal{C}_0^n \oplus \mathcal{C}^r)$ with $\mathcal{S}(\theta_0, .)$ self adjoint. Define $\mathcal{M}_0 = \{(x_0, w) \in \mathcal{C}_0^n \oplus \mathcal{C}^r; x_0 = 0\}$ and $\mathcal{M} = \mathcal{M}_0 + m$ with $m = (x_0, 0) \in \mathcal{C}_0^n \oplus \mathcal{C}^r$. $\mathcal{M}_0$ is a closed subspace of $\mathcal{C}_0^n \oplus \mathcal{C}^r$ and $\mathcal{M}$ a translation of $\mathcal{M}_0$ by the element m. From (7.16),

$$\inf_{\zeta \in \mathcal{M}_0} \left\{ \frac{-\langle \mathcal{S}\zeta; \zeta \rangle}{\|\zeta\|_2^2} \right\} = \inf_{w \in \mathcal{C}^r} \left\{ \frac{\delta^2 \|w\|_2^2 - \left\| \widetilde{Z}(\theta_0, 0, w) \right\|_2^2}{\|w\|_2^2} \right\} \geq \alpha^2 > 0$$

and invoking Lemma 7.10, we obtain that there exists a unique element $\widehat{\zeta} \in \mathcal{M}$ such that

$$-\widetilde{\mathcal{J}}_\delta(\theta_0, \widehat{\zeta}) = \inf_{\zeta \in \mathcal{M}} -\widetilde{\mathcal{J}}(\theta_0, \zeta),$$

where $\widehat{\zeta} = p + m$, $p = \mathcal{W}'(\theta_0, m)$ for some $\mathcal{W}'(\theta_0, .) \in \mathbb{B}(\mathcal{C}_0^n \oplus \mathcal{C}^r)$. Therefore,

$$\mathcal{W}'(\theta_0, m) = \mathcal{W}'(\theta_0, x_0, 0) = (0, \widehat{w})$$

and for some $\mathcal{W} \in \mathbb{B}(\mathcal{C}_0^n, \mathcal{C}^r)$, $\widehat{w} = \mathcal{W}(\theta_0, x_0)$. □

Now, for $\theta_0 \in \Theta_0$ fixed define the operators

$$\widehat{X}(\theta_0, .) \in \mathbb{B}(\mathcal{C}_0^0, \mathcal{C}^n), \quad \widehat{Z}(\theta_0, .) \in \mathbb{B}(\mathcal{C}_0^n, \mathcal{C}^r) \tag{7.17}$$

as:

$$\widehat{X}(\theta_0, x_0) \triangleq \widetilde{X}(\theta_0, x_0, \mathcal{W}(\theta_0, x_0)) = X_{\mathcal{F}(X)}(\theta_0, x_0, \mathcal{W}(\theta_0, x_0), \mathcal{Q}(\theta_0, \mathcal{W}(\theta_0, x_0)))$$
$$= (x_0, \widehat{x}(1), \ldots) = \widehat{x}$$
$$\widehat{Z}(\theta_0, x_0) \triangleq \widetilde{Z}(\theta_0, x_0, \mathcal{W}(\theta_0, x_0)) = Z_{\mathcal{F}(X)}(\theta_0, x_0, \mathcal{W}(\theta_0, x_0), \mathcal{Q}(\theta_0, \mathcal{W}(\theta_0, x_0)))$$
$$= (z_0, \widehat{z}(1), \ldots) = \widehat{z}$$

so that

$$\widehat{\mathcal{J}}_\delta(\theta_0, x_0) = \sup_{w \in \mathcal{C}^r} \inf_{q \in \mathcal{C}^m} \mathcal{J}_\delta(\theta_0, x_0, w, q) = \left\| \widehat{Z}(\theta_0, x_0) \right\|_2^2 - \delta^2 \|\mathcal{W}(\theta_0, x_0)\|_2^2$$
$$= \left\langle \widehat{Z}(\theta_0, x_0); \widehat{Z}(\theta_0, x_0) \right\rangle - \delta^2 \left\langle \mathcal{W}(\theta_0, x_0); \mathcal{W}(\theta_0, x_0) \right\rangle$$
$$= \left\langle (\widehat{Z}^* \widehat{Z} - \delta^2 \mathcal{W}^* \mathcal{W})(\theta_0, x_0); x_0 \right\rangle = \langle \mathfrak{B}(\theta_0, x_0); x_0 \rangle$$

where $\mathfrak{B}(\theta_0, .) \in \mathbb{B}(\mathcal{C}_0^n)$ is defined as $\mathfrak{B}(\theta_0, .) = (\widehat{Z}^* \widehat{Z} - \delta^2 \mathcal{W}^* \mathcal{W})(\theta_0, .)$. Since for any $x_0 \in \mathcal{C}_0^n$,

$$\widehat{\mathcal{J}}_\delta(\theta_0, x_0) = \sup_{w \in \mathcal{C}^r} \inf_{q \in \mathcal{C}^m} \mathcal{J}_\delta(\theta_0, x_0, w, q)$$
$$= \sup_{w \in \mathcal{C}^r} \left\{ \left\| \widetilde{Z}(\theta_0, x_0, w) \right\|_2^2 - \delta^2 \|w\|_2^2 \right\} \geq \left\| \widetilde{Z}(\theta_0, x_0, 0) \right\|_2^2 \geq 0$$

it follows that $\mathfrak{B}(\theta_0, .) \geq 0$. For each $i \in \mathbb{N}$ and $x_0 \in \mathbb{C}^n$, define $P_i \in \mathbb{B}(\mathbb{C}^n)$ such that

$$P_i x_0 = E(\mathfrak{B}(i, x_0)).$$

For every $x_0 \in \mathbb{C}^n$,

$$\widehat{\mathcal{J}}_\delta(i, x_0) = \langle \mathfrak{B}(i, x_0); x_0 \rangle = E(\langle \mathfrak{B}(i, x_0); x_0 \rangle)$$
$$= \langle E(\mathfrak{B}(i, x_0)); x_0 \rangle = \langle P_i x_0; x_0 \rangle = x_0^* P_i x_0 \geq 0$$

so that $P_i \geq 0$.

In order to prove that $P = (P_1, \ldots, P_N)$, as defined before, satisfies the conditions of the theorem, we consider a truncated minimax problem:

$$\widehat{\mathcal{J}}_\delta^k(\theta_0, x_0) = \sup_{w \in \mathcal{C}^{k,r}} \inf_{q \in \mathcal{C}^m} \mathcal{J}_\delta(\theta_0, x_0, w, q) \tag{7.18}$$

where for ι integer,

$$\mathcal{C}^{k,\iota} = \{s = (s(0), s(1), \ldots) \in \mathcal{C}^\iota; \ s(i) = 0 \text{ for } i \geq k\}.$$

Moreover, setting $\varphi^k \in \mathbb{B}(\mathcal{C}^r, \mathcal{C}^{k,r})$ as

$$\varphi^k(w) = (w(0), \ldots, w(k-1), 0, 0, \ldots),$$

where $w = (w(0), w(1), , \ldots) \in \mathcal{C}^r$, we get from (7.12) and (7.18) that

$$\widetilde{\mathcal{J}}_\delta(\theta_0, x_0, \varphi^k(\mathcal{W}(\theta_0, x_0))) \leq \widehat{\mathcal{J}}_\delta^k(\theta_0, x_0) \leq \widehat{\mathcal{J}}_\delta(\theta_0, x_0).$$

Since $\varphi^k(\mathcal{W}(\theta_0, x_0)) \to \mathcal{W}(\theta_0, x_0)$ as $k \to \infty$ we obtain, from continuity of $\widetilde{\mathcal{J}}_\delta(\theta_0, x_0, .)$, that

$$\widehat{\mathcal{J}}_\delta^k(\theta_0, x_0) \to \widehat{\mathcal{J}}_\delta(\theta_0, x_0) \text{ as } k \to \infty. \tag{7.19}$$

Furthermore, notice that (7.18) can be rewritten as:

$$\widehat{\mathcal{J}}_\delta^k(\theta_0, x_0) = \sup_{w \in \mathcal{C}^{k,r}} \inf_{q \in \mathcal{C}^{k,m}} \left\{ \sum_{l=0}^{k-1} \left(\left\| Q_{\theta(l)}^{1/2} x(l) \right\|_2^2 + \left\| F_{\theta(l)}(X) x(l) + q(l) \right\|_2^2 \right. \right.$$
$$\left. \left. - \delta^2 \|w(l)\|_2^2 \right) + \left\| X_{\theta(k)}^{1/2} x(k) \right\|_2^2 \right\} \tag{7.20}$$

We shall now obtain a solution for (7.18) in a recursive way. Define the sequences $P^k = (P_1^k, \ldots, P_N^k)$, $P_i^k \in \mathbb{B}(\mathbb{C}^n)$, $F^k = (F_1^k, \ldots, F_N^k)$, $F_i^k \in \mathbb{B}(\mathbb{C}^n, \mathbb{C}^m)$, and $U^k = (U_1^k, \ldots, U_N^k)$, $U_i^k \in \mathbb{B}(\mathbb{C}^n, \mathbb{C}^r)$ as:

$$P^0 = (P_1^0, \ldots, P_N^0) \triangleq X = (X_1, \ldots, X_N)$$

(where X is the mean square stabilizing solution of the CARE (4.35)), and

$$P_i^{k+1} \triangleq C_i^* C_i + A_i^* \mathcal{E}_i(P^k) A_i - A_i^* \mathcal{E}_i(P^k) \left[B_i \ \tfrac{1}{\delta} G_i \right]$$
$$\times \left(\begin{bmatrix} I & 0 \\ 0 & -I \end{bmatrix} + \begin{bmatrix} B_i^* \\ \tfrac{1}{\delta} G_i^* \end{bmatrix} \mathcal{E}_i(P^k) \left[B_i \ \tfrac{1}{\delta} G_i \right] \right)^{-1} \begin{bmatrix} B_i^* \\ \tfrac{1}{\delta} G_i^* \end{bmatrix} \mathcal{E}_i(P^k) A_i$$
$$= C_i^* C_i + (A_i + B_i F_i^{k+1} + \tfrac{1}{\delta} G_i U_i^{k+1})^* \mathcal{E}_i(P^k)(A_i + B_i F_i^{k+1} + \tfrac{1}{\delta} G_i U_i^{k+1})$$
$$+ (F_i^{k+1})^* F_i^{k+1} - (U_i^{k+1})^* U_i^{k+1}$$

for $k \geq 0$ and $i \in \mathbb{N}$, with

$$
\begin{aligned}
F_i^{k+1} \triangleq & - (I + B_i^* \mathcal{E}_i(P^k) B_i \\
& + \frac{1}{\delta^2} B_i^* \mathcal{E}_i(P^k) G_i (I - \frac{1}{\delta^2} G_i^* \mathcal{E}_i(P^k) G_i)^{-1} G_i^* \mathcal{E}_i(P^k) B_i)^{-1} \\
& \times (B_i^*(I + \frac{1}{\delta^2} \mathcal{E}_i(P^k) G_i (I - \frac{1}{\delta^2} G_i^* \mathcal{E}_i(P^k) G_i)^{-1} G_i^*) \mathcal{E}_i(P^k) A_i),
\end{aligned}
$$

$$
\begin{aligned}
U_i^{k+1} \triangleq & (I - \frac{1}{\delta^2} G_i^* \mathcal{E}_i(P^k) G_i \\
& + \frac{1}{\delta^2} G_i^* \mathcal{E}_i(P^k) B_i (I + B_i^* \mathcal{E}_i(P^k) B_i)^{-1} B_i^* \mathcal{E}_i(P^k) G_i)^{-1} \\
& \times (\frac{1}{\delta} G_i^*(I - \mathcal{E}_i(P^k) B_i (I + B_i^* \mathcal{E}_i(P^k) B_i)^{-1} B_i^*) \mathcal{E}_i(P^k) A_i).
\end{aligned}
$$

The existence of the above inverses will be established in the proof of the proposition below. First we define, for $V = (V_1, \ldots, V_N) \in \mathbb{H}^{n+}$ such that $I - \frac{1}{\delta^2} G_i^* \mathcal{E}_i(V) G_i > 0$, $i \in \mathbb{N}$,

$$
\begin{aligned}
\mathfrak{R}(V) &\triangleq (\mathfrak{R}_1(V), \ldots, \mathfrak{R}_N(V)), \\
\mathfrak{D}(V) &\triangleq (\mathfrak{D}_1(V), \ldots, \mathfrak{D}_N(V)), \\
\mathfrak{M}(V) &\triangleq (\mathfrak{M}_1(V), \ldots, \mathfrak{M}_N(V)), \\
\mathfrak{U}(V) &\triangleq (\mathfrak{U}_1(V), \ldots, \mathfrak{U}_N(V))
\end{aligned}
$$

as

$$
\begin{aligned}
\mathfrak{R}_i(V) \triangleq & C_i^* C_i + A_i^* \mathcal{E}_i(V) A_i - A_i^* \mathcal{E}_i(V) \left[B_i \ \tfrac{1}{\delta} G_i \right] \\
& \times \left(\begin{bmatrix} I & 0 \\ 0 & -I \end{bmatrix} + \begin{bmatrix} B_i^* \\ \tfrac{1}{\delta} G_i^* \end{bmatrix} \mathcal{E}_i(V) \left[B_i \ \tfrac{1}{\delta} G_i \right] \right)^{-1} \begin{bmatrix} B_i^* \\ \tfrac{1}{\delta} G_i^* \end{bmatrix} \mathcal{E}_i(V) A_i, \\
\mathfrak{M}_i(V) \triangleq & I + B_i^* \mathcal{E}_i(V) B_i, \\
\mathfrak{D}_i(V) \triangleq & \delta^2 (I - \frac{1}{\delta^2} G_i^* \mathcal{E}_i(V) G_i + \frac{1}{\delta^2} G_i^* \mathcal{E}_i(V) B_i (I + B_i^* \mathcal{E}_i(V) B_i)^{-1} B_i^* \mathcal{E}_i(V) G_i),
\end{aligned}
$$

and

$$
\mathfrak{U}_i(V) \triangleq G_i^*(I - \mathcal{E}_i(V) B_i (I + B_i^* \mathcal{E}_i(V) B_i)^{-1} B_i^*) \mathcal{E}_i(V) A_i \qquad (7.21)
$$

for $i \in \mathbb{N}$.

Proposition 7.12. *Consider $\theta_0 \in \Theta_0$ fixed. Under the hypothesis of Lemma 7.9, for each $k \geq 0$, we have that*

1. *$P^k = (P_1^k, \ldots, P_N^k) \in \mathbb{H}^{n+}$.*
2. *$\delta^2 I - G_i^* \mathcal{E}_i(P^k) G_i \geq \alpha^2 I$ for all $i \in \mathbb{N}$ and $\alpha > 0$ as in (7.16).*

3. $\widehat{\mathcal{J}}_\delta^k(\theta_0, x_0) = \mathcal{J}_\delta(\theta_0, x_0, \widehat{w}^k, \widehat{q}^k) = E(x_0^* P_{\theta_0} x_0),$ where

$$\widehat{w}^k = (\widehat{w}^k(0), \ldots, \widehat{w}^k(k-1), 0, 0, \ldots),$$
$$\widehat{q}^k = (\widehat{q}^k(0), \ldots, \widehat{q}^k(k-1), 0, 0,, \ldots),$$

with

$$\widehat{w}^k(l) = \frac{1}{\delta} U_{\theta(l)}^{k-l} \widehat{x}^k(l)$$
$$\widehat{q}^k(l) = (-\mathcal{F}_{\theta(l)}(X) + F_{\theta(l)}^{k-l}) \widehat{x}^k(l), \quad l = 0, \ldots, k-1,$$
$$\widehat{x}^k = (\widehat{x}^k(0), \widehat{x}^k(1), \ldots,) = X_{\mathcal{F}(X)}(\theta_0, x_0, \widehat{w}^k, \widehat{q}^k).$$

Proof. Let us apply induction on k. From Theorem 4.5, we have that 1) and 3) are clearly true for $k = 0$. Let us prove now that 2) is satisfied for $k = 0$. Fix $\theta_0 = i \in \mathbb{N}$ and consider $w(0) \in \mathbb{C}^r$, $w = (w(0), 0, \ldots) \in \mathcal{C}^r$. Then, from (7.15),

$$\alpha^2 \|w\|_2^2 \leq \delta^2 \|w\|_2^2 - \|\bar{Z}(i, 0, w)\|_2^2 \leq \delta^2 \|w\|_2^2 - \left\|\widetilde{Z}(i, 0, w)\right\|_2^2.$$

But

$$\left\|\widetilde{Z}(i, 0, w)\right\|_2^2 = \left\|Z_{\mathcal{F}(X)}(\theta(1), G_{\theta(1)} w(0), 0)\right\|_2^2 = w(0)^* G_i^* \mathcal{E}_i(X) G_i w(0)$$

and $\|w\|_2^2 = w(0)^* w(0)$, so that, from (7.15),

$$w(0)^* \left(\delta^2 I - G_i^* \mathcal{E}_i(X) G_i - \alpha^2 I\right) w(0) \geq 0$$

and since i and $w(0)$ are arbitrary, the result is proved for $k = 0$. Suppose now that the proposition holds for k. For $x_0 \in \mathcal{C}_0^n$ and $\theta_0 \in \Theta_0$, we define

$$\beta^{k+1}(\theta_0, x_0) = \sup_{w_0 \in \mathcal{C}_0^r} \inf_{q_0 \in \mathcal{C}_0^m} E\Big(\left\|C_{\theta(0)} x(0)\right\|^2$$
$$+ \left\|\mathcal{F}_{\theta(0)}(X) x(0) + q(0)\right\|^2 - \delta^2 \|w(0)\|^2 + x(1)^* P_{\theta(1)}^k x(1) \Big) \tag{7.22}$$

where

$$x(1) = A_{\theta_0} x_0 + B_{\theta_0} u_0 + G_{\theta_0} w_0 \quad \text{and} \quad u_0 = \mathcal{F}_{\theta_0}(X) x_0 + q_0.$$

Straightforward but somewhat lengthy algebraic calculations show that for any $i \in \mathbb{N}$, $x_0 \in \mathbb{C}^n$, $u_0 \in \mathbb{C}^m$, $w_0 \in \mathbb{C}^r$,

$$\|C_i x_0\|^2 + \|u_0\|^2 - \delta^2 \|w_0\|^2$$
$$+ (A_i x_0 + B_i u_0 + G_i w_0)^* \mathcal{E}_i(V)(A_i x_0 + B_i u_0 + G_i w_0)$$
$$= x_0^* \mathfrak{R}_i(V) x_0 + \left\|\mathfrak{M}_i(V)^{1/2}(u_0 + \widehat{u}_0(V, i, x_0, w_0))\right\|^2$$
$$- \left\|\mathfrak{D}_i(V)^{1/2}(w_0 - \widehat{w}(V, i, x_0))\right\|^2 \tag{7.23}$$

where

$$\widehat{w}(V, i, x_0) = \mathfrak{D}_i(V)^{-1}\mathfrak{U}_i(V)x_0$$
$$\widehat{u}_0(V, i, x_0, w_0) = \mathfrak{M}_i(V)^{-1}(B_i^*\mathcal{E}_i(V)A_i x_0 + B_i^*\mathcal{E}_i(V)G_i w_0),$$

which shows that the solution of (7.22) is given by

$$u_0 = \mathcal{F}_{\theta_0}(X)x_0 + q_0$$
$$= -\widehat{u}(P^k, i, x_0, w_0)$$
$$= -(I + B_{\theta_0}^*\mathcal{E}_{\theta_0}(P^k)B_{\theta_0})^{-1}(B_{\theta_0}\mathcal{E}_{\theta_0}(P^k)A_{\theta_0}x_0 + B_{\theta_0}^*\mathcal{E}_{\theta_0}(P^k)G_{\theta_0}w_0),$$
$$w_0 = \frac{1}{\delta}U_{\theta_0}^{k+1}x_0,$$

and therefore,

$$q_0 = -\mathcal{F}_{\theta_0}(X)x_0 + (I + B_{\theta_0}^*\mathcal{E}_{\theta_0}(P^k)B_{\theta_0})^{-1}B_{\theta_0}\mathcal{E}_{\theta_0}(P^k)(A_{\theta_0} + \frac{1}{\delta}G_{\theta_0}U_{\theta_0}^{k+1})x_0$$
$$= (-\mathcal{F}_{\theta_0}(X) + F_{\theta_0}^{k+1})x_0.$$

We have from (7.21), (7.22), (7.23) and above that

$$\beta^{k+1}(\theta_0, x_0) = E(x_0^*(P_{\theta_0}^{k+1})x_0) \geq 0,$$

and thus, $P_i^{k+1} \geq 0$, showing 1). Notice now that, by definition, $\widehat{\mathcal{J}}_\delta^{k+1}(\theta_0, x_0) \leq \beta^{k+1}(\theta_0, x_0)$. On the other hand, consider, for any $q \in \mathcal{C}^{k+1,m}$, $w(q) = (w(q)(0), \ldots, w(q)(k), 0, 0, \ldots) \in \mathcal{C}^{k+1,r}$ as

$$w(q)(l) = \frac{1}{\delta}U_{\theta(l)}^{k+1-l}x(q)(l), \ l = 0, \ldots, k,$$

with

$$x(q) = (x(q)(0), x(q)(1), \ldots) = X_{\mathcal{F}(X)}(\theta_0, x_0, w(q), q).$$

We get that

$$\beta^{k+1}(\theta_0, x_0) = \inf_{q \in \mathcal{C}^{k+1,m}} \left\{ \sum_{l=0}^{k} \left(\|C_{\theta(l)}x(q)(l)\|_2^2 + \|\mathcal{F}_{\theta(l)}(X)x(q)(l) + q(l)\|_2^2 \right. \right.$$
$$\left. \left. -\delta^2 \|w(q)(l)\|_2^2 \right) + \left\| X_{\theta(k)}^{1/2}x(k+1) \right\|_2^2 \right\}.$$

Taking the supremum over $w \in \mathcal{C}^{k+1,r}$ we get from (7.20) that $\beta^{k+1}(\theta_0, x_0) \leq \widehat{\mathcal{J}}_\delta^{k+1}(\theta_0, x_0)$, showing 3), that is,

$$\widehat{\mathcal{J}}_\delta^{k+1}(\theta_0, x_0) = \beta^{k+1}(\theta_0, x_0) = E(x_0^*P_{\theta_0}^{k+1}x_0).$$

Finally, let us show 2). Consider

$$w = (w(0), \ldots, w(k+1), 0, 0, \ldots) \in \mathcal{C}^{k+2,r},$$
$$\widehat{w}^{k+1} = (\widehat{w}^{k+1}(0), \ldots, \widehat{w}^{k+1}(k), 0, 0, \ldots) \in \mathcal{C}^{k+1,r},$$

and

$$q = (q(0), \ldots, q(k+1), 0, 0, \ldots) \in \mathcal{C}^{k+2,m},$$
$$\widehat{q}^{k+1} = (\widehat{q}^{k+1}(0), \ldots, \widehat{q}^{k+1}(k), 0, 0, \ldots) \in \mathcal{C}^{k+1,m},$$

with

$$w(0) = w_0 \in \mathbb{C}^r, \quad w(l) = \widehat{w}^{k+1}(l-1) = \frac{1}{\delta} U_{\theta(l)}^{k+2-l} x(l), \ l = 1, \ldots, k,$$

$$q(0) = 0, \ q(l) = \widehat{q}^{k+1}(l-1) = (-\mathcal{F}_{\theta(l)}(X) + F_{\theta(l)}^{k+2-l}) x(l), \ l = 1, \ldots, k$$

where $x = (x(0), x(1), \ldots) = X_{\mathcal{F}(X)}(\theta_0, x_0, w, q)$. Then, for $\theta_0 = i$ and $x_0 = 0$,

$$\left\| Z_{\mathcal{F}(X)}(i, 0, w, q) \right\|_2^2 - \delta^2 \left\| w \right\|_2^2$$
$$= \left\| \widetilde{Z}(\theta(1), G_i w_0, \widehat{w}^{k+1}) \right\|_2^2 - \delta^2 \left\| \widehat{w}^{k+1} \right\|_2^2 - \delta^2 \left\| w_0 \right\|^2$$
$$= \widehat{\mathcal{J}}_\delta^{k+1}(\theta(1), G_i w_0) - \delta^2 \left\| w_0 \right\|^2$$
$$= E(w_0^* G_i^* P_{\theta(1)}^{k+1} G_i w_0) - \delta^2 \left\| w_0 \right\|^2$$
$$= w_0^* G_i^* \mathcal{E}_i(P^{k+1}) G_i w_0 - \delta^2 \left\| w_0 \right\|^2 .$$

But, from (7.13) and (7.15),

$$\left\| \widetilde{Z}(\theta(1), G_i w_0, \widehat{w}^{k+1}) \right\|_2^2 - \delta^2 \left\| \widehat{w}^{k+1} \right\|_2^2$$
$$\leq \left\| \bar{Z}(\theta(1), G_i w_0, \widehat{w}^{k+1}) \right\|_2^2 - \delta^2 \left\| \widehat{w}^{k+1} \right\|_2^2$$
$$= \left\| \bar{Z}(i, 0, w) \right\|_2^2 - \delta^2 \left\| w \right\|_2^2 + \delta^2 \left\| w_0 \right\|^2$$
$$< -\alpha^2 \left\| w_0 \right\|^2 + \delta^2 \left\| w_0 \right\|^2$$

that is, for every $w_0 \in \mathbb{C}^r$

$$w_0^* (\delta^2 I - G_i^* \mathcal{E}_i(P^{k+1}) G_i) w_0 > \alpha^2 \left\| w_0 \right\|^2$$

showing that for every i,

$$\delta^2 I - G_i^* \mathcal{E}_i(P^{k+1}) G_i \geq \alpha^2 I.$$

□

We can now proceed to the proof of Lemma 7.9.

Proof (of Lemma 7.9). Let us show that P_i as defined above satisfies 1), 2) and 3) of Theorem 7.3. Since for every $x_0 \in \mathbb{C}^n$ and i

$$x_0^* P_i^k x_0 = \widehat{\mathcal{J}}_\delta^k(\theta_0, x_0) \uparrow \widehat{\mathcal{J}}_\delta(\theta_0, x_0) = x_0^* P_i x_0 \text{ as } k \to \infty,$$

we get that $P_i^k \uparrow P_i$ as $k \uparrow \infty$, and thus $P \in \mathbb{H}^{n+}$. Taking the limit as $k \to \infty$ in Proposition 7.12, we get that P satisfies 1) and 2) of Theorem 7.3. Moreover from uniqueness of $\widehat{w}$ established in Proposition 7.11, and that $\widehat{w}^k$ is a maximizing sequence for $\widehat{\mathcal{J}}_\delta(\theta_0, x_0)$ (see (7.10)), we can conclude, using the same arguments as in the proof of Proposition 3 in [209], that $\widehat{w}^k \to \widehat{w}$ as $k \to \infty$. Continuity of $\widetilde{X}(\theta_0, x_0, .)$ implies that $\widetilde{X}(\theta_0, x_0, \widehat{w}^k) \to \widetilde{X}(\theta_0, x_0, \widehat{w})$ as $k \to \infty$, and thus

$$\widehat{x}^k = (\widehat{x}^k(0), \widehat{x}^k(1), \ldots) = \widetilde{X}(\theta_0, x_0, \widehat{w}^k) \to \widetilde{X}(\theta_0, x_0, \widehat{w}) = (\widehat{x}(0), \widehat{x}(1), \ldots) = \widehat{x}.$$

Therefore, for each $l = 0, 1, \ldots,$

$$\widehat{w}^k(l) = \frac{1}{\delta} U_{\theta(l)}^{k-l} \widehat{x}^k(l) \to \frac{1}{\delta} U_{\theta(l)} \widehat{x}(l) \text{ as } k \to \infty.$$

Similarly,

$$\mathcal{Q}(\theta_0, \widehat{w}^k) = q^k = (\widehat{q}^k(0), \widehat{q}^k(1), \ldots)$$
$$\to \mathcal{Q}(\theta_0, \widehat{w}) = \widehat{q} = (\widehat{q}(0), \widehat{q}(1), , \ldots) \text{ as } k \to \infty$$

so that for each $l = 0, 1, \ldots$

$$\widehat{q}^k(l) = (-\mathcal{F}_{\theta(l)}(X) + F_{\theta(l)}^{k-l}) \widehat{x}^k(l) \to (-\mathcal{F}_{\theta(l)}(X) + F_{\theta(l)}) \widehat{x}(l) \text{ as } k \to \infty.$$

This shows that

$$\widehat{q} = (\widehat{q}(0), \widehat{q}(1), \ldots) = \mathcal{Q}(\theta_0, \mathcal{W}(\theta_0, x_0))$$
$$= ((-\mathcal{F}_{\theta_0}(X) + F_{\theta_0}) x_0, \ (-\mathcal{F}_{\theta(1)}(X) + F_{\theta(1)}) \widehat{x}(1), \ldots)$$

$$\widehat{w} = (\widehat{w}(0), \widehat{w}(1), \ldots) = \mathcal{W}(\theta_0, x_0) = (\frac{1}{\delta} U_{\theta_0} x_0, \frac{1}{\delta} U_{\theta(1)} \widehat{x}(1), \frac{1}{\delta} U_{\theta(2)} \widehat{x}(2), , \ldots)$$

and thus

$$\widehat{X}(\theta_0, x_0) = (x_0, \widehat{x}(1), \widehat{x}(2), \ldots)$$

where

$$\widehat{x}(k+1) = (A_{\theta(k)} + B_{\theta(k)} F_{\theta(k)} + \frac{1}{\delta} G_{\theta(k)} U_{\theta(k)}) \widehat{x}(k), \ \widehat{x}(0) = x_0, \ \theta(0) = \theta_0.$$

Since $\widehat{X}(\theta_0, .) \in \mathbb{B}(\mathcal{C}_0^n, \mathcal{C}^n)$ (as seen in (7.17)), we get that $\widehat{X}(\theta_0, x_0) \in \mathcal{C}^n$ for any $\theta_0 \in \Theta_0$ and $x_0 \in \mathcal{C}_0^n$ which implies, from Theorem 3.34, that $r_\sigma(\mathcal{L}) < 1$.

$\square$

Remark 7.13. The condition that (C, A) is mean square detectable (together with mean square stabilizability of (A, B)) is only used to guarantee the existence of the mean square stabilizing solution to the CARE (see Corollary A.16)

$$X_i = C_i^* C_i + A_i^* \mathcal{E}_i(X) A_i - A_i^* \mathcal{E}_i(X) B_i (I + B_i^* \mathcal{E}_i(X) B_i)^{-1} B_i^* \mathcal{E}_i(X) A_i.$$

In view of Theorem A.20 and Remark A.22, (C, A) mean square detectability could be replaced by the following condition. For each $i \in \mathbb{N}$, one of the following conditions below is satisfied:

1. $(C_i, p_{ii} A_i)$ has no unobservable modes inside the closed unitary complex disk, or
2. $(C_i, p_{ii} A_i)$ has no unobservable modes over the unitary complex disk, 0 is not an unobservable mode of $(C_i, p_{ii} A_i)$ and $\zeta(i) < \infty$, where $\zeta(i)$ is as in Theorem A.20.

7.4 Recursive Algorithm for the H_∞-control CARE

The proof of Lemma 7.9 suggests a recursive algorithm for obtaining a solution for the H_∞ problem. This algorithm can be seen as an adaptation of the algorithm originally presented in [204] from the deterministic to the Markov jump case. The next theorem is the main result of this section.

Theorem 7.14. *Suppose that either (C, A) is mean square detectable or the conditions 1 and 2 of Remark 7.13 are satisfied. Suppose also that there exists $P = (P_1, \ldots, P_N) \in \mathbb{H}^{n+}$ satisfying conditions 1), 2) and 3) of Theorem 7.3. Set for $\kappa = 0, 1, \ldots$, $P^{\kappa+1} = \mathfrak{R}(P^\kappa)$, where $P^0 = X$ is the maximal solution of (7.21). Then P^κ converges to P exponentially fast as κ goes to infinity.*

Proof. First of all notice that from Proposition 7.5, $F = (F_1, \ldots, F_N)$ as defined in Theorem 7.3 stabilizes (A, B) in the mean square sense and thus (A, B) is mean square stabilizable. From mean square detectability or conditions 1 and 2 of Remark 7.13 we get that there exists the mean square stabilizing solution $P^0 = X$ to Equation (4.35), which coincides with the maximal solution. Moreover, after some algebraic manipulation, we get that

$$(P_i - P_i^0) - (A_i + B_i F_i)^* \mathcal{E}_i (P - P^0)(A_i + B_i F_i)$$
$$= U_i^* (I - \frac{1}{\delta^2} G_i^* \mathcal{E}_i(P) G_i) U_i + (F_i - F_i^0)^* (I + B_i^* \mathcal{E}_i(P^0) B_i)(F_i - F_i^0)$$

where $F^0 = (F_1^0, \ldots, F_N^0)$ is given by

$$F_i^0 = -(I + B_i^* \mathcal{E}_i(P^0) B_i)^{-1} B_i^* \mathcal{E}_i(P^0) A_i = \mathcal{F}_i(X), \ i \in \mathbb{N}.$$

Since F stabilizes (A, B) in the mean square sense, we get from Proposition 3.20 that $P - P^0 \geq 0$. This also shows that $I - \frac{1}{\delta^2} G_i^* \mathcal{E}_i(P^0) G_i \geq I - \frac{1}{\delta^2} G_i^* \mathcal{E}_i(P) G_i > 0, i \in \mathbb{N}$.

Let us show by induction on κ that $P^0 \leq P^\kappa \leq P^{\kappa+1} \leq P$ (and thus $I - \frac{1}{\delta^2} G_i^* \mathcal{E}_i(P^\kappa) G_i \geq I - \frac{1}{\delta^2} G_i^* \mathcal{E}_i(P) G_i > 0$). From the proof of Lemma 7.9, $P^0 \leq P^\kappa \leq P^{\kappa+1}$ for all $\kappa \geq 0$. For $\kappa = 0$, we have already shown that $P^0 \leq P$. Setting

$$x(1) = A_i x_0 + B_i u_0 + G_i w_0$$

we have from (7.22) that

$$x_0^* P_i^1 x_0 = \sup_{w_0} \inf_{u_0} \left\{ \|C_i x_0\|^2 + \|u_0\|^2 - \delta^2 \|w_0\|^2 + x(1)^* \mathcal{E}_i(P^0) x(1) \right\}$$

$$\leq \sup_{w_0} \inf_{u_0} \left\{ \|C_i x_0\|^2 + \|u_0\|^2 - \delta^2 \|w_0\|^2 + x(1)^* \mathcal{E}_i(P) x(1) \right\}$$

$$= x_0^* \mathcal{R}_i(P) x_0 = x_0^* P_i x_0$$

since that $\mathcal{R}(P) = P$. Therefore, $P^1 \leq P$.

Suppose now that $P^{\kappa-1} \leq P$. Let us show that $P^\kappa \leq P$. Indeed, (7.22) yields to

$$x_0^* P_i^\kappa x_0 = \sup_{w_0} \inf_{u_0} \left\{ \|C_i x_0\|^2 + \|u_0\|^2 - \delta^2 \|w_0\|^2 + x(1)^* \mathcal{E}_i(P^\kappa) x(1) \right\}$$

$$\leq \sup_{w_0} \inf_{u_0} \{ \|C_i x_0\|^2 + \|u_0\|^2 - \delta^2 \|w_0\|^2 + x(1)^* \mathcal{E}_i(P) x(1) \} = x_0^* P_i x_0$$

showing that indeed $P^\kappa \leq P$. Therefore, from [201], p. 79, there exists $\widetilde{P} \in \mathbb{H}^{n+}$ such that P^κ converges to $\widetilde{P}$ as κ goes to infinity. For arbitrary $(w, q) \in \mathcal{C}^r \oplus \mathcal{C}^m$ consider (see (7.3)) $X_{\mathcal{F}(X)}(\theta_0, x_0, w, q) = x = (x_0, x(1), \ldots)$.

Recalling that $u(k) = \mathcal{F}_{\theta(k)}(X) x(k) + q(k)$, we have from (7.21) and (7.23) that

$$E(x_0^* P_{\theta_0} x_0) - E(x(\nu + 1)^* P_{\theta(\nu+1)} x(\nu + 1))$$

$$= \sum_{k=0}^{\nu} E\left(x(k)^* P_{\theta(k)} x(k) - x(k+1)^* P_{\theta(k+1)} x(k+1) \right)$$

$$= \sum_{k=0}^{\nu} E\left(E\left(x(k)^* P_{\theta(k)} x(k) - x(k+1)^* P_{\theta(k+1)} x(k+1) \right) \mid \mathfrak{F}_k \right)$$

$$= \sum_{k=0}^{\nu} E\left(x(k)^* P_{\theta(k)} x(k) - x(k+1)^* \mathcal{E}_{\theta(k)}(P) x(k+1) \right)$$

$$= \sum_{k=0}^{\nu} E\left(\|C_{\theta(k)} x(k)\|^2 + \|u(k)\|^2 - \delta^2 \|w(k)\|^2 \right)$$

$$- \sum_{k=0}^{\nu} E\left(\left\| \mathfrak{M}_{\theta(k)}(P)^{1/2} (u(k) + \widehat{u}(P, \theta(k), x(k), w(k))) \right\|^2 \right)$$

$$+ \sum_{k=0}^{\nu} E\left(\left\| \mathfrak{D}_{\theta(k)}(P)^{1/2} (w(k) - \widehat{w}(P, \theta(k), x(k))) \right\|^2 \right)$$

and thus

$$\sum_{k=0}^{\infty} E\left(\left\|C_{\theta(k)}x(k)\right\|^2 + \|u(k)\|^2 - \delta^2 \|w(k)\|^2\right)$$

$$= E(x_0^* P_{\theta_0} x_0) + \sum_{k=0}^{\infty} E\left(\left\|\mathfrak{M}_{\theta(k)}(P)^{1/2}\left(u(k) + \widehat{u}(P,\theta(k),x(k),w(k))\right)\right\|^2\right)$$

$$+ \sum_{k=0}^{\infty} E\left(\left\|\mathfrak{D}_{\theta(k)}(P)^{1/2}\left(w(k) - \widehat{w}(P,\theta(k),x(k))\right)\right\|^2\right)$$

and it is clear from above that

$$E(x_0^* P_{\theta_0} x_0) = \widehat{\mathcal{J}}_\delta(\theta_0,x_0) = \left\|Z_{\mathcal{F}(X)}(\theta_0,x_0,\widehat{w},\mathcal{Q}(\theta_0,\widehat{w}))\right\|_2^2 - \delta^2 \|\widehat{w}\|_2^2 \quad (7.24)$$

where

$$\widehat{w} = (\widehat{w}(0),\widehat{w}(1),\ldots), \quad \widehat{w}(k) = \frac{1}{\delta}U_{\theta(k)}\widehat{x}(k),$$

and

$$\widehat{x}(k+1) = (A_{\theta(k)} + BF_{\theta(k)} + \frac{1}{\delta}G_{\theta(k)}U_{\theta(k)})\widehat{x}(k), \widehat{x}(0) = x_0, \theta(0) = \theta_0.$$

Recall also from the previous section that

$$\widehat{\mathcal{J}}_\delta^k(\theta_0,x_0) = \beta^k(\theta_0,x_0) = E(x_0^* P_{\theta_0}^k x_0)$$

$$= \left\|Z_{\mathcal{F}(X)}(\theta_0,x_0,w^\kappa,\mathcal{Q}(\theta_0,w^\kappa))\right\|_2^2 - \delta^2 \|w^\kappa\|_2$$

$$= \left\|Z_{\mathcal{F}(X)}(\theta_0,x_0,w^\kappa,q^\kappa)\right\|_2^2 - \delta^2 \|w^\kappa\|_2^2 \quad (7.25)$$

where

$$w^\kappa = (w^\kappa(0),\ldots,w^\kappa(\kappa-1),0,\ldots),$$

$$w^\kappa(k) = \mathfrak{D}_{\theta(k)}(P^{\kappa-k-1})^{-1}\mathfrak{B}_{\theta(k)}(P^{\kappa-k-1})x(k),$$

$$q^\kappa = (q^\kappa(0),q^\kappa(1),\ldots),$$

$$q^\kappa(k) = -\mathcal{F}_{\theta(k)}(X)x(k) + u^\kappa(k),$$

and

$$u^\kappa(k) = \mathfrak{M}_{\theta(k)}(P^{\kappa-k-1})^{-1}B_{\theta(k)}^*\mathcal{E}_{\theta(k)}(P^{\kappa-k-1})(A_{\theta(k)}$$

$$+ G_{\theta(k)}\mathfrak{D}_{\theta(k)}(P^{\kappa-k-1})^{-1}\mathfrak{B}_{\theta(k)}(P^{\kappa-k-1}))x(k)$$

for $k = 0,\ldots,\kappa-1$ and

$$u^\kappa(k) = \mathfrak{M}_{\theta(k)}(P^0)^{-1}B_{\theta(k)}^*\mathcal{E}_{\theta(k)}(P^0)A_{\theta(k)}x(k)$$

for $k \geq \kappa$. Set $\widehat{w}^\kappa = (\widehat{w}^\kappa(0),\ldots,\widehat{w}^\kappa(\kappa-1),0,\ldots,0), \widehat{w}^\kappa(k) = \widehat{w}(k)$ for $k = 0,\ldots,\kappa-1$. From the fact that $r_\sigma(\mathcal{T}) < 1$ (condition 3) of Theorem 7.3), we can find $a > 0$, $0 < b < 1$, such that $\left\|\mathcal{T}^k\right\| \le ab^k$. Moreover, we have from Proposition 3.1 that

$$\|\widehat{w}(k)\|_2^2 = E\left(\frac{1}{\delta}\left\|U_{\theta(k)}\widehat{x}(k)\right\|^2\right) \le c_1 E(\|\widehat{x}(k)\|^2) \le c_2\left\|\mathcal{T}^k\right\|\|x_0\|_2^2 \le c_3 b^k\|x_0\|_2^2$$

for appropriate constants c_1, c_2 and c_3. Therefore

$$\|w - \widehat{w}^\kappa\|_2^2 = \sum_{k=\kappa}^\infty \|\widehat{w}(k)\|_2^2 \le \frac{1}{1-b}c_3 b^k\|x_0\|_2^2. \tag{7.26}$$

Similarly we can show that $\left\|Z_{\mathcal{F}(X)}(\theta_0, x_0, \widehat{w}, \mathcal{Q}(\theta_0, \widehat{w}))\right\|_2^2 \le c_3'\|x_0\|_2^2$ for some appropriate constant $c_3' > 0$. From (7.24) with $x_0 = 0$, we have that

$$0 = \sup_{w \in \mathcal{C}^r}\left\{\left\|Z_{\mathcal{F}(X)}(\theta_0, 0, \widehat{w}, \mathcal{Q}(\theta_0, \widehat{w}))\right\|_2^2 - \delta^2\|w\|_2^2\right\}$$

$$\ge \left\|Z_{\mathcal{F}(X)}(\theta_0, 0, \widehat{w} - \widehat{w}^\kappa, \mathcal{Q}(\theta_0, \widehat{w} - \widehat{w}^\kappa))\right\|_2^2 - \delta^2\|\widehat{w} - \widehat{w}^\kappa\|_2^2. \tag{7.27}$$

Equations (7.24), (7.25), (7.26) and (7.27) with $x_0 \in \mathbb{C}^n$ and $\theta_0 = i$ yield

$$x_0^* P_i^\kappa x_0 \ge \left\|Z_{\mathcal{F}(X)}(i, x_0, \widehat{w}^\kappa, \mathcal{Q}(\theta_0, \widehat{w}^\kappa))\right\|_2^2 - \delta^2\|\widehat{w}^\kappa\|_2$$

$$\ge \left\|Z_{\mathcal{F}(X)}(i, x_0, \widehat{w}, \mathcal{Q}(i, \widehat{w})) - Z_{\mathcal{F}(X)}(i, 0, \widehat{w} - \widehat{w}^\kappa, \mathcal{Q}(i, \widehat{w} - \widehat{w}^\kappa))\right\|_2^2$$
$$\quad - \delta^2\|\widehat{w}\|_2^2$$

$$\ge \left\|Z_{\mathcal{F}(X)}(i, x_0, \widehat{w}, \mathcal{Q}(i, \widehat{w}))\right\|_2^2 - \delta^2\|\widehat{w}\|_2^2$$
$$\quad - 2\left|\langle Z_{\mathcal{F}(X)}(i, x_0, \widehat{w}, \mathcal{Q}(i, \widehat{w})); Z_{\mathcal{F}(X)}(i, 0, \widehat{w} - \widehat{w}^\kappa, \mathcal{Q}(i, \widehat{w} - \widehat{w}^\kappa))\rangle\right|$$

$$\ge x_0^* P_i x_0 - 2\left\|Z_{\mathcal{F}(X)}(i, x_0, \widehat{w}, \mathcal{Q}(i, \widehat{w}))\right\|_2$$
$$\quad \times \left\|Z_{\mathcal{F}(X)}(i, 0, \widehat{w} - \widehat{w}^\kappa, \mathcal{Q}(i, \widehat{w} - \widehat{w}^\kappa))\right\|_2$$

$$\ge x_0^* P_i x_0 - c_4\|x_0\|_2\|\widehat{w} - \widehat{w}^\kappa\|_2 \ge x_0^* P_i x_0 - c_5\|x_0\|_2^2 b^{\kappa/2}$$

for appropriate positive constants c_4 and c_5. Thus

$$0 \le \frac{1}{\|x_0\|^2}x_0^*(P_i - P_i^\kappa)x_0 \le c_5 b^{\kappa/2}$$

which shows that P^κ converges to P exponentially fast as κ goes to infinity.

□

In summary, we have the following procedure for deriving the stabilizing solution P satisfying 1), 2), and 3) of Theorem 7.3, whenever it exists.

Algorithm 7.15 (H_∞-control Recursive Algorithm) *The following steps produce a stabilizing solution for the H_∞-control problem.*

1. *Solve the convex programming problem given by (A.17a)–(A.17c). Set the solution of this problem as P^0.*

2. *Determine P^κ through the following iterations: $P^{\kappa+1} = \mathfrak{M}(P^\kappa)$.*
3. *If P^κ converges to P as κ goes to infinity, then check if $r_\sigma(\mathcal{T}) < 1$, where $\mathcal{T}$ is given by Equation 3) of Theorem 7.3. If it is, then P is the desired solution. Otherwise there is no solution for the given δ.*

7.5 Historical Remarks

A fundamental problem in control theory is to design controllers which give satisfactory performance in the presence of uncertainties such as unknown model parameters and disturbances which affect the system dynamics. The H_∞-control theory originated in an effort to codify classical control methods where frequency response functions are shaped to meet certain performance objectives. It was originally formulated as a linear design problem in the frequency domain (see, for instance, [126] and [225]). The linear H_∞-control theory has developed extensively since the early 1980s. By the late 1980s this theory had achieved a highly satisfactory degree of completeness. In a state-space formulation (cf. [12], [96], [136], [203], [209], [228]), the H_∞-control problem consists in obtaining a controller that stabilizes a linear system and ensures that the ℓ_2-induced norm from the additive input disturbance to the output is less than a pre-specified attenuation value. The H_∞-control analysis has been extended to comprise non-linear systems (cf. [140] [141], [212]), infinite dimensional linear systems (see, e.g. [151]) and the LQG problem (cf. [153]). The min-max nature of the H_∞ problem led easily to connections with game theory (cf. [12], [96], [162] and [209]). The connection between H_∞-control, risk-sensitive, differential games and entropy can be found, for instance, in [110], [133] and [142].

Regarding MJLS, [85], [123] and [186] seem to be the first works dealing with H_∞-control. The technique in [85] and [123] provides sufficient conditions for a solution via coupled Riccati equations. In [186] the differential game interpretation for the problem is employed. Necessary and sufficient conditions, which extend the results in [123] for the discrete-time case, are obtained in [63] and include the case in which the state space of the Markov chain is infinite countable. The results described in this book comprise the results obtained in [63], when restricted to the case in which the state space of $\theta(k)$ is finite. Results for the H_∞-control of continuous-time MJLS with infinite countable Markov chain were obtained in [124]. Many papers have come out in the last few years (see, for instance, [38], [39], [40], [50], [82], [128], [167], [187], [197], [199]), and this topic continues to deserve a great deal of interest, since it includes mixed variants such as robust H_∞-control for MJLS with time-delay, etc.

8

Design Techniques and Examples

This chapter presents and discusses some applications of the theoretical results introduced earlier. Also, some design-oriented techniques, especially those making use of linear matrix inequalities, are presented here. This final chapter is intended to conclude the book assembling some problems in the Markov jump context and the tools to solve them.

8.1 Some Applications

The examples in the previous chapters were very simple and straightforward. The applications presented in this section, on the other hand, refer to more realistic problems with some engineering concerns in mind. Also, greater attention is given to the problem descriptions, in order not only to justify the option for a Markov jump approach, but mainly to better express how it can be done and how to benefit from the use of such models.

Some additional very interesting applications not included in this chapter can be found in [89], [135], [152] and [202].

8.1.1 Optimal Control for a Solar Thermal Receiver

We now return to the solar thermal receiver presented in Chapter 1. Sworder and Rogers in [208] proposed a Markov jump controller for the Solar One energy plant, located in Dagget, California.

The plant is basically composed of a set of adjustable mirrors, the heliostats, surrounding a tower with a boiler (see Figure 8.1). The attitude of the heliostats is controlled in order to keep sunlight focused onto the boiler. The heat transferred can make the plant generate nearly 10 MWe under favorable weather conditions.

Our main interest is to control the feedwater flow rate to the boiler in order to maintain the outlet steam temperature within adequate boundaries.

Fig. 8.1. Solar One energy plant. Courtesy of the National Renewable Energy Laboratory; photo credit Sandia National Laboratory

The stochastic nature of this control problem arises because the system dynamics is heavily dependent on the instantaneous insolation. Cloud movement over the heliostats can cause sudden changes in insolation and can be treated, for practical purposes, as a stochastic process.

From insolation data collected at the site, it was established that the mean duration of a cloudy period was approximately 2.3 min, while the mean interval of direct insolation was 4.3 min. Based on this information, two operation modes were defined: (1) sunny; and (2) cloudy.

With these modes and the mean interval durations associated to them, a transition probability matrix was obtained, which is, for a sample time of 6 s,

$$P = \begin{bmatrix} 0.9767 & 0.0233 \\ 0.0435 & 0.9565 \end{bmatrix}. \tag{8.1}$$

The thermal receiver is described by the following simplified version of (4.25), given by

$$\mathcal{G} = \begin{cases} x(k+1) = A_{\theta(k)}x(k) + B_{\theta(k)}u(k) \\ z(k) = C_{\theta(k)}x(k) + D_{\theta(k)}u(k) \end{cases}$$

with the parameters given in Table 8.1 for $\theta(k) \in \{1, 2\}$.

Table 8.1. Parameters for the solar thermal receiver model

		Sunny	Cloudy
Plant		$A_1 = 0.8353$	$A_2 = 0.9646$
		$B_1 = 0.0915$	$B_2 = 0.0982$
Weights		$C_1^* C_1 = 0.0355$	$C_2^* C_2 = 0.0355$
		$D_1^* D_1 = 1$	$D_2^* D_2 = 1$

Solving the convex programming problem (see Appendix A, Problem A.11 and Theorem A.12), we obtain the results of Table 8.2. Notice that for the closed loop system, we have $r_\sigma(\mathcal{L}) = 0.8871$; therefore it is stable in the mean square sense.

Table 8.2. Optimal control for the solar thermal receiver

		Sunny	Cloudy
CARE solution		$X_1 = 0.1295$	$X_2 = 0.3603$
Optimal controller		$F_1 = -0.0103$	$F_2 = -0.0331$

Monte Carlo simulations of the closed loop system are presented in Figure 8.2. The figure contains 2000 possible trajectories for initial condition $x(0) = 1$. The thick line in the figure is the expected trajectory. As can be seen, an unfavorable sequence of states in the Markov chain may lead to a poor performance (the upper trajectories in the figure), but no other controller with the same structure will present better expected performance.

8.1.2 Optimal Policy for the National Income with a Multiplier–Accelerator Model

Samuelson's multiplier–accelerator model, published in 1939 [196], is possibly the first dynamic model based on economic theories to address the problem of income determination and the business cycle.

A very interesting application of MJLS to economic modeling employing the multiplier–accelerator model is presented in [28] and here is slightly adapted to fit our framework.

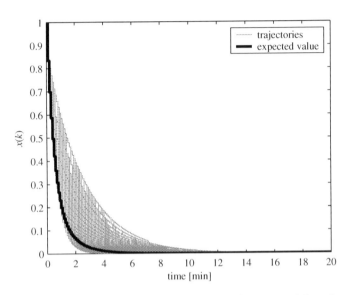

Fig. 8.2. Monte Carlo simulations for the solar thermal receiver with optimal control

A state-space version of the multiplier–accelerator model is employed, literally,

$$x(k+1) = \begin{bmatrix} 0 & 1 \\ -\alpha & 1-s+\alpha \end{bmatrix} x(k) + \begin{bmatrix} 0 \\ 1 \end{bmatrix} u(k) \qquad (8.2)$$

where $x_2(k)$ stands for the *national income* ($x_1(k)$ differs from $x_2(k)$ only by a one-step lag), s is the *marginal propensity to save* ($1/s$ is the so-called multiplier), α is an *accelerator coefficient* and $u(k)$ is the *government expenditure*.

Based on historical data of the United States Department of Commerce, the parameters s and α were grouped in three classes, according to Table 8.3.

Table 8.3. Multiplier–accelerator modes (from [28])

Mode	Name	Description
1	Norm	s and α in mid-range
2	Boom	s in low range (or α in high)
3	Slump	s in high range (or α in low)

The transition probability matrix connecting these modes is

$$P = \begin{bmatrix} 0.67 & 0.17 & 0.16 \\ 0.30 & 0.47 & 0.23 \\ 0.26 & 0.10 & 0.64 \end{bmatrix}, \tag{8.3}$$

while the parameters for the model are given in Table 8.4.

Table 8.4. Parameters for the multiplier–accelerator model

	Norm	Boom	Slump
Model	$A_1 = \begin{bmatrix} 0 & 1 \\ -2.5 & 3.2 \end{bmatrix}$	$A_2 = \begin{bmatrix} 0 & 1 \\ -43.7 & 45.4 \end{bmatrix}$	$A_3 = \begin{bmatrix} 0 & 1 \\ 5.3 & -5.2 \end{bmatrix}$
	$B_1 = \begin{bmatrix} 0 \\ 1 \end{bmatrix}$	$B_2 = \begin{bmatrix} 0 \\ 1 \end{bmatrix}$	$B_3 = \begin{bmatrix} 0 \\ 1 \end{bmatrix}$
Weights	$C_1^* C_1 = \begin{bmatrix} 3.6 & -3.8 \\ -3.8 & 4.87 \end{bmatrix}$	$C_2^* C_2 = \begin{bmatrix} 10 & -3 \\ -3 & 8 \end{bmatrix}$	$C_3^* C_3 = \begin{bmatrix} 5 & -4.5 \\ -4.5 & 4.5 \end{bmatrix}$
	$D_1^* D_1 = 2.6$	$D_2^* D_2 = 1.165$	$D_3^* D_3 = 1.111$

The optimal policy, like in the thermal solar receiver example, can be determined by solving a convex programming problem (see Appendix A, Problem A.11 and Theorem A.12), which yields the results of Table 8.5.

Table 8.5. Optimal control for the multiplier–accelerator model

	CARE solution	Optimal controller
Norm	$X_1 = \begin{bmatrix} 18.6616 & -18.9560 \\ -18.9560 & 28.1085 \end{bmatrix}$	$F_1 = \begin{bmatrix} 2.3172 & -2.3317 \end{bmatrix}$
Boom	$X_2 = \begin{bmatrix} 30.8818 & -21.6010 \\ -21.6010 & 36.2739 \end{bmatrix}$	$F_2 = \begin{bmatrix} 4.1684 & -3.7131 \end{bmatrix}$
Slump	$X_3 = \begin{bmatrix} 35.4175 & -38.6129 \\ -38.6129 & 36.2739 \end{bmatrix}$	$F_3 = \begin{bmatrix} -5.1657 & 5.7933 \end{bmatrix}$

Figure 8.3 presents Monte Carlo simulations of the multiplier–accelerator model with the optimal policy given by the controller of Table 8.5. The figure presents 4000 randomly picked trajectories for the national income $x_2(k)$ with initial conditions given by $x_1(0) = x_2(0) = 1$. The thick line is the expected value for $x_2(k)$. Notice that for this controller we have $r_\sigma(\mathcal{L}) = 0.4609 < 1$.

8.1.3 Adding Noise to the Solar Thermal Receiver problem

In Subsection 8.1.1 we considered the optimal control problem for a solar thermal receiver. In that framework, it was assumed that both the Markov

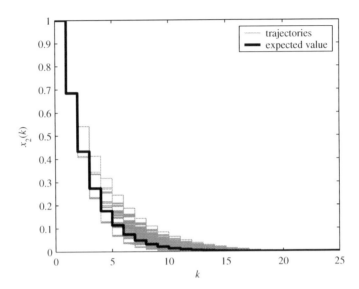

Fig. 8.3. Monte Carlo simulations for the multiplier–accelerator model with optimal policy

chain and the system states were perfectly known. In Chapter 6 a framework was developed that could account for partial noisy observations of the state vector. It was shown there that we could design independently the optimal controller and an optimal filter, benefiting from a separation principle.

Here we will assume that the thermal receiver is subject to a noise as described in Chapter 6. We will use the same controller of Subsection 8.1.1 in series with a stationary Markov filter. System and control data are the ones given in Tables 8.1 and 8.2 respectively. Table 8.6 presents the parameters for this filter.

Table 8.6. Optimal filter for the solar thermal receiver

	Sunny	Cloudy
Noise	$G_1 = \begin{bmatrix} 0.05 & 0 \end{bmatrix}$	$G_2 = \begin{bmatrix} 0.03 & 0 \end{bmatrix}$
	$H_1 = \begin{bmatrix} 0 & 0.2 \end{bmatrix}$	$H_2 = \begin{bmatrix} 0 & 0.1 \end{bmatrix}$
CARE solution	$Y_1 = 0.0040$	$Y_2 = 0.0012$
Optimal filter	$M_1 = 0.1113$	$M_2 = 0.2436$

Monte Carlo simulations of the closed loop system, similar to those presented in Figure 8.2 are presented in Figure 8.4. The figure contains 2000 possible trajectories for initial condition $x(0) = 1$, where the thick line is the expected trajectory.

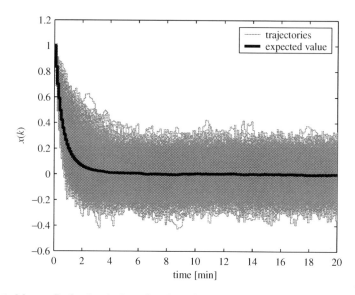

Fig. 8.4. Monte Carlo simulations for the solar thermal receiver with partial observations

8.2 Robust Control via LMI Approximations

The use of LMI techniques to obtain numerical solutions to various control and filtering problems has become a very strong trend in the area since the development of very efficient and reliable algorithms (see for instance [182] and [19]).

The basic idea is to obtain a convex programming problem (which may be in the form of a LMI or may be converted to one) that is equivalent in some sense to a certain control or filtering problem, and then solve the optimization problem instead of the original one. This has been done in Chapter 4 for optimal and H_2-control, and the solutions to the examples given in the previous section were obtained using LMIs.

The general subject of LMIs is too wide to be treated here, and we invite the reader who looks for an introduction on the matter to study some of the many excellent works available, for instance [42], [131] and also [7].

There are enough reasons that contribute to make LMI techniques such an attractive tool for design. To mention a few:

1. There are efficient algorithms to solve LMIs. Once a problem is expressed as an LMI, it is usually simple to get a numerical solution to it. In an area still as widely unexplored as MJLS, a general straightforward way to solve general (and eventually new) problems is highly valuable.
2. Some control and filtering problems involving MJLS can only be solved, with our current knowledge on the area, via LMI approximations.
3. The LMI framework allows for the convenient introduction of several useful enhancements in control or filter design, like the inclusion of robustness, uncertainties, special structures for the filters or controllers, restrictions on variables, etc, that would be very difficult, or even impossible, to account for using other techniques.

In this section we present a small sample of basic LMI tricks and tools involving problems considered earlier in this book or complementing them. A comprehensive description of MJLS design problems benefiting from LMI techniques would be far beyond the scope of this book, and necessarily incomplete, for the literature on the subject grows wider every year.

Since there is no perfect design technique, the reader must beware of intrinsic limitations on the use of LMIs, in control and filter design in general and, specifically when concerning MJLS.

First, LMI techniques are just a tool to obtain solutions for some control and filtering problems. They do not give further insight or improve knowledge on the problems themselves.

Second, many problems are not intrinsically convex, so their description as LMIs involves some degree of approximation. The effects are usually the necessity of using sometimes unnatural or not very useful models in order to keep the problems convex or, which may be more serious, conservativeness of the attained solution. In the following this will be put in evidence for the presented problems.

Even considering these (and other) disadvantages, LMI techniques are still very attractive design tools, and for many problems, the current state-of-the-art approach is to use LMIs. Also, for some of them, it is the only known effective way of obtaining a solution.

8.2.1 Robust H_2-control

The results presented earlier in Chapter 4 regarding the H_2-control of MJLS that lead to the equivalence of this control problem to a convex programming problem can be modified in order to include uncertainties, both in the transition probability matrix and in the system matrices. This is possible by including these uncertainties as restrictions in the associated optimization problem.

In order to keep the convexity of the optimization problem, the uncertainties must be described in an adequate manner, with structures that can be inserted in the problem formulation without affecting its basic properties. Therefore for this section we will consider the following version of System (1.3) with uncertainties,

$$\mathcal{G} = \begin{cases} x(k+1) = (A_{\theta(k)} + \Delta A_{\theta(k)})x(k) + (B_{\theta(k)} + \Delta B_{\theta(k)})u(k) \\ \qquad\qquad + G_{\theta(k)}w(k) \\ z(k) = C_{\theta(k)}x(k) + D_{\theta(k)}u(k) \\ x_0 \in \mathcal{C}_0^n, \theta(0) = \theta_0 \in \Theta_0 \end{cases} \tag{8.4}$$

for $k = 0, 1, \ldots$, where the uncertainties given by ΔA_i, ΔB_i satisfy the following norm bounded conditions for $i \in \mathbb{N}$ and appropriate matrices Al_i, Ar_i, Bl_i, Br_i,

$$\Delta A_i = Al_i \Delta_i Ar_i$$
$$\Delta B_i = Bl_i \Delta_i Br_i$$
$$\Delta_i \Delta_i^* \le I .$$

The use of norm bounded expressions is a clever trick to include relatively realistic and reasonable uncertainties while still maintaining the problem convex.

Also, the transition probability matrix associated with the Markov chain P is assumed to be not exactly known, but to belong to a polytope defined by

$$\mathbb{P} = \{P; P = \sum_{t=1}^{\rho} \alpha^t P^t, \ \alpha^t \ge 0, \sum_{t=1}^{\rho} \alpha^t = 1\} \tag{8.5}$$

where the ρ vertices P^t, $t = 1, \ldots, \rho$ are all known transition probability matrices.

The replacement of a given specified parameter in a problem by a convex set, as in the case of the polytope described above, makes it possible to encompass a large variety of uncertainties while still keeping the problem in the LMI framework.

We recall the convex programming problem of Subsect. 4.4.4. Its cost function is given by

$$\lambda(W) = \sum_{j=1}^{N} \text{tr}\left(\begin{bmatrix} C_i^* C_i & 0 \\ 0 & D_i D_i^* \end{bmatrix} \begin{bmatrix} W_{i1} & W_{i2} \\ W_{i2}^* & W_{i3} \end{bmatrix} \right),$$

while the restrictions given by the convex set Ψ are redefined. The new restrictions are expressed in terms of a new convex set Ψ, given by

$$\Psi = \bigcap_{t=1}^{\rho} \Psi^t$$

where each Ψ^t, $t = 1, \ldots, \rho$, is defined as

$$\Psi^t = \{ \ W = (W_1, \ldots, W_N); \ \text{for } i \in \mathbb{N},$$

$$W_i = \begin{bmatrix} W_{i1} & W_{i2} \\ W_{i2}^* & W_{i3} \end{bmatrix} \geq 0, \ W_{i1} > 0, \ \mathcal{H}_i^t(W) \geq 0 \ \}$$

where for $i, j \in \mathbb{N}$,

$$\mathcal{M}_i(W) = \begin{bmatrix} Ar_i W_{i1} Ar_i^* & Ar_i W_{i2} Br_i^* \\ Br_i W_{i2}^* Ar_i^* & Br_i W_{i3} Br_i^* \end{bmatrix}$$

$$\mathcal{X}_{ij}^t(W) = \left[\sqrt{p_{ij}^t}(A_i W_{i1} + B_i W_{i2}^*) Ar_i^* \ \ \sqrt{p_{ij}^t}(A_i W_{i2} + B_i W_{i3}) Br_i^* \right]$$

$$\mathcal{Z}_j^t(W) = \sum_{i=1}^N p_{ij}^t (A_i W_{i1} A_i^* + B_i W_{i2}^* A_i^* + A_i W_{i2} B_i^*$$

$$+ B_i W_{i3} B_i^* + Al_i Al_i^* + Bl_i Bl_i^* + \upsilon_i G_i G_i^*)$$

$$\mathcal{H}_j^t(W) = \begin{bmatrix} W_{j1} - \mathcal{Z}_j^t(W) & \mathcal{X}_{1j}^t(W) & \cdots & \mathcal{X}_{Nj}^t(W) \\ \mathcal{X}_{1j}^t(W)^* & I - \mathcal{M}_1(W) & \cdots & 0 \\ \vdots & \vdots & \ddots & \vdots \\ \mathcal{X}_{Nj}^t(W)^* & 0 & \cdots & I - \mathcal{M}_N(W) \end{bmatrix}.$$

with all matrices assumed real.

Proposition 8.1 presents an auxiliary result that will be needed in Theorem 8.2, given in the following, which is the main result in this subsection.

Proposition 8.1. *Suppose that $W \in \Psi$. Then for $j \in \mathbb{N}$,*

$$W_{j1} \geq \sum_{i=1}^N p_{ij}((A_i + B_i F_i + \Delta A_i + \Delta B_i F_i) W_{i1} (A_i + B_i F_i + \Delta A_i + \Delta B_i F_i)^*$$

$$+ \upsilon_i G_i G_i^*)$$

where $\mathrm{P} = [p_{ij}] \in \mathbb{P}$, $\Delta A_i = Al_i \Delta_i Ar_i$, $\Delta B_i = Bl_i \Delta_i Br_i$, Δ_i *are such that* $\Delta_i \Delta_i^* \leq I$ *and* $F_i = W_{i2}^* W_{i1}^{-1}$ *for* $i \in \mathbb{N}$.

Proof. If $W \in \Psi$ then $W \in \Psi^t$ for each $t = 1, \ldots, \rho$. From the Schur complement (see Lemma 2.23), $\mathcal{H}_j^t(W) \geq 0$ if and only if

$$0 \leq \begin{bmatrix} I - \mathcal{M}_1(W) & \cdots & 0 \\ \vdots & \ddots & \vdots \\ 0 & \cdots & I - \mathcal{M}_N(W) \end{bmatrix},$$

$$0 \leq W_{j1} - \mathcal{Z}_j^t(W)$$

$$- \left[\mathcal{X}_{1j}^t(W) \ \ldots \ \mathcal{X}_{Nj}^t(W) \right] \begin{bmatrix} I - \mathcal{M}_1(W) & \cdots & 0 \\ \vdots & \ddots & \vdots \\ 0 & \cdots & I - \mathcal{M}_N(W) \end{bmatrix}^\dagger \begin{bmatrix} \mathcal{X}_{1j}^t(W)^* \\ \vdots \\ \mathcal{X}_{Nj}^t(W)^* \end{bmatrix}$$

$$(8.6)$$

and

$$\left[\mathcal{X}_{1j}^t(W) \ldots \mathcal{X}_{Nj}^t(W)\right] = \left[\mathcal{X}_{1j}^t(W) \ldots \mathcal{X}_{Nj}^t(W)\right]$$
$$\begin{bmatrix} I - \mathcal{M}_1(W) \ldots & & 0 \\ \vdots & \ddots & \vdots \\ 0 & \ldots I - \mathcal{M}_N(W) \end{bmatrix}^\dagger \begin{bmatrix} I - \mathcal{M}_1(W) \ldots & & 0 \\ \vdots & \ddots & \vdots \\ 0 & \ldots I - \mathcal{M}_N(W) \end{bmatrix} \quad (8.7)$$

for $j \in \mathbb{N}$. Equations (8.6) and (8.7) can be rewritten as

$$0 \le W_{j1} - \sum_{i=1}^N p_{ij}^t (A_i W_{i1} A_i^* + B_i W_{i2}^* A_i^* + A_i W_{i2} B_i^* + B_i W_{i3} B_i^*$$
$$+ Al_i Al_i^* + Br_i Bl_i^* + v_i G_i G_i^* + \sigma_i(W)) \quad (8.8)$$

where

$$\sigma_i(W) = \left[(A_i W_{i1} + B_i W_{i2}^*) Ar_i^* \; (A_i W_{i2} + B_i W_{i3}) Br_i^*\right]$$
$$\times \left(I - \begin{bmatrix} Ar_i W_{i1} Ar_i^* & Ar_i W_{i2} Br_i^* \\ Br_i W_{i2}^* Ar_i^* & Br_i W_{i3} Br_i^* \end{bmatrix}\right)^\dagger \begin{bmatrix} Ar_i (A_i W_{i1} + B_i W_{i2}^*)^* \\ Br_i (A_i W_{i2} + B_i W_{i3})^* \end{bmatrix}$$

and

$$\left[(A_i W_{i1} + B_i W_{i2}^*) Ar_i^* \; (A_i W_{i2} + B_i W_{i3}) Br_i^*\right]$$
$$= \left[(A_i W_{i1} + B_i W_{i2}^*) Ar_i^* \; (A_i W_{i2} + B_i W_{i3}) Br_i^*\right]$$
$$\left(I - \begin{bmatrix} Ar_i W_{i1} Ar_i^* & Ar_i W_{i2} Br_i^* \\ Br_i W_{i2}^* Ar_i^* & Br_i W_{i3} Br_i^* \end{bmatrix}\right)^\dagger \left(I - \begin{bmatrix} Ar_i W_{i1} Ar_i^* & Ar_i W_{i2} Br_i^* \\ Br_i W_{i2}^* Ar_i^* & Br_i W_{i3} Br_i^* \end{bmatrix}\right) \quad (8.9)$$

for $j \in \mathbb{N}$. Write now

$$\tau_i(W) = \left[(A_i W_{i1} + B_i W_{i2}^*) Ar_i^* \; (A_i W_{i2} + B_i W_{i3}) Br_i^*\right]$$
$$\times \left(\left(I - \begin{bmatrix} Ar_i W_{i1} Ar_i^* & Ar_i W_{i2} Br_i^* \\ Br_i W_{i2}^* Ar_i^* & Br_i W_{i3} Br_i^* \end{bmatrix}\right)^\dagger\right)^{1/2}$$
$$- \left[Al_i \; Bl_i\right] \Delta_i \left(I - \begin{bmatrix} Ar_i W_{i1} Ar_i^* & Ar_i W_{i2} Br_i^* \\ Br_i W_{i2}^* Ar_i^* & Br_i W_{i3} Br_i^* \end{bmatrix}\right)^{1/2}$$

for $i \in \mathbb{N}$. Then, from (8.9) and the properties of the generalized inverse,

$$0 \le \tau_i(W) \tau_i(W)^*$$
$$= \sigma_i(W) + \left[Al_i \; Bl_i\right] \Delta_i \left(I - \begin{bmatrix} Ar_i W_{i1} Ar_i^* & Ar_i W_{i2} Br_i^* \\ Br_i W_{i2}^* Ar_i^* & Br_i W_{i3} Br_i^* \end{bmatrix}\right) \Delta_i^* \begin{bmatrix} Al_i^* \\ Bl_i^* \end{bmatrix}$$
$$- \left[(A_i W_{i1} + B_i W_{i2}^*) Ar_i^* \; (A_i W_{i2} + B_i W_{i3}) Br_i^*\right] \Delta_i^* \begin{bmatrix} Al_i^* \\ Bl_i^* \end{bmatrix}$$
$$- \left[Al_i \; Bl_i\right] \Delta_i \begin{bmatrix} Ar_i (A_i W_{i1} + B_i W_{i2}^*)^* \\ Br_i (A_i W_{i2} + B_i W_{i3})^* \end{bmatrix} \quad (8.10)$$

for $i \in \mathbb{N}$. From (8.10) and the fact that $\Delta_i \Delta_i^* \leq I$ for $i \in \mathbb{N}$, we have

$$\begin{bmatrix} Al_i & Bl_i \end{bmatrix} \Delta_i \Delta_i^* \begin{bmatrix} Al_i^* \\ Bl_i^* \end{bmatrix} \leq Al_i Al_i^* + Bl_i Bl_i \tag{8.11}$$

for $i \in \mathbb{N}$. From (8.10) and (8.11),

$$\sigma_i(W) + Al_i Al_i^* + Bl_i Bl_i^*$$
$$\geq \begin{bmatrix} Al_i & Bl_i \end{bmatrix} \Delta_i \begin{bmatrix} Ar_i W_{i1} Ar_i^* & Ar_i W_{i2} Br_i^* \\ Br_i W_{i2}^* Ar_i^* & Br_i W_{i3}^* Br_i^* \end{bmatrix} \Delta_i^* \begin{bmatrix} Al_i^* \\ Bl_i^* \end{bmatrix}$$
$$+ \begin{bmatrix} (A_i W_{i1} + B_i W_{i2}^*) Ar_i^* & (A_i W_{i2} + Br_i W_{i3}) N_i^* \end{bmatrix} \Delta_i^* \begin{bmatrix} Al_i^* \\ Bl_i^* \end{bmatrix}$$
$$+ \begin{bmatrix} Al_i & Bl_i \end{bmatrix} \Delta_i \begin{bmatrix} Ar_i (A_i W_{i1} + B_i W_{i2}^*)^* \\ Br_i (A_i W_{i2} + B_i W_{i3})^* \end{bmatrix} \tag{8.12}$$

for $i \in \mathbb{N}$. From (8.8) and (8.12) and recalling that $W_{i3} \geq W_{i2}^* W_{i1}^{-1} W_{i2}$ for $i \in \mathbb{N}$, we get

$$W_{j1} \geq \sum_{i=1}^{N} p_{ij}^t \Big(A_i W_{i1} A_i^* + B_i W_{i2}^* A_i^* + A_i W_{i2} B_i^*$$
$$+ B_i W_{i3} B_i^* + Al_i Al_i^* + Bl_i Bl_i^* + \sigma_i(W) + v_i G_i G_i^* \Big)$$
$$\geq \sum_{i=1}^{N} p_{ij}^t \Big(A_i W_{i1} A_i^* + B_i W_{i2}^* A_i^* + A_i W_{i2} B_i^* + B_i W_{i3} B_i^*$$
$$+ \begin{bmatrix} Al_i & Bl_i \end{bmatrix} \Delta_i \begin{bmatrix} Ar_i W_{i1} Ar_i^* & Ar_i W_{i2} Br_i^* \\ Br_i W_{i2}^* Ar_i^* & Br_i W_{i3}^* Br_i^* \end{bmatrix} \Delta_i^* \begin{bmatrix} Al_i^* \\ Bl_i^* \end{bmatrix}$$
$$+ \begin{bmatrix} (A_i W_{i1} + B_i W_{i2}^*) Ar_i^* & (A_i W_{i2} + B_i W_{i3}) Br_i^* \end{bmatrix} \Delta_i^* \begin{bmatrix} Al_i^* \\ Bl_i^* \end{bmatrix}$$
$$+ \begin{bmatrix} Al_i & Bl_i \end{bmatrix} \Delta_i \begin{bmatrix} Ar_i (A_i W_{i1} + B_i W_{i2}^*)^* \\ Br_i (A_i W_{i2} + B_i W_{i3})^* \end{bmatrix} + v_i G_i G_i^* \Big)$$
$$= \sum_{i=1}^{N} p_{ij}^t \Big((A_i + \Delta A_i) W_{i1} (A_i + \Delta A_i)^* + (B_i + \Delta B_i) W_{i3} (B_i + \Delta B_i)^*$$
$$+ (A_i + \Delta A_i) W_{i2} (B_i + \Delta B_i)^* + (B_i + \Delta B_i) W_{i2}^* (A_i + \Delta A_i)^*$$
$$+ v_i G_i G_i^* \Big)$$
$$\geq \sum_{i=1}^{N} p_{ij}^t ((A_i + B_i F_i + \Delta A_i + \Delta B_i F_i) W_{i1} (A_i$$
$$+ B_i F_i + \Delta A_i + \Delta B_i F_i)^* + v_i G_i G_i^*) \tag{8.13}$$

for $j \in \mathbb{N}$. Since $p_{ij} = \sum_{t=1}^{\rho} \alpha^t p_{ij}^t$ for some $\alpha^t \geq 0$, $\sum_{t=1}^{\rho} \alpha^t = 1$, and (8.13) is satisfied for every $t = 1, \ldots, \rho$ (recall that $W \in \bigcap_{t=1}^{\rho} \Psi^t$) we have, after multiplying by α^t and taking the sum over $t = 1, \ldots, \rho$, the desired result.   □

Notice that for the case in which there are uncertainties neither on A nor on B, nor on the transition probability P (that is, $Al_i = 0$, $Bl_i = 0$, $Ar_i = 0$, $Br_i = 0$, $\mathbb{P} = \{P\}$), the restriction $\mathcal{H}_j^1(W) \geq 0$ reduces to

$$0 \leq W_{j1} - \mathcal{Z}_j^1(W)$$

$$= W_{j1} - \sum_{i=1}^{N} p_{ij}(A_i W_{i1} A_i^* + B_i W_{i2}^* A_i^* + A_i W_{i2} B_i^* + B_i W_{i3} B_i^* + v_i G_i G_i^*)$$

and thus the set Ψ coincides with Ψ of Subsection 4.4.4.

Finally, the next theorem gives us the means to solve the robust H_2-control problem.

Theorem 8.2. *Suppose there exists $W = (W_1, \ldots, W_N) \in \Psi$ such that*

$$\lambda(W) = min\{\lambda(V); \, V \in \Psi\}.$$

Then for $F = (F_1, \ldots, F_N)$ with $F_i = W_{i2}^* W_{i1}^{-1}$, $i \in \mathbb{N}$, we have that System $\mathcal{G}_F$ is MSS and

$$\|\mathcal{G}_F\|_2^2 \leq \lambda(W).$$

Proof. Let us denote $\Delta = (\Delta_1, \ldots, \Delta_N)$, and for any Δ_i satisfying $\Delta_i \Delta_i^* \leq I$ for $i \in \mathbb{N}$ and $V = (V_1, \ldots, V_N)$, let $\mathcal{T}_\Delta(V) = (\mathcal{T}_{\Delta 1}(V), \ldots, \mathcal{T}_{\Delta N}(V))$, where

$$\mathcal{T}_{\Delta_j}(V) \triangleq \sum_{i=1}^{N} p_{ij}((A_i + B_i F_i + \Delta A_i + \Delta B_i F_i)V_i(A_i + B_i F_i + \Delta A_i + \Delta B_i F_i)^*$$
$$+ v_i G_i G_i^*)$$

for $j \in \mathbb{N}$. Let us write $W_1 = (W_{11}, \ldots, W_{N1})$. From Proposition 8.1 we have

$$W_1 \geq \mathcal{T}_\Delta(W_1)$$

and from Theorem 3.9 we get that System $\mathcal{G}_F$ is MSS. From Proposition 4.9 we have $\mathcal{K}\mathcal{Y}(W) \leq W$, and finally from Proposition 4.8,

$$\|\mathcal{G}_F\|_2^2 = \lambda(\mathcal{K}\mathcal{Y}(W)) \leq \lambda(W)$$

completing the proof of the theorem.   □

Theorem 8.2 gives us an easily implementable approach to solve the robust H_2-control problem, that is, find W such that $\lambda(W) = min\{\lambda(V); V \in \Psi\}$ and use the correspondent controller, as given by the theorem. The optimization problem is convex and so a numerical solution can be easily found. Note that this solution is not necessarily optimal, since the theorem presents only sufficient conditions (notice that Theorem 4.10 presents necessary and sufficient conditions for the case without uncertainties). However, mean square stability for the closed loop system is guaranteed.

Example 8.3 (including robustness on H_2-control design). This simple example, borrowed from [72], illustrates how the LMI framework presented above
can be used to include reasonable uncertainties in the control design. Consider
the system with parameters given by Table 8.7.

Table 8.7. Parameters for the robut H_2-control design

Mode 1	Mode 2	Mode 3
$A_1 = \begin{bmatrix} 0 & 1 \\ -2.2308+\sigma & 2.5462+\sigma \end{bmatrix}$	$A_2 = \begin{bmatrix} 0 & 1 \\ -38.9103+\sigma & 2.5462+\sigma \end{bmatrix}$	$A_3 = \begin{bmatrix} 0 & 1 \\ 4.6384+\sigma & -4.7455+\sigma \end{bmatrix}$
$B_1 = \begin{bmatrix} 0 \\ 1 \end{bmatrix}$	$B_2 = \begin{bmatrix} 0 \\ 1 \end{bmatrix}$	$B_3 = \begin{bmatrix} 0 \\ 1 \end{bmatrix}$
$C_1 = \begin{bmatrix} 1.5049 & -1.0709 \\ -1.0709 & 1.6160 \\ 0 & 0 \end{bmatrix}$	$C_2 = \begin{bmatrix} 10.2036 & -10.3952 \\ -10.3952 & 11.2819 \\ 0 & 0 \end{bmatrix}$	$C_3 = \begin{bmatrix} 1.7335 & -1.2255 \\ -1.2255 & 1.6639 \\ 0 & 0 \end{bmatrix}$
$D_1 = \begin{bmatrix} 0 \\ 0 \\ 1.6125 \end{bmatrix}$	$D_2 = \begin{bmatrix} 0 \\ 0 \\ 1.0794 \end{bmatrix}$	$D_3 = \begin{bmatrix} 0 \\ 0 \\ 1.0540 \end{bmatrix}$
$G_1 = \begin{bmatrix} 1 & 0 \\ 0 & 1 \end{bmatrix}$	$G_2 = \begin{bmatrix} 1 & 0 \\ 0 & 1 \end{bmatrix}$	$G_3 = \begin{bmatrix} 1 & 0 \\ 0 & 1 \end{bmatrix}$

We consider three cases, according to the level of uncertainties included in
the problem.

1. No uncertainties
   This case reduces to the H_2-control problem described in Chapter 4. We
   assume that $\sigma = 0$, which implies that $\Delta A_i = 0$ and $\Delta B_i = 0$ for $i =
   1, \ldots, 4$. We also assume that the probability transition matrix is exactly
   known, given by

$$P = \begin{bmatrix} 0.67 & 0.17 & 0.16 \\ 0.30 & 0.47 & 0.23 \\ 0.26 & 0.10 & 0.64 \end{bmatrix}.$$

The optimal controller is given by

$$F_1 = \begin{bmatrix} 2.2153 & -1.5909 \end{bmatrix}$$
$$F_2 = \begin{bmatrix} 38.8637 & -38.8864 \end{bmatrix}$$
$$F_3 = \begin{bmatrix} -4.6176 & 5.6267 \end{bmatrix}$$

and

$$|\mathcal{G}_F\|_2^2 = 4124.$$

Since this problem does not consider uncertainties and therefore is in the
framework of Chapter 4, the solution above is non-conservative.

2. Uncertain transition probability matrix
   For this case we assume that matrix P is not precisely known, but is such
   that $P \in \mathbb{P}$, where $\mathbb{P}$ is the polytope with vertices

$$P^1 = \begin{bmatrix} 0.51\ 0.25\ 0.24 \\ 0.14\ 0.55\ 0.31 \\ 0.10\ 0.18\ 0.72 \end{bmatrix}$$

$$P^2 = \begin{bmatrix} 0.83\ 0.09\ 0.08 \\ 0.46\ 0.39\ 0.15 \\ 0.42\ 0.02\ 0.56 \end{bmatrix}$$

$$P^3 = \begin{bmatrix} 0.50\ 0.25\ 0.25 \\ 0.20\ 0.50\ 0.30 \\ 0.30\ 0.30\ 0.40 \end{bmatrix}$$

$$P^4 = \begin{bmatrix} 1\ 0\ 0 \\ 0\ 1\ 0 \\ 0\ 0\ 1 \end{bmatrix}.$$

Notice that the transition probability matrix used in Case 1 also belongs
to this polytope. The robust controller is given by

$$F_1 = \begin{bmatrix} 2.2167\ -1.7979 \end{bmatrix}$$
$$F_2 = \begin{bmatrix} 38.8861\ -39.1083 \end{bmatrix}$$
$$F_3 = \begin{bmatrix} -4.6279\ 5.4857 \end{bmatrix}$$

with

$$\|\mathcal{G}_F\|_2^2 = 4577.$$

As expected, this controller presents a higher H_2-norm when compared
with the optimal controller of Case 1, but this is a common tradeoff between performance and robustness. With this controller, mean square stability is guaranteed for any $P \in \mathbb{P}$, including the system of Case 1.
Since we are in the framework of Theorem 8.2, this solution is possibly
conservative, i.e. there may be another controller F, different from the one
obtained here, that also guarantees mean square stability and presents a
smaller H_2-norm.

3. Including parameter σ
   We consider P uncertain as in the previous case and also that $\sigma \in$
   $[-0.1, 0.1]$. To include uncertainty σ in our problem we pick matrices

$$Al_1 = Al_2 = Al_3 = Al_4 = \begin{bmatrix} 0 \\ 0.3162 \end{bmatrix}$$
$$Ar_1 = Ar_2 = Ar_3 = Ar_4 = \begin{bmatrix} 0.3162\ 0.3162 \end{bmatrix}$$
$$Bl_1 = Bl_2 = Bl_3 = Bl_4 = \begin{bmatrix} 0 \\ 0 \end{bmatrix}$$
$$Br_1 = Br_2 = Br_3 = Br_4 = 0,$$

which yield σ as in Table 8.7 for the system given by (8.4). The robust controller obtained is

$$F_1 = \begin{bmatrix} 2.2281 & -2.4440 \end{bmatrix}$$
$$F_2 = \begin{bmatrix} 38.8998 & -39.7265 \end{bmatrix}$$
$$F_3 = \begin{bmatrix} -4.6360 & 4.8930 \end{bmatrix}$$

with

$$\|\mathcal{G}_F\|_2^2 = 6840.$$

Due to the higher degree of uncertainty, the H_2-norm is much higher than in the previous cases.

8.2.2 Robust Mixed H_2/H_∞-control

The robust control for the H_2-control problem, presented in the previous subsection, was obtained as an extension to the results of Chapter 4. This subsection presents an approximation for the mixed H_2/H_∞-control problem, also based on convex programming techniques, but with an independent development.

The results presented here can be seen, when restricted to the deterministic case, as analogous to those of [130].

We consider the following version of System (1.3):

$$\mathcal{G} = \begin{cases} x(k+1) = A_{\theta(k)}x(k) + B_{\theta(k)}u(k) + Gw(k) \\ z(k) = C_{\theta(k)}x(k) + D_{\theta(k)}u(k) \\ x(0) = 0, \theta(0) = \theta_0 \in \Theta_0 \end{cases} \quad (8.14)$$

which is slightly different from the one considered in Subsection 4.3.1. Here we will not consider uncertainties on the system matrices, like in Subsection 8.2.1, and most importantly, in the present case, matrix G is assumed constant over the values taken by the Markov chain. This restriction is bundled into Theorem 8.5 proof development, as will be seen in this subsection. We will also assume (see Remark 4.1) that

1. $D_i^* D_i > 0$ and
2. $C_i^* D_i = 0$ for $i \in \mathbb{N}$.

As before, we will denote by $\mathcal{G}_F$ the system above when $u(k) = F_{\theta(k)}x(k)$. As in the previous subsection all matrices will be assumed real. The following proposition, presented in [59], will be useful for further developments.

Proposition 8.4. *Suppose* (C, A) *is mean square detectable and* $X = (X_1, \ldots, X_N) \in \mathbb{H}^n$ *and* $F = (F_1, \ldots, F_N) \in \mathbb{H}^{n,m}$ *satisfy*

$$-X_i + (A_i + B_iF_i)^*\mathcal{E}_i(X)(A_i + B_iF_i) + (C_i + D_iF_i)^*(C_i + D_iF_i) \le 0 \quad (8.15)$$

for $i \in \mathbb{N}$. *Then* F *stabilizes* (A, B) *in the mean square sense.*

Proof. From the hypothesis that (C, A) is mean square detectable (see Definition 3.41), there is $J = (J_1, \ldots, J_N) \in \mathbb{H}^{q,n}$ such that $r_\sigma(\mathcal{L}) < 1$ for the operator $\mathcal{L}$ defined as in (3.7) with $\Gamma_i = A_i + J_i C_i$. We define

$$\bar{B}_i = \left[B_i (D_i^* D_i)^{-1/2} \ J_i \right]$$

and

$$\bar{K}_i = \begin{bmatrix} 0 \\ C_i \end{bmatrix}.$$

Clearly $\bar{B}_i \bar{K}_i = J_i C_i$. Define also

$$\bar{F}_i = \begin{bmatrix} (D_i^* D_i)^{1/2} F_i \\ 0 \end{bmatrix}$$

so that $\bar{B}_i \bar{F}_i = B_i F_i$. Then

$$\bar{K}_i - \bar{F}_i = \begin{bmatrix} (D_i^* D_i)^{1/2} F_i \\ C_i \end{bmatrix}$$

and (recalling that $C_i^* D_i = 0$)

$$(\bar{K}_i - \bar{F}_i)^* (\bar{K}_i - \bar{F}_i) = (C_i + D_i F_i)^* (C_i + D_i F_i)$$

and from (8.15),

$$X_i - (A_i + \bar{B}_i \bar{F}_i)^* \mathcal{E}_i(X)(A_i + \bar{B}_i \bar{F}_i) \geq (\bar{K}_i - \bar{F}_i)^* (\bar{K}_i - \bar{F}_i).$$

From Lemma A.8, $r_\sigma(\bar{\mathcal{L}}) < 1$ where the operator $\bar{\mathcal{L}}$ is defined as in (3.7) replacing Γ_i by Λ_i, with $\Lambda_i = A_i + B_i F_i$. Thus F stabilizes (A, B).   □

Mixed H_2/H_∞-control deals simultaneously with optimality and robustness. The objective is to obtain a controller that stabilizes System $\mathcal{G}$, as given by (8.14), in the mean square sense such that an upper bound for the H_2-norm of the closed loop system is minimized under the restriction that the output gain to ℓ_2-additive disturbance sequences is smaller than a pre-specified value δ.

The next theorem presents sufficient conditions for the solution of the mixed H_2/H_∞-control problem, and will constitute the basis for deriving the LMI approach to obtain a solution for the problem.

Theorem 8.5. *Suppose (C, A) is mean square detectable and $\delta > 0$ a fixed real number. If there exist $0 \leq X = (X_1, \ldots, X_N) \in \mathbb{H}^n$ and $F = (F_1, \ldots, F_N) \in \mathbb{H}^{n,m}$ such that for $i \in \mathbb{N}$*

$$-X_i + (A_i + B_i F_i)^* \mathcal{E}_i(X)(A_i + B_i F_i)$$
$$+ (C_i + D_i F_i)^* (C_i + D_i F_i) + \delta^{-2} X_i GG^* X_i \leq 0, \quad (8.16)$$

then F stabilizes (A, B) in the mean square sense and

$$\sup_{\theta_0 \in \Theta_0} \left\| Z_F^0(\theta_0, .) \right\|^2 \le \delta^2(1 - \nu) \le \delta^2$$

where operator Z_F^0 is as defined in (7.5) and $\nu \in (0, \delta^{-2} \sum_{i=1}^N \mathrm{tr}\{G^* X_i G\})$. Moreover

$$\|\mathcal{G}_F\|_2^2 \le \sum_{i=1}^N v_i \, \mathrm{tr}\{G^* \mathcal{E}_i(X) G\}.$$

Proof. Comparing (8.15) and (8.16) it is immediate from Proposition 8.4 that F stabilizes (A, B) in the mean square sense. Set $\Gamma_i = A_i + B_i F_i$ and $\mathcal{O}_i(F) = C_i^* C_i + F_i^* D_i^* D_i F_i$ for $i \in \mathbb{N}$. From Theorem 3.34, $x = (0, x(1), \dots) \in \mathcal{C}^n$ for every $w = (w(0), w(1), \dots) \in \mathcal{C}^r$. So we get from (8.16) that

$$\begin{aligned}
& E(x(k+1)^* X_{\theta(k+1)} x(k+1)) \\
&= E(x(k+1)^* E(X_{\theta(k+1)} | \mathfrak{F}_k) x(k+1)) \\
&= E((\Gamma_{\theta(k)} x(k) + Gw(k))^* \mathcal{E}_{\theta(k)}(X)(\Gamma_{\theta(k)} x(k) + Gw(k))) \\
&\le E(x(k)^* (X_{\theta(k)} - \mathcal{O}_{\theta(k)}(F) - \delta^{-2} X_{\theta(k)} GG^* X_{\theta(k)}) x(k) \\
&\quad + w(k)^* G^* \mathcal{E}_{\theta(k)}(X) \Gamma_{\theta(k)} x(k) + x(k)^* \Gamma_{\theta(k)}^* \mathcal{E}_{\theta(k)}(X) Gw(k) \\
&\quad + w(k)^* G^* \mathcal{E}_{\theta(k)}(X) Gw(k))
\end{aligned}$$

so that

$$\begin{aligned}
& \left\| X_{\theta(k+1)}^{1/2} x(k+1) \right\|_2^2 - \left\| X_{\theta(k)}^{1/2} x(k) \right\|_2^2 + \|z(k)\|_2^2 \\
&\le -\delta^{-2} \left\| G^* X_{\theta(k)} x(k) \right\|_2^2 + E(w(k)^* G^* \mathcal{E}_{\theta(k)}(X) \Gamma_{\theta(k)} x(k) \\
&\quad + E(x(k)^* \Gamma_{\theta(k)}^* \mathcal{E}_{\theta(k)}(X) Gw(k)) + \left\| \mathcal{E}_{\theta(k)}(X)^{1/2} Gw(k) \right\|_2^2 \\
&= -\delta^{-2} \left\| G^* X_{\theta(k+1)} x(k+1) \right\|_2^2 + \delta^{-2} \left\| G^* X_{\theta(k+1)} x(k+1) \right\|_2^2 \\
&\quad - \delta^{-2} \left\| G^* X_{\theta(k)} x(k) \right\|_2^2 + 2E(w(k)^* G^* \mathcal{E}_{\theta(k)}(X)(\Gamma_{\theta(k)} x(k) + Gw(k))) \\
&\quad - E(w(k)^* G^* \mathcal{E}_{\theta(k)}(X) Gw(k)).
\end{aligned}$$

Thus

$$\begin{aligned}
& \left\| X_{\theta(k+1)}^{1/2} x(k+1) \right\|_2^2 - \left\| X_{\theta(k)}^{1/2} x(k) \right\|_2^2 + \|z(k)\|_2^2 \\
&\quad + \delta^{-2} \left\| G^* X_{\theta(k)} x(k) \right\|_2^2 - \delta^{-2} \left\| G^* X_{\theta(k+1)} x(k+1) \right\|_2^2 \\
&\le -\delta^{-2} \left\| G^* X_{\theta(k+1)} x(k+1) \right\|_2^2 + 2E(w(k)^* G^* X_{\theta(k+1)} x(k+1)) \\
&\quad - \delta^2 \|w(k)\|_2^2 + E(w(k)^* (\nu^2 I - G^* \mathcal{E}_{\theta(k)}(X) G) w(k)) \\
&= \left\| \delta^{-1} G^* X_{\theta(k+1)} x(k+1) - \delta w(k) \right\|_2^2 + E(w(k)^* (\delta^2 I - G^* \mathcal{E}_{\theta(k)}(X) G) w(k)) \\
&\le E(w(k)^* (\delta^2 I - G^* \mathcal{E}_{\theta(k)}(X) G) w(k)).
\end{aligned}$$

Notice that this argument is valid only if matrix G does not depend on $\theta(k)$. Taking the sum from $k = 0$ to ∞, and recalling that $x(0) = 0$ and that $\|x(k)\|_2 \to 0$ as $k \to \infty$, we get for $z = (z(0), z(1), \dots)$,

$$\|z\|_2^2 \leq \delta^2 \sum_{k=0}^{\infty} E(w(k)^*(I - \delta^{-2}G^*\mathcal{E}_{\theta(k)}(X)G)w(k)) \leq \delta^2(1 - \nu)\|w\|_2^2$$

where $\nu \in (0, \delta^{-2}\sum_{i=1}^{N} \operatorname{tr}\{G^*X_iG\})$. Thus

$$\sup_{\theta_0 \in \Theta_0} \left\|Z_F^0(\theta_0, .)\right\|^2 \leq \delta^2(1 - \nu) \leq \delta^2.$$

Notice from Proposition 4.8 that

$$\|\mathcal{G}_F\|_2^2 = \sum_{i=1}^{N} \upsilon_i \operatorname{tr}\{G^*\mathcal{E}_i(\mathcal{S}o)G\},$$

where

$$\mathcal{S}o_i = \Gamma_i^*\mathcal{E}_i(\mathcal{S}o)\Gamma_i + \mathcal{O}_i(F)$$

for $i \in \mathbb{N}$. From (8.16), we have for some $V = (V_1, \dots, V_N) \geq 0$

$$X_i = \Gamma_i^*\mathcal{E}_i(X)\Gamma_i + \mathcal{O}_i(F) + \delta^{-2}X_iGG^*X_i + V_i^*V_i$$

for $i \in \mathbb{N}$, so that from Theorem 3.9, $X \geq \mathcal{S}o$. This implies that

$$\|\mathcal{G}_F\|_2^2 = \sum_{i=1}^{N} \upsilon_i \operatorname{tr}\{G^*\mathcal{E}_i(\mathcal{S}o)G\} \leq \sum_{i=1}^{N} \upsilon_i \operatorname{tr}\{G^*\mathcal{E}_i(X)G\},$$

completing the proof of the theorem.   □

The theorem above suggests the following approach to solve the mixed H_2/H_∞-control problem: for $\delta > 0$ fixed, find $X = (X_1, \dots, X_N) \geq 0$ and $F = (F_1, \dots, F_N)$ that minimize $\sum_{i=1}^{N} \upsilon_i \operatorname{tr}\{G^*\mathcal{E}_i(X)G\}$ subject to (8.16) and such that the closed loop system is mean square stable.

As will be shown with the next results, indeed a convex approximation for this problem can be obtained, and the resulting LMI optimization problem has an adequate structure for the inclusion of uncertainties. We will consider uncertainties in the transition probability matrix P in the same manner of Subsection 8.2.1, that is, we will assume that P is not exactly known, but that $P \in \mathbb{P}$, where $\mathbb{P}$ is as given in (8.5). We define now the following problem:

Problem 8.6. Set $\mu = \delta^2$. Find real $0 < X = (X_1, \dots, X_N) \in \mathbb{H}^n$, $0 < Q = (Q_1, \dots, Q_N) \in \mathbb{H}^n$, $0 < L = (L_1, \dots, L_N) \in \mathbb{H}^n$ and $Y = (Y_1, \dots, Y_N) \in \mathbb{H}^{n,m}$ such that

$$\xi = \min \sum_{i=1}^{N} \upsilon_i \operatorname{tr}\{G^*\mathcal{E}_i(X)G\} \tag{8.17a}$$

subject to

$$
\begin{bmatrix}
Q_i & Q_iA_i^* + Y_i^*B_i^* & Q_iC_i^* & Y_i^*D_i^* & G \\
A_iQ_i + B_iY_i & L_i & 0 & 0 & 0 \\
C_iQ_i & 0 & I & 0 & 0 \\
D_iY_i & 0 & 0 & I & 0 \\
G^* & 0 & 0 & 0 & \mu I
\end{bmatrix} \geq 0 \text{ for } i \in \mathbb{N} \qquad (8.17b)
$$

$$
\begin{bmatrix}
L_i & L_i\sqrt{p_{i1}^t} & \cdots & L_i\sqrt{p_{iN}^t} \\
\sqrt{p_{i1}^t}\,L_i & Q_1 & 0 & 0 \\
\vdots & 0 & \ddots & 0 \\
\sqrt{p_{iN}^t}\,L_i & 0 & 0 & Q_N
\end{bmatrix} \geq 0 \text{ for } i \in \mathbb{N},\, t = 1,\ldots,\rho \qquad (8.17c)
$$

$$
\begin{bmatrix} X_i & I \\ I & Q_i \end{bmatrix} \geq 0 \text{ for } i \in \mathbb{N}. \qquad (8.17d)
$$

The following theorem makes the connection between Optimization Problem 8.6 and the control problem.

Theorem 8.7. *Suppose Problem 8.6 has a solution given by* (P, Q, L, Y) *and set* $F = (F_1, \ldots, F_N) \in \mathbb{H}^{n,m}$ *with* $F_i = Y_iQ_i^{-1}$ *for* $i \in \mathbb{N}$. *Then System* $\mathcal{G}_F$ *is mean square stable,*

$$
\|\mathcal{G}_F\|_2 \leq \xi^{1/2}
$$

and

$$
\sup_{\theta_0 \in \Theta_0} \|Z_F^0(\theta_0, .)\| \leq \delta
$$

for every $\mathrm{P} \in \mathbb{P}$.

Proof. First of all notice that (8.17b)–(8.17d) are equivalent (from the Schur complement, Lemma 2.23) to

$$
Q_i \geq Q_i(A_i + B_iF_i)^*L_i^{-1}(A_i + B_iF_i)Q_i
$$
$$
+ Q_i(C_i + D_iF_i)^*(C_i + D_iF_i)Q_i + \mu^{-1}Q_i(Q_i^{-1})G^*G(Q_i^{-1})Q_i \qquad (8.18a)
$$

$$
L_i \geq L_i\left(\sum_{j=1}^{N} p_{ij}^t Q_j^{-1}\right) L_i \text{ for } t = 1,\ldots,\rho \qquad (8.18b)
$$

$$
X_i \geq Q_i^{-1} \qquad (8.18c)
$$

for $i \in \mathbb{N}$. Since we are minimizing $\sum_{i=1}^{N} v_i \operatorname{tr}\{G^*\mathcal{E}_i(X)G\}$ we must have from (8.18c) that $X_i = Q_i^{-1}$ is an optimal solution. Consider any $\mathrm{P} \in \mathbb{P}$. By definition we have $p_{ij} = \sum_{t=1}^{r} \alpha^t p_{ij}^t$ for some $\alpha^t \geq 0$, $\sum_{t=1}^{r} \alpha^t = 1$. Thus from (8.18b) we get

$$
L_i^{-1} \geq \sum_{j=1}^{N} p_{ij}Q_j^{-1} = \mathcal{E}_i(X)
$$

for $i \in \mathbb{N}$. From (8.18a) we have

$$
\begin{aligned}
X_i &= Q_i^{-1} \\
&\geq (A_i + B_i F_i)^* L_i^{-1}(A_i + B_i F_i) + (C_i + D_i F_i)^*(C_i + D_i F_i) \\
&\quad + \delta^{-1} Q_i G^* G Q_i \\
&\geq (A_i + B_i F_i)^* \mathcal{E}_i(X)(A_i + B_i F_i) + (C_i + D_i F_i)^*(C_i + D_i F_i) \\
&\quad + \delta^{-1} Q_i G^* G Q_i.
\end{aligned}
$$

Applying Theorems 8.5 and 3.9, we obtain the desired result. □

Notice that the theorem above presents only a sufficient condition, and so any solution to the control problem obtained with it might be a conservative one, although mean square stability is guaranteed. Also, if there is no solution to the optimization problem, it does not mean that there is no solution to the control problem.

8.2.3 Robust H_∞-control

The Optimization Problem 8.6 considered in the previous subsection can be directly adapted to the H_∞ case. In fact, given the results above, it is a simpler problem, as presented below.

Problem 8.8. Find $\mu \in \mathbb{R}^+$, $0 < Q = (Q_1, \ldots, Q_N) \in \mathbb{H}^n$ real, $0 < L = (L_1, \ldots, L_N) \in \mathbb{H}^n$ real, and $Y = (Y_1, \ldots, Y_N) \in \mathbb{H}^{n,m}$ real, such that

$$
\zeta = \min\{\mu\} \tag{8.19a}
$$

subject to

$$
\begin{bmatrix}
Q_i & Q_i A_i^* + Y_i^* B_i^* & Q_i C_i^* & Y_i^* D_i^* & G \\
A_i Q_i + B_i Y_i & L_i & 0 & 0 & 0 \\
C_i Q_i & 0 & I & 0 & 0 \\
D_i Y_i & 0 & 0 & I & 0 \\
G^* & 0 & 0 & 0 & \mu I
\end{bmatrix} \geq 0 \text{ for } i \in \mathbb{N} \tag{8.19b}
$$

$$
\begin{bmatrix}
L_i & L_i \sqrt{p_{i1}^t} & \cdots & L_i \sqrt{p_{iN}^t} \\
\sqrt{p_{i1}^t} L_i & Q_1 & 0 & 0 \\
\vdots & 0 & \ddots & 0 \\
\sqrt{p_{iN}^t} L_i & 0 & 0 & Q_N
\end{bmatrix} \geq 0 \text{ for } i \in \mathbb{N}, \, t = 1, \ldots, \rho. \tag{8.19c}
$$

Since we are not concerned with the value of the H_2-norm, X was withdrawn from the problem and so we have no equivalent to the LMI (8.17d) of Problem 8.6. It is easy to see that this problem also leads to a mean square stabilizing solution (just notice that only LMIs (8.17b) and (8.17c) are related to stability in Problem 8.6).

Although the solution might be conservative, like in the previous subsection, we are concerned with finding a minimal upper bound for the disturbance gain, while in Chapter 7, we were only interested in finding a solution for a fixed value of the upper bound. Also, using this LMI approach, it is possible to include uncertainties in the problem, as seen.

8.3 Achieving Optimal H_∞-control

8.3.1 Algorithm

In this book there are two approaches for the H_∞-control problem. These approaches differ basically by what we may refer to as optimality and suboptimality of H_∞-control.

In Chapter 7 we presented necessary and sufficient conditions for the existence of a solution as well as an iterative algorithm to solve the H_∞ CARE. Using this approach, it is possible, for a given fixed real number $\delta > 0$, find an adequate mean square stabilizing controller that guarantees that the closed loop system will have a gain smaller than δ for any ℓ_2-additive disturbance sequence, if such a controller exists. It is not possible however to know directly what is the minimum value of δ for which a solution still exists.

In Subsection 8.2.2, an LMI optimization problem was presented, whose solution lead to a mean square stabilizing controller with minimal δ. However, this solution may be conservative, and the δ found may not be the real minimum. On the other hand, the approach allows the inclusion of uncertainties on the problem and can make use of very efficient numerical methods.

It is possible, by means of a very simple iterative method, to combine both approaches to find the real optimal solution. Since we will deal simultaneously with the two methods, we will restrict ourselves to systems whose structure is compatible with both, that is, we will assume that there are no uncertainties on the system (as in Chapter 7) and that G is constant over the states of the Markov chain (as in Subsection 8.2.2). The procedure is as follows.

Algorithm 8.9 (Bisection)   *1. Solve Problem 8.8 and find an initial upper bound δ^0 (this is a possibly conservative minimum for the problem). Thus we have that the desired value lies in the interval $(0, \delta^0]$. Alternatively one could consider a very large initial value for δ^0 instead of solving Problem 8.8.*

2. Bisect the interval obtained in the previous step and use the procedure of Chapter 7 to check if the problem is feasible in the mean point of the interval, $\delta^1 = \delta^0/2$. If it is feasible, the optimal solution lies in the interval $(0, \delta^1]$. Otherwise it lies in the interval $(\delta^1, \delta^0]$.

3. Keep bisecting the feasible interval obtaining δ^κ, $\kappa = 2, 3, \ldots$ until the length of the feasible interval is reduced to an acceptable value ϵ.

This procedure, although with a somewhat heavy computational burden, guarantees that the maximum error ϵ (given by the length of the last interval) for the optimal value is obtained after η iterations, where

$$\eta = \min\left\{n \text{ integer}; \, n \geq \frac{\ln(\delta^0/\epsilon)}{\ln(2)}\right\}.$$

8.3.2 H_∞-control for the UarmII Manipulator

Underactuated manipulators are robotic systems with less actuators than degrees of freedom. Usually, when dealing with a robotic arm, each joint of the arm has its own manipulator (what we call an *active joint*). In an underactuated manipulator however, there may be joints without actuators (called *passive joints*). Even though we cannot directly control passive joints, we can use the dynamic coupling between them and the other active joints to do so.

The main interest in the study of underactuated manipulators lies in the operation of normal manipulators that are subject to faults while performing critical operations in environments where it is not convenient or possible to perform maintenance, such as in space vehicles, deep sea underwater structures, nuclear containments, etc.

When there is a fault in a given joint, it becomes a passive one and the control system should be able to position the manipulator using only the remaining active joints. Basically, the control strategy is to swing the robotic arm with the faulty joints unlocked until the arm reaches a certain adequate position. Then brakes are applied on the passive joints and the arm is moved to the final desired position. For more details on this technique see [8].

The Uarm II (Underactuated Arm II) research manipulator, described in [200] (see Figure 8.5), is a planar robot especially built for the study of control algorithms for underactuated manipulators. It is comprised of three special-purpose joints with DC motors and pneumatic brakes, capable of simulating a large number of faulty conditions. We denote these conditions by three letter acronyms. AAA stands for (active-active-active), meaning that the robot is in normal operation, while PAA (passive-active-active) means that joint 1 is passive and so on.

Siqueira and Terra, in [200], employed a Markov jump approach for the control of the Uarm II manipulator, reported in the following.

Modeling of manipulators is a somewhat complex task, usually yielding nonlinear expressions. In order to keep the models linear, the attainable angular excursion of 20° for each joint was divided in two regions centered on 5° and 15°. Since we have three joints, the total number of regions is eight.

For simplicity, only faults on joint 2 are considered, yielding three possible configurations.

AAA : normal operation; all joints active.
$\text{AP}_\text{u}\text{A}$: faulty operation; joint 2 passive and unlocked.

Fig. 8.5. Uarm II manipulator

AP_lA : faulty operation; joint 2 passive and locked.

With eight regions for linearization and three configurations, we have 24 operation modes, given by Table 8.8, where q_i stands for the position of joint i.

Table 8.8. Operation modes for the Uarm II manipulator (from [200])

Operation modes			Linearization points					
AAA	AP_uA	AP_lA	q_1	q_2	q_3	$\dot{q}_1$	$\dot{q}_2$	$\dot{q}_3$
1	9	17	5°	5°	5°	0	0	0
2	10	18	15°	5°	5°	0	0	0
3	11	19	5°	15°	5°	0	0	0
4	12	20	15°	15°	5°	0	0	0
5	13	21	5°	5°	15°	0	0	0
6	14	22	15°	5°	15°	0	0	0
7	15	23	5°	15°	15°	0	0	0
8	16	24	15°	15°	15°	0	0	0

The probability transition matrix connecting the operation modes is given by

$$P = \left[\begin{array}{c|c|c} P_{AAA} & P_f & P_0 \\ \hline P_0 & P_{AP_uA} & P_s \\ \hline P_0 & P_s & P_{AP_lA} \end{array}\right].$$
(8.20)

The matrix was partitioned in 8×8 blocks according to the possible configurations. The blocks are given by

$$P_{AAA} = \begin{bmatrix} 0.89 & 0.10 & 0 & 0 & 0 & 0 & 0 & 0 \\ 0.10 & 0.79 & 0.10 & 0 & 0 & 0 & 0 & 0 \\ 0 & 0.10 & 0.79 & 0.10 & 0 & 0 & 0 & 0 \\ 0 & 0 & 0.10 & 0.79 & 0.10 & 0 & 0 & 0 \\ 0 & 0 & 0 & 0.10 & 0.79 & 0.10 & 0 & 0 \\ 0 & 0 & 0 & 0 & 0.10 & 0.79 & 0.10 & 0 \\ 0 & 0 & 0 & 0 & 0 & 0.10 & 0.79 & 0.10 \\ 0 & 0 & 0 & 0 & 0 & 0 & 0.10 & 0.89 \end{bmatrix}$$

$$P_f = \begin{bmatrix} 0.01 & 0 & 0 & 0 & 0 & 0 & 0 & 0 \\ 0 & 0.01 & 0 & 0 & 0 & 0 & 0 & 0 \\ 0 & 0 & 0.01 & 0 & 0 & 0 & 0 & 0 \\ 0 & 0 & 0 & 0.01 & 0 & 0 & 0 & 0 \\ 0 & 0 & 0 & 0 & 0.01 & 0 & 0 & 0 \\ 0 & 0 & 0 & 0 & 0 & 0.01 & 0 & 0 \\ 0 & 0 & 0 & 0 & 0 & 0 & 0.01 & 0 \\ 0 & 0 & 0 & 0 & 0 & 0 & 0 & 0.01 \end{bmatrix}$$

$$P_0 = \begin{bmatrix} 0 & 0 & 0 & 0 & 0 & 0 & 0 & 0 \\ 0 & 0 & 0 & 0 & 0 & 0 & 0 & 0 \\ 0 & 0 & 0 & 0 & 0 & 0 & 0 & 0 \\ 0 & 0 & 0 & 0 & 0 & 0 & 0 & 0 \\ 0 & 0 & 0 & 0 & 0 & 0 & 0 & 0 \\ 0 & 0 & 0 & 0 & 0 & 0 & 0 & 0 \\ 0 & 0 & 0 & 0 & 0 & 0 & 0 & 0 \\ 0 & 0 & 0 & 0 & 0 & 0 & 0 & 0 \end{bmatrix}$$

$$P_{AP_uA} = \begin{bmatrix} 0.78 & 0.20 & 0 & 0 & 0 & 0 & 0 & 0 \\ 0.10 & 0.78 & 0.10 & 0 & 0 & 0 & 0 & 0 \\ 0 & 0.10 & 0.78 & 0.10 & 0 & 0 & 0 & 0 \\ 0 & 0 & 0.10 & 0.78 & 0.10 & 0 & 0 & 0 \\ 0 & 0 & 0 & 0.10 & 0.78 & 0.10 & 0 & 0 \\ 0 & 0 & 0 & 0 & 0.10 & 0.78 & 0.10 & 0 \\ 0 & 0 & 0 & 0 & 0 & 0.10 & 0.78 & 0.10 \\ 0 & 0 & 0 & 0 & 0 & 0 & 0.2 & 0.78 \end{bmatrix}$$

with

$$P_{AP_lA} = P_{AP_uA} \qquad \text{and} \qquad P_s = 2P_f .$$

The blocks are arranged such that P_{AAA} presents the transition probabilities between operation modes for normal operation and P_f presents the probability of faults occurring while in normal operation. If a fault occurs, the system may only move to configuration AP_uA and then it may move between AP_uA and AP_1A according to the probabilities given by P_s.

Rigorously speaking, the switching between AP_uA and AP_1A is not intrinsically stochastic, but since it is quite complex to assess this operation sequence a priori on a general basis, it is not at all unreasonable to assume it is a Markov chain.

The models for the 24 operation modes are presented in Tables 8.9, 8.10 and 8.11. It is assumed that $G_i = B_i$ for $i = 1, \ldots, 24$. The system state is formed by the angular position plus the angular velocity of each joint.

The obtained H_∞-controller is presented in Table 8.12. For this controller, $\delta = 27.6$.

Figure 8.6 presents the system response to a 20° step on the setpoint for the three joints. On $t = 4\,\text{s}$, a fault on joint 2 is artificially induced, changing the configuration from AAA to AP_uA. The associated sequence of the Markov chain is presented in Figure 8.7.

In Figure 8.8, the estimated applied torque on each joint is shown. [200] reports that the performance of this controller compares favorably with other control designs employed in similar experiments.

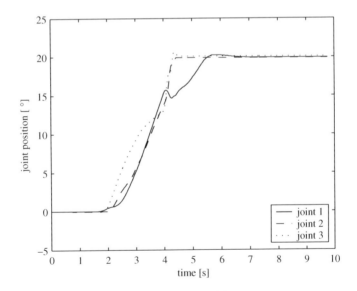

Fig. 8.6. Joint positions

Table 8.9. Uarm II model for configuration AAA

i	A_i	B_i	C_i	D_i
1	$\begin{bmatrix} 1.01 & -0.01 & 0.00 & 0.06 & 0.00 \\ -0.03 & 1.04 & -0.02 & -0.01 & -0.01 \\ 0.02 & -0.03 & 1.02 & -0.03 & 0.06 \\ 0.50 & -0.56 & 0.20 & 1.26 & 0.08 \\ -1.13 & 1.44 & -0.64 & -0.46 & -0.27 \\ 0.79 & -1.28 & 1.00 & 2.19 & 1.42 \end{bmatrix}$	$\begin{bmatrix} 1.26 & -2.81 & 1.98 \\ -2.81 & 7.22 & -6.39 \\ 1.98 & -6.39 & 10.00 \\ 50.26 & -112.50 & 79.18 \\ -112.50 & 288.86 & -255.68 \\ 79.18 & -255.68 & 399.92 \end{bmatrix}$	$\begin{bmatrix} 0.05 & -0.00 & 0.00 & -0.00 & 0.00 & 0.00 \\ -0.00 & 0.05 & -0.00 & 0.00 & -0.00 & -0.00 \\ 0.00 & -0.00 & 0.05 & 0 & 0.00 & 0.00 \\ 0 & 0 & 0 & 1 & 0 & 0 \\ 0 & 0 & 0 & 0 & 1 & 0 \end{bmatrix}$	$\begin{bmatrix} 0.03 & -0.07 & 0.05 \\ -0.07 & 0.18 & -0.16 \\ 0.05 & -0.16 & 0.25 \\ 1 & 0 & 0 \\ 0 & 1 & 0 \end{bmatrix}$
2	$\begin{bmatrix} 1.01 & -0.01 & 0.00 & 0.06 & 0.00 \\ -0.03 & 1.04 & -0.02 & -0.01 & -0.01 \\ 0.03 & -0.03 & 1.03 & 0.01 & 0.06 \\ 3.50 & -0.56 & 0.20 & 1.26 & 0.08 \\ -1.13 & 1.44 & -0.64 & -0.46 & -0.27 \\ 0.79 & -1.28 & 1.00 & 2.19 & 1.42 \end{bmatrix}$	$\begin{bmatrix} 1.26 & -2.81 & 1.98 \\ -2.81 & 7.22 & -6.39 \\ 1.98 & -6.39 & 10.00 \\ 50.26 & -112.50 & 79.18 \\ -112.50 & 288.86 & -255.68 \\ 79.18 & -255.68 & 399.92 \end{bmatrix}$	$\begin{bmatrix} 0.05 & -0.00 & 0.00 & -0.00 & 0.00 & 0.00 \\ -0.00 & 0.05 & -0.00 & 0.00 & -0.00 & -0.00 \\ 0.00 & -0.00 & 0.05 & 0 & 0.00 & 0.00 \\ 0 & 0 & 0 & 1 & 0 & 0 \\ 0 & 0 & 0 & 0 & 1 & 0 \end{bmatrix}$	$\begin{bmatrix} 0.03 & -0.07 & 0.05 \\ -0.07 & 0.18 & -0.16 \\ 0.05 & -0.16 & 0.25 \\ 1 & 0 & 0 \\ 0 & 1 & 0 \end{bmatrix}$
3	$\begin{bmatrix} 1.01 & -0.01 & 0.00 & 0.06 & 0.00 \\ -0.02 & 1.03 & -0.01 & -0.01 & -0.01 \\ 0.02 & -0.03 & 1.02 & 0.01 & 0.06 \\ 0.44 & -0.49 & 0.17 & 1.23 & 0.07 \\ -0.97 & 1.26 & -0.58 & -0.40 & -0.24 \\ 0.69 & -1.15 & 0.96 & 2.04 & 1.41 \end{bmatrix}$	$\begin{bmatrix} 1.10 & -2.43 & 1.72 \\ -2.43 & 6.29 & -5.76 \\ 1.72 & -5.76 & 9.56 \\ 43.99 & -97.20 & 68.71 \\ -97.20 & 251.78 & -230.24 \\ 68.71 & -230.24 & 382.48 \end{bmatrix}$	$\begin{bmatrix} 0.05 & -0.00 & 0.00 & -0.00 & 0.00 & 0.00 \\ -0.00 & 0.05 & -0.00 & 0.00 & -0.00 & -0.00 \\ 0.00 & -0.00 & 0.05 & 0 & 0.00 & 0.00 \\ 0 & 0 & 0 & 1 & 0 & 0 \\ 0 & 0 & 0 & 0 & 1 & 0 \end{bmatrix}$	$\begin{bmatrix} 0.03 & -0.06 & 0.04 \\ -0.06 & 0.16 & -0.14 \\ 0.04 & -0.14 & 0.24 \\ 1 & 0 & 0 \\ 0 & 1 & 0 \end{bmatrix}$
4	$\begin{bmatrix} 1.01 & -0.01 & 0.00 & 0.06 & 0.00 \\ -0.02 & 1.03 & -0.01 & 0.08 & -0.01 \\ 0.02 & -0.03 & 1.02 & 0.01 & 0.06 \\ 0.44 & -0.49 & 0.17 & 1.23 & 0.07 \\ -0.97 & 1.26 & -0.58 & -0.40 & -0.24 \\ 0.69 & -1.15 & 0.96 & 2.04 & 1.41 \end{bmatrix}$	$\begin{bmatrix} 1.70 & -2.43 & -5.76 \\ 1.72 & 6.29 & -5.76 \\ 1.72 & -5.76 & 9.56 \\ 43.99 & -97.20 & 68.71 \\ -97.20 & 251.78 & -230.24 \\ 68.71 & -230.24 & 382.48 \end{bmatrix}$	$\begin{bmatrix} 0.05 & -0.00 & 0.00 & -0.00 & 0.00 & 0.00 \\ -0.00 & 0.05 & -0.00 & 0.00 & -0.00 & -0.00 \\ 0.00 & -0.00 & 0.05 & 0 & 0.00 & 0.00 \\ 0 & 0 & 0 & 1 & 0 & 0 \\ 0 & 0 & 0 & 0 & 1 & 0 \end{bmatrix}$	$\begin{bmatrix} 0.03 & -0.06 & 0.04 \\ -0.06 & 0.16 & -0.14 \\ 0.04 & -0.14 & 0.24 \\ 1 & 0 & 0 \\ 0 & 1 & 0 \end{bmatrix}$
5	$\begin{bmatrix} 1.01 & -0.01 & 0.00 & 0.06 & 0.00 \\ -0.03 & 1.04 & -0.02 & -0.01 & -0.01 \\ 0.02 & -0.03 & 1.02 & 0.01 & 0.06 \\ 0.50 & -0.56 & 0.20 & 1.26 & 0.08 \\ -1.13 & 1.44 & -0.63 & -0.47 & -0.27 \\ 0.80 & -1.27 & 0.98 & 2.19 & 1.42 \end{bmatrix}$	$\begin{bmatrix} 1.26 & -2.82 & 1.99 \\ -2.82 & 7.21 & -6.35 \\ 1.99 & -6.35 & 9.81 \\ 50.38 & -112.76 & 79.72 \\ -112.76 & 288.48 & -253.82 \\ 79.72 & -253.82 & 392.49 \end{bmatrix}$	$\begin{bmatrix} 0.05 & -0.00 & 0.00 & -0.00 & 0.00 & 0.00 \\ -0.00 & 0.05 & -0.00 & 0.00 & -0.00 & -0.00 \\ 0.00 & -0.00 & 0.05 & 0 & 0.00 & 0.00 \\ 0 & 0 & 0 & 1 & 0 & 0 \\ 0 & 0 & 0 & 0 & 1 & 0 \end{bmatrix}$	$\begin{bmatrix} 0.03 & -0.07 & 0.05 \\ -0.07 & 0.18 & -0.16 \\ 0.05 & -0.16 & 0.25 \\ 1 & 0 & 0 \\ 0 & 1 & 0 \end{bmatrix}$
6	$\begin{bmatrix} 1.01 & -0.01 & 0.00 & 0.06 & 0.00 \\ -0.03 & 1.04 & -0.02 & -0.01 & -0.01 \\ 0.02 & -0.03 & 1.02 & 0.01 & 0.06 \\ 0.50 & -0.56 & 0.20 & 1.26 & 0.08 \\ -1.13 & 1.44 & -0.63 & -0.47 & -0.27 \\ 0.80 & -1.27 & 0.98 & 2.19 & 1.42 \end{bmatrix}$	$\begin{bmatrix} 1.26 & -2.82 & 1.99 \\ -2.82 & 7.21 & -6.35 \\ 1.99 & -6.35 & 9.81 \\ 50.38 & -112.76 & 79.72 \\ -112.76 & 288.48 & -253.82 \\ 79.72 & -253.82 & 392.49 \end{bmatrix}$	$\begin{bmatrix} 0.05 & -0.00 & 0.00 & -0.00 & 0.00 & 0.00 \\ -0.00 & 0.05 & -0.00 & 0.00 & -0.00 & -0.00 \\ 0.00 & -0.00 & 0.05 & 0 & 0.00 & 0.00 \\ 0 & 0 & 0 & 1 & 0 & 0 \\ 0 & 0 & 0 & 0 & 1 & 0 \end{bmatrix}$	$\begin{bmatrix} 0.03 & -0.07 & 0.05 \\ -0.07 & 0.18 & -0.16 \\ 0.05 & -0.16 & 0.25 \\ 1 & 0 & 0 \\ 0 & 1 & 0 \end{bmatrix}$
7	$\begin{bmatrix} 1.01 & -0.01 & 0.00 & 0.06 & 0.00 \\ -0.02 & 1.03 & -0.01 & -0.01 & -0.01 \\ 0.02 & -0.03 & 1.02 & 0.01 & 0.06 \\ 0.45 & -0.49 & 0.18 & 1.23 & 0.08 \\ -0.99 & 1.28 & -0.59 & -0.41 & -0.25 \\ 0.72 & -1.18 & 0.96 & 2.05 & 1.41 \end{bmatrix}$	$\begin{bmatrix} 1.81 & -2.47 & -5.90 \\ -2.47 & 6.39 & 9.59 \\ 1.81 & -5.90 & 72.27 \\ 44.53 & -98.85 & -235.85 \\ -98.85 & 255.57 & 383.65 \\ 72.27 & -235.85 & \end{bmatrix}$	$\begin{bmatrix} 0.05 & -0.00 & 0.00 & -0.00 & 0.00 & 0.00 \\ -0.00 & 0.05 & -0.00 & 0.00 & -0.00 & -0.00 \\ 0.00 & -0.00 & 0.05 & 0 & 0.00 & 0.00 \\ 0 & 0 & 0 & 1 & 0 & 0 \\ 0 & 0 & 0 & 0 & 1 & 0 \end{bmatrix}$	$\begin{bmatrix} 0.03 & -0.06 & 0.05 \\ -0.06 & 0.16 & -0.15 \\ 0.05 & -0.15 & 0.24 \\ 1 & 0 & 0 \\ 0 & 1 & 0 \end{bmatrix}$
8	$\begin{bmatrix} 1.01 & -0.01 & 0.00 & 0.06 & 0.00 \\ -0.02 & 1.03 & -0.01 & -0.01 & -0.01 \\ 0.02 & -0.03 & 1.02 & 0.01 & 0.06 \\ 0.45 & -0.49 & 0.18 & 1.23 & 0.08 \\ -0.99 & 1.28 & -0.59 & -0.52 & -0.25 \\ 0.72 & -1.18 & 0.96 & 0.38 & 1.41 \end{bmatrix}$	$\begin{bmatrix} 1.81 & -2.47 & -5.90 \\ -2.47 & 6.39 & 9.59 \\ 1.81 & -5.90 & 72.27 \\ 44.53 & -98.85 & -235.85 \\ -98.85 & 255.57 & 383.65 \\ 72.27 & -235.85 & \end{bmatrix}$	$\begin{bmatrix} 0.05 & -0.00 & 0.00 & -0.00 & 0.00 & 0.00 \\ -0.00 & 0.05 & -0.00 & 0.00 & -0.00 & -0.00 \\ 0.00 & -0.00 & 0.05 & 0 & 0.00 & 0.00 \\ 0 & 0 & 0 & 1 & 0 & 0 \\ 0 & 0 & 0 & 0 & 1 & 0 \end{bmatrix}$	$\begin{bmatrix} 0.03 & -0.06 & 0.05 \\ -0.06 & 0.16 & -0.15 \\ 0.05 & -0.15 & 0.24 \\ 1 & 0 & 0 \\ 0 & 1 & 0 \end{bmatrix}$

Table 8.10. Uarm II model for configuration AP_uA

Configuration 9:
```
⎡ 1.14  -0.18   0    0.09  -0.04 ⎤   ⎡ -3.48    0   -11.29  ⎤
⎢-0.18   1.33   0   -0.06   0.12 ⎥   ⎢  4.05    0    20.72  ⎥
⎢  0      0     1    0      0    ⎥   ⎢   0      0     0      ⎥
⎢ 5.60  -7.15   0    2.64  -1.59 ⎥   ⎢-139.25   0  -451.67  ⎥
⎣-7.17  13.27   0   -2.59   3.93 ⎦   ⎣ 161.95   0   828.72  ⎦

⎡ 0.05  -0.00   0    0.00  -0.00 ⎤   ⎡ -0.09   0   -0.28 ⎤
⎢-0.00   0.06   0    0.00  -0.00 ⎥   ⎢  0.10   0    0.52 ⎥
⎢  0      0     0    0      0    ⎥   ⎢   0     0     0   ⎥
⎢-0.00   0.00   0    0.00  -0.00 ⎥   ⎢   1     0     0   ⎥
⎣  0      0     0    0      0    ⎦   ⎣   0     0     1   ⎦
```

Configuration 10:
```
⎡ 1.14  -0.18   0    0.09  -0.04 ⎤   ⎡ -3.48    0   -11.29  ⎤
⎢-0.18   1.33   0   -0.06   0.12 ⎥   ⎢  4.05    0    20.72  ⎥
⎢  0      0     1    0      0    ⎥   ⎢   0      0     0      ⎥
⎢ 5.60  -7.15   0    2.64  -1.59 ⎥   ⎢-139.25   0  -451.67  ⎥
⎣-7.17  13.27   0   -2.59   3.93 ⎦   ⎣ 161.95   0   828.72  ⎦

⎡ 0.05  -0.00   0    0.00  -0.00 ⎤   ⎡ -0.07   0   -0.25 ⎤
⎢-0.00   0.06   0    0.00  -0.00 ⎥   ⎢  0.09   0    0.48 ⎥
⎢  0      0     0    0      0    ⎥   ⎢   0     0     0   ⎥
⎢-0.00   0.00   0    0.00  -0.00 ⎥   ⎢   1     0     0   ⎥
⎣  0      0     0    0      0    ⎦   ⎣   0     0     1   ⎦
```

Configuration 11:
```
⎡ 1.12  -0.16   0    0.09  -0.04 ⎤   ⎡ -2.98    0   -10.07  ⎤
⎢-0.16   1.31   0   -0.06   0.12 ⎥   ⎢  3.48    0    19.32  ⎥
⎢  0      0     1    0      0    ⎥   ⎢   0      0     0      ⎥
⎢ 4.83  -6.39   0    2.44  -1.42 ⎥   ⎢-119.30   0  -402.80  ⎥
⎣-6.29  12.40   0   -2.36   3.74 ⎦   ⎣ 139.19   0   772.92  ⎦

⎡ 0.05  -0.00   0    0.00  -0.00 ⎤   ⎡ -0.07   0   -0.25 ⎤
⎢-0.00   0.06   0    0.00  -0.00 ⎥   ⎢  0.09   0    0.48 ⎥
⎢  0      0     0    0      0    ⎥   ⎢   0     0     0   ⎥
⎢-0.00   0.00   0    0.00  -0.00 ⎥   ⎢   1     0     0   ⎥
⎣  0      0     0    0      0    ⎦   ⎣   0     0     1   ⎦
```

Configuration 12:
```
⎡ 1.12  -0.16   0    0.09  -0.04 ⎤   ⎡ -3.45    0   -10.95  ⎤
⎢-0.16   1.31   0   -0.06   0.12 ⎥   ⎢  3.97    0    19.85  ⎥
⎢  0      0     1    0      0    ⎥   ⎢   0      0     0      ⎥
⎢ 4.83  -6.39   0    2.44  -1.42 ⎥   ⎢-138.07   0  -437.91  ⎥
⎣-6.29  12.40   0   -2.36   3.74 ⎦   ⎣ 158.80   0   794.02  ⎦

⎡ 0.05  -0.00   0    0.00  -0.00 ⎤   ⎡ -0.08   0   -0.26 ⎤
⎢-0.00   0.06   0    0.00  -0.00 ⎥   ⎢  0.09   0    0.48 ⎥
⎢  0      0     0    0      0    ⎥   ⎢   0     0     0   ⎥
⎢-0.00   0.00   0    0.00  -0.00 ⎥   ⎢   1     0     0   ⎥
⎣  0      0     0    0      0    ⎦   ⎣   0     0     1   ⎦
```

Configuration 13:
```
⎡ 1.14  -0.17   0    0.09  -0.04 ⎤   ⎡ -3.45    0   -10.95  ⎤
⎢-0.17   1.32   0   -0.06   0.12 ⎥   ⎢  3.97    0    19.85  ⎥
⎢  0      0     1    0      0    ⎥   ⎢   0      0     0      ⎥
⎢ 5.53  -6.93   0    2.60  -1.54 ⎥   ⎢-138.07   0  -437.91  ⎥
⎣-6.99  12.71   0   -2.50   3.81 ⎦   ⎣ 158.80   0   794.02  ⎦

⎡ 0.05  -0.00   0    0.00  -0.00 ⎤   ⎡ -0.08   0   -0.26 ⎤
⎢-0.00   0.06   0    0.00  -0.00 ⎥   ⎢  0.09   0    0.48 ⎥
⎢  0      0     0    0      0    ⎥   ⎢   0     0     0   ⎥
⎢-0.00   0.00   0    0.00  -0.00 ⎥   ⎢   1     0     0   ⎥
⎣  0      0     0    0      0    ⎦   ⎣   0     0     1   ⎦
```

Configuration 14:
```
⎡ 1.14  -0.17   0    0.09  -0.04 ⎤   ⎡ -3.08    0   -10.30  ⎤
⎢-0.17   1.32   0   -0.06   0.12 ⎥   ⎢  3.63    0    19.32  ⎥
⎢  0      0     1    0      0    ⎥   ⎢   0      0     0      ⎥
⎢ 5.53  -6.93   0    2.60  -1.54 ⎥   ⎢-138.07   0  -437.91  ⎥
⎣-6.99  12.71   0   -2.50   3.81 ⎦   ⎣ 158.80   0   794.02  ⎦

⎡ 0.05  -0.00   0    0.00  -0.00 ⎤   ⎡ -0.09   0   -0.27 ⎤
⎢-0.00   0.06   0    0.00  -0.00 ⎥   ⎢  0.10   0    0.50 ⎥
⎢  0      0     0    0      0    ⎥   ⎢   0     0     0   ⎥
⎢-0.00   0.00   0    0.00  -0.00 ⎥   ⎢   1     0     0   ⎥
⎣  0      0     0    0      0    ⎦   ⎣   0     0     1   ⎦
```

Configuration 15:
```
⎡ 1.12  -0.16   0    0.09  -0.04 ⎤   ⎡ -3.08    0   -10.30  ⎤
⎢-0.16   1.31   0   -0.06   0.12 ⎥   ⎢  3.63    0    19.32  ⎥
⎢  0      0     1    0      0    ⎥   ⎢   0      0     0      ⎥
⎢ 4.98  -6.53   0    2.48  -1.45 ⎥   ⎢-123.13   0  -412.03  ⎥
⎣-6.49  12.39   0   -2.39   3.74 ⎦   ⎣ 145.17   0   772.79  ⎦

⎡ 0.05  -0.00   0    0.00  -0.00 ⎤   ⎡ -0.09   0   -0.27 ⎤
⎢-0.00   0.06   0    0.00  -0.00 ⎥   ⎢  0.10   0    0.50 ⎥
⎢  0      0     0    0      0    ⎥   ⎢   0     0     0   ⎥
⎢-0.00   0.00   0    0.00  -0.00 ⎥   ⎢   1     0     0   ⎥
⎣  0      0     0    0      0    ⎦   ⎣   0     0     1   ⎦
```

Configuration 16:
```
⎡ 1.12  -0.16   0    0.09  -0.04 ⎤   ⎡ -3.08    0   -10.30  ⎤
⎢-0.16   1.31   0   -0.06   0.12 ⎥   ⎢  3.63    0    19.32  ⎥
⎢  0      0     1    0      0    ⎥   ⎢   0      0     0      ⎥
⎢ 4.98  -6.53   0    2.48  -1.45 ⎥   ⎢-123.13   0  -412.03  ⎥
⎣-6.49  12.39   0   -2.39   3.74 ⎦   ⎣ 145.17   0   772.79  ⎦

⎡ 0.05  -0.00   0    0.00  -0.00 ⎤   ⎡ -0.08   0   -0.26 ⎤
⎢-0.00   0.06   0    0.00  -0.00 ⎥   ⎢  0.09   0    0.48 ⎥
⎢  0      0     0    0      0    ⎥   ⎢   0     0     0   ⎥
⎢-0.00   0.00   0    0.00  -0.00 ⎥   ⎢   1     0     0   ⎥
⎣  0      0     0    0      0    ⎦   ⎣   0     0     1   ⎦
```

Table 8.11. Uarm II model for configuration AP$_1$A

17

$$\begin{bmatrix} 1.01 & 0 & -0.01 & 0.05 & 0 & -0.00 \\ 0 & 1 & 0 & 0 & 0 & 0.05 \\ -0.02 & 0 & 1.06 & -0.00 & 0 & -0.01 \\ 0.34 & 0 & -0.22 & 1.08 & 0 & 0 \\ 0 & 0 & 0 & 0 & 1 & 1.11 \\ -0.73 & 0 & 2.23 & -0.17 & 0 & 0 \end{bmatrix} \quad \begin{bmatrix} 0.16 & 0 & -0.49 \\ 0 & 0 & 0 \\ -0.50 & 0 & 4.20 \\ 6.45 & 0 & -19.68 \\ 0 & 0 & 0 \\ -19.88 & 0 & 168.02 \end{bmatrix} \quad \begin{bmatrix} 0.05 & 0 & -0.00 & 0.00 & 0 & -0.00 \\ 0 & 0 & 0 & 0 & 0 & 0 \\ -0.00 & 0 & 0.05 & -0.00 & 0 & 0.00 \\ 0 & 0 & 0 & 0 & 0 & 0 \\ 0 & 0 & 0 & 0 & 0 & 0 \end{bmatrix} \quad \begin{bmatrix} 0.00 & 0 & -0.01 \\ 0 & 0 & 0 \\ -0.01 & 0 & 0.11 \\ 0 & 0 & 0 \\ 1 & 0 & 0 \\ 0 & 0 & 1 \end{bmatrix}$$

18

$$\begin{bmatrix} 1.01 & 0 & -0.01 & 0.05 & 0 & -0.00 \\ 0 & 1 & 0 & 0 & 0 & 0.05 \\ -0.02 & 0 & 1.06 & -0.00 & 0 & -0.01 \\ 0.34 & 0 & -0.22 & 1.08 & 0 & 0 \\ 0 & 0 & 0 & 0 & 1 & 1.11 \\ -0.73 & 0 & 2.23 & -0.17 & 0 & 0 \end{bmatrix} \quad \begin{bmatrix} 0.16 & 0 & -0.49 \\ 0 & 0 & 0 \\ -0.50 & 0 & 4.20 \\ 6.45 & 0 & -19.68 \\ 0 & 0 & 0 \\ -19.88 & 0 & 168.02 \end{bmatrix} \quad \begin{bmatrix} 0.05 & 0 & -0.00 & 0.00 & 0 & -0.00 \\ 0 & 0 & 0 & 0 & 0 & 0 \\ -0.00 & 0 & 0.05 & -0.00 & 0 & 0.00 \\ 0 & 0 & 0 & 0 & 0 & 0 \\ 0 & 0 & 0 & 0 & 0 & 0 \end{bmatrix} \quad \begin{bmatrix} 0.00 & 0 & -0.01 \\ 0 & 0 & 0 \\ -0.01 & 0 & 0.11 \\ 0 & 0 & 0 \\ 1 & 0 & 0 \\ 0 & 0 & 1 \end{bmatrix}$$

19

$$\begin{bmatrix} 1.01 & 0 & -0.01 & 0.05 & 0 & -0.00 \\ 0 & 1 & 0 & 0 & 0 & 0.05 \\ -0.02 & 0 & 1.06 & -0.00 & 0 & -0.01 \\ 0.34 & 0 & -0.22 & 1.08 & 0 & 0 \\ 0 & 0 & 0 & 0 & 1 & 1.11 \\ -0.72 & 0 & 2.21 & -0.17 & 0 & 0 \end{bmatrix} \quad \begin{bmatrix} 0.16 & 0 & -0.49 \\ 0 & 0 & 0 \\ -0.49 & 0 & 4.16 \\ 6.48 & 0 & -19.47 \\ 0 & 0 & 0 \\ -19.67 & 0 & 166.45 \end{bmatrix} \quad \begin{bmatrix} 0.05 & 0 & -0.00 & 0.00 & 0 & -0.00 \\ 0 & 0 & 0 & 0 & 0 & 0 \\ -0.00 & 0 & 0.05 & -0.00 & 0 & 0.00 \\ 0 & 0 & 0 & 0 & 0 & 0 \\ 0 & 0 & 0 & 0 & 0 & 0 \end{bmatrix} \quad \begin{bmatrix} 0.00 & 0 & -0.01 \\ 0 & 0 & 0 \\ -0.01 & 0 & 0.10 \\ 0 & 0 & 0 \\ 1 & 0 & 0 \\ 0 & 0 & 1 \end{bmatrix}$$

20

$$\begin{bmatrix} 1.01 & 0 & -0.01 & 0.05 & 0 & -0.00 \\ 0 & 1 & 0 & 0 & 0 & 0.05 \\ -0.02 & 0 & 1.06 & -0.00 & 0 & -0.01 \\ 0.34 & 0 & -0.22 & 1.08 & 0 & 0 \\ 0 & 0 & 0 & 0 & 1 & 1.11 \\ -0.72 & 0 & 2.21 & -0.17 & 0 & 0 \end{bmatrix} \quad \begin{bmatrix} 0.16 & 0 & -0.47 \\ 0 & 0 & 0 \\ -0.49 & 0 & 4.16 \\ 6.48 & 0 & -18.80 \\ 0 & 0 & 0 \\ -19.67 & 0 & 166.45 \end{bmatrix} \quad \begin{bmatrix} 0.05 & 0 & -0.00 & 0.00 & 0 & -0.00 \\ 0 & 0 & 0 & 0 & 0 & 0 \\ -0.00 & 0 & 0.05 & -0.00 & 0 & 0.00 \\ 0 & 0 & 0 & 0 & 0 & 0 \\ 0 & 0 & 0 & 0 & 0 & 0 \end{bmatrix} \quad \begin{bmatrix} 0.00 & 0 & -0.01 \\ 0 & 0 & 0 \\ -0.01 & 0 & 0.10 \\ 0 & 0 & 0 \\ 1 & 0 & 0 \\ 0 & 0 & 1 \end{bmatrix}$$

21

$$\begin{bmatrix} 1.01 & 0 & -0.01 & 0.05 & 0 & -0.00 \\ 0 & 1 & 0 & 0 & 0 & 0.05 \\ -0.02 & 0 & 1.05 & -0.00 & 0 & -0.01 \\ 0.34 & 0 & -0.21 & 1.08 & 0 & 0 \\ 0 & 0 & 0 & 0 & 1 & 1.10 \\ -0.69 & 0 & 2.18 & -0.16 & 0 & 0 \end{bmatrix} \quad \begin{bmatrix} 0.16 & 0 & -0.47 \\ 0 & 0 & 0 \\ -0.48 & 0 & 4.10 \\ 6.33 & 0 & -18.80 \\ 0 & 0 & 0 \\ -19.00 & 0 & 163.84 \end{bmatrix} \quad \begin{bmatrix} 0.05 & 0 & -0.00 & 0.00 & 0 & -0.00 \\ 0 & 0 & 0 & 0 & 0 & 0 \\ -0.00 & 0 & 0.05 & -0.00 & 0 & 0.00 \\ 0 & 0 & 0 & 0 & 0 & 0 \\ 0 & 0 & 0 & 0 & 0 & 0 \end{bmatrix} \quad \begin{bmatrix} 0.00 & 0 & -0.01 \\ 0 & 0 & 0 \\ -0.01 & 0 & 0.10 \\ 0 & 0 & 0 \\ 1 & 0 & 0 \\ 0 & 0 & 1 \end{bmatrix}$$

22

$$\begin{bmatrix} 1.01 & 0 & -0.01 & 0.05 & 0 & -0.00 \\ 0 & 1 & 0 & 0 & 0 & 0.05 \\ -0.02 & 0 & 1.05 & -0.00 & 0 & -0.01 \\ 0.34 & 0 & -0.21 & 1.08 & 0 & 0 \\ 0 & 0 & 0 & 0 & 1 & 1.10 \\ -0.69 & 0 & 2.18 & -0.16 & 0 & 0 \end{bmatrix} \quad \begin{bmatrix} 0.16 & 0 & -0.47 \\ 0 & 0 & 0 \\ -0.48 & 0 & 4.10 \\ 6.33 & 0 & -18.80 \\ 0 & 0 & 0 \\ -19.00 & 0 & 163.84 \end{bmatrix} \quad \begin{bmatrix} 0.05 & 0 & -0.00 & 0.00 & 0 & -0.00 \\ 0 & 0 & 0 & 0 & 0 & 0 \\ -0.00 & 0 & 0.05 & -0.00 & 0 & 0.00 \\ 0 & 0 & 0 & 0 & 0 & 0 \\ 0 & 0 & 0 & 0 & 0 & 0 \end{bmatrix} \quad \begin{bmatrix} 0.00 & 0 & -0.01 \\ 0 & 0 & 0 \\ -0.01 & 0 & 0.10 \\ 0 & 0 & 0 \\ 1 & 0 & 0 \\ 0 & 0 & 1 \end{bmatrix}$$

23

$$\begin{bmatrix} 1.01 & 0 & -0.01 & 0.05 & 0 & -0.00 \\ 0 & 1 & 0 & 0 & 0 & 0.05 \\ -0.02 & 0 & 1.05 & -0.00 & 0 & -0.01 \\ 0.34 & 0 & -0.20 & 1.08 & 0 & 0 \\ 0 & 0 & 0 & 0 & 1 & 1.10 \\ -0.67 & 0 & 2.14 & -0.16 & 0 & 0 \end{bmatrix} \quad \begin{bmatrix} 0.16 & 0 & -0.46 \\ 0 & 0 & 0 \\ -0.46 & 0 & 4.02 \\ 6.32 & 0 & -18.29 \\ 0 & 0 & 0 \\ -18.49 & 0 & 160.88 \end{bmatrix} \quad \begin{bmatrix} 0.05 & 0 & -0.00 & 0.00 & 0 & -0.00 \\ 0 & 0 & 0 & 0 & 0 & 0 \\ -0.00 & 0 & 0.05 & -0.00 & 0 & 0.00 \\ 0 & 0 & 0 & 0 & 0 & 0 \\ 0 & 0 & 0 & 0 & 0 & 0 \end{bmatrix} \quad \begin{bmatrix} 0.00 & 0 & -0.01 \\ 0 & 0 & 0 \\ -0.01 & 0 & 0.10 \\ 0 & 0 & 0 \\ 1 & 0 & 0 \\ 0 & 0 & 1 \end{bmatrix}$$

24

$$\begin{bmatrix} 1.01 & 0 & -0.01 & 0.05 & 0 & -0.00 \\ 0 & 1 & 0 & 0 & 0 & 0.05 \\ -0.02 & 0 & 1.05 & -0.00 & 0 & -0.01 \\ 0.34 & 0 & -0.20 & 1.08 & 0 & 0 \\ 0 & 0 & 0 & 0 & 1 & 1.10 \\ -0.67 & 0 & 2.14 & -0.16 & 0 & 0 \end{bmatrix} \quad \begin{bmatrix} 0.16 & 0 & -0.46 \\ 0 & 0 & 0 \\ -0.46 & 0 & 4.02 \\ 6.32 & 0 & -18.29 \\ 0 & 0 & 0 \\ -18.49 & 0 & 160.88 \end{bmatrix} \quad \begin{bmatrix} 0.05 & 0 & -0.00 & 0.00 & 0 & -0.00 \\ 0 & 0 & 0 & 0 & 0 & 0 \\ -0.00 & 0 & 0.05 & -0.00 & 0 & 0.00 \\ 0 & 0 & 0 & 0 & 0 & 0 \\ 0 & 0 & 0 & 0 & 0 & 0 \end{bmatrix} \quad \begin{bmatrix} 0.00 & 0 & -0.01 \\ 0 & 0 & 0 \\ -0.01 & 0 & 0.10 \\ 0 & 0 & 0 \\ 1 & 0 & 0 \\ 0 & 0 & 1 \end{bmatrix}$$

Table 8.12. $H\infty$-controller for Uarm II

i	F_i
1	$\begin{bmatrix} -0.052 & -0.017 & 0.003 & -0.040 & -0.016 & -0.003 \\ -0.008 & -0.028 & -0.014 & -0.013 & -0.015 & -0.004 \\ -0.001 & -0.004 & -0.027 & -0.002 & -0.003 & -0.005 \end{bmatrix}$
2	$\begin{bmatrix} -0.052 & -0.017 & 0.003 & -0.040 & -0.016 & -0.003 \\ -0.008 & -0.028 & -0.014 & -0.013 & -0.015 & -0.004 \\ -0.001 & -0.004 & -0.027 & -0.002 & -0.003 & -0.005 \end{bmatrix}$
3	$\begin{bmatrix} -0.053 & -0.016 & 0.002 & -0.040 & -0.016 & -0.003 \\ -0.007 & -0.029 & -0.014 & -0.013 & -0.015 & -0.004 \\ -0.001 & -0.004 & -0.027 & -0.002 & -0.003 & -0.005 \end{bmatrix}$
4	$\begin{bmatrix} -0.053 & -0.016 & 0.002 & -0.040 & -0.016 & -0.003 \\ -0.007 & -0.029 & -0.014 & -0.013 & -0.015 & -0.004 \\ -0.001 & -0.004 & -0.027 & -0.002 & -0.003 & -0.005 \end{bmatrix}$
5	$\begin{bmatrix} -0.052 & -0.017 & 0.003 & -0.040 & -0.016 & -0.003 \\ -0.008 & -0.028 & -0.015 & -0.013 & -0.015 & -0.004 \\ -0.001 & -0.004 & -0.027 & -0.002 & -0.003 & -0.005 \end{bmatrix}$
6	$\begin{bmatrix} -0.052 & -0.017 & 0.003 & -0.040 & -0.016 & -0.003 \\ -0.008 & -0.028 & -0.015 & -0.013 & -0.015 & -0.004 \\ -0.001 & -0.004 & -0.027 & -0.002 & -0.003 & -0.005 \end{bmatrix}$
7	$\begin{bmatrix} -0.053 & -0.016 & 0.003 & -0.040 & -0.015 & -0.003 \\ -0.007 & -0.029 & -0.015 & -0.013 & -0.015 & -0.004 \\ -0.001 & -0.004 & -0.027 & -0.002 & -0.003 & -0.005 \end{bmatrix}$
8	$\begin{bmatrix} -0.053 & -0.016 & 0.003 & -0.040 & -0.015 & -0.003 \\ -0.007 & -0.029 & -0.015 & -0.013 & -0.015 & -0.004 \\ -0.001 & -0.004 & -0.027 & -0.002 & -0.003 & -0.005 \end{bmatrix}$

i	F_i
9	$\begin{bmatrix} 0.002 & 0.066 & 0.024 & 0.001 & 0.014 & 0.006 \\ 0 & 0 & 0 & 0 & 0 & 0 \\ -0.001 & -0.001 & -0.028 & -0.000 & 0.001 & -0.006 \end{bmatrix}$
10	$\begin{bmatrix} 0.002 & 0.066 & 0.024 & 0.001 & 0.014 & 0.006 \\ 0 & 0 & 0 & 0 & 0 & 0 \\ -0.001 & -0.001 & -0.028 & -0.000 & 0.001 & -0.006 \end{bmatrix}$
11	$\begin{bmatrix} 0.002 & 0.067 & 0.023 & 0.001 & 0.015 & 0.006 \\ 0 & 0 & 0 & 0 & 0 & 0 \\ -0.001 & -0.000 & -0.028 & -0.000 & 0.001 & -0.006 \end{bmatrix}$
12	$\begin{bmatrix} 0.002 & 0.067 & 0.023 & 0.001 & 0.015 & 0.006 \\ 0 & 0 & 0 & 0 & 0 & 0 \\ -0.001 & -0.000 & -0.028 & -0.000 & 0.001 & -0.006 \end{bmatrix}$
13	$\begin{bmatrix} 0.002 & 0.066 & 0.024 & 0.001 & 0.014 & 0.006 \\ 0 & 0 & 0 & 0 & 0 & 0 \\ -0.001 & -0.001 & -0.028 & -0.000 & 0.001 & -0.006 \end{bmatrix}$
14	$\begin{bmatrix} 0.002 & 0.066 & 0.024 & 0.001 & 0.014 & 0.006 \\ 0 & 0 & 0 & 0 & 0 & 0 \\ -0.001 & -0.001 & -0.028 & -0.000 & 0.001 & -0.006 \end{bmatrix}$
15	$\begin{bmatrix} 0.002 & 0.067 & 0.024 & 0.001 & 0.015 & 0.006 \\ 0 & 0 & 0 & 0 & 0 & 0 \\ -0.001 & -0.000 & -0.028 & -0.000 & 0.001 & -0.006 \end{bmatrix}$
16	$\begin{bmatrix} 0.002 & 0.067 & 0.024 & 0.001 & 0.015 & 0.006 \\ 0 & 0 & 0 & 0 & 0 & 0 \\ -0.001 & -0.000 & -0.028 & -0.000 & 0.001 & -0.006 \end{bmatrix}$

i	F_i
17	$\begin{bmatrix} -0.131 & -0.000 & -0.016 & -0.069 & -0.000 & -0.008 \\ 0 & 0 & 0 & 0 & 0 & 0 \\ -0.006 & -0.001 & -0.045 & -0.006 & -0.000 & -0.006 \end{bmatrix}$
18	$\begin{bmatrix} -0.131 & -0.000 & -0.016 & -0.069 & -0.000 & -0.008 \\ 0 & 0 & 0 & 0 & 0 & 0 \\ -0.006 & -0.001 & -0.045 & -0.006 & -0.000 & -0.006 \end{bmatrix}$
19	$\begin{bmatrix} -0.131 & -0.000 & -0.016 & -0.069 & -0.000 & -0.008 \\ 0 & 0 & 0 & 0 & 0 & 0 \\ -0.006 & -0.001 & -0.045 & -0.006 & -0.000 & -0.006 \end{bmatrix}$
20	$\begin{bmatrix} -0.131 & -0.000 & -0.016 & -0.069 & -0.000 & -0.008 \\ 0 & 0 & 0 & 0 & 0 & 0 \\ -0.006 & -0.001 & -0.045 & -0.006 & -0.000 & -0.006 \end{bmatrix}$
21	$\begin{bmatrix} -0.131 & -0.000 & -0.015 & -0.069 & -0.000 & -0.007 \\ 0 & 0 & 0 & 0 & 0 & 0 \\ -0.006 & -0.001 & -0.045 & -0.006 & -0.000 & -0.006 \end{bmatrix}$
22	$\begin{bmatrix} -0.131 & -0.000 & -0.015 & -0.069 & -0.000 & -0.007 \\ 0 & 0 & 0 & 0 & 0 & 0 \\ -0.006 & -0.001 & -0.045 & -0.006 & -0.000 & -0.006 \end{bmatrix}$
23	$\begin{bmatrix} -0.131 & -0.000 & -0.015 & -0.069 & -0.000 & -0.007 \\ 0 & 0 & 0 & 0 & 0 & 0 \\ -0.006 & -0.001 & -0.045 & -0.006 & -0.000 & -0.006 \end{bmatrix}$
24	$\begin{bmatrix} -0.131 & -0.000 & -0.015 & -0.069 & -0.000 & -0.007 \\ 0 & 0 & 0 & 0 & 0 & 0 \\ -0.006 & -0.001 & -0.045 & -0.006 & -0.000 & -0.006 \end{bmatrix}$

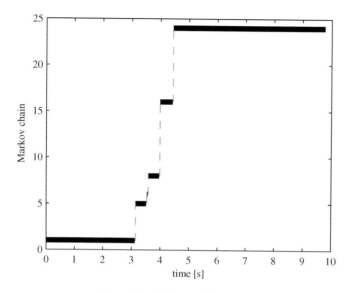

Fig. 8.7. Markov chain state

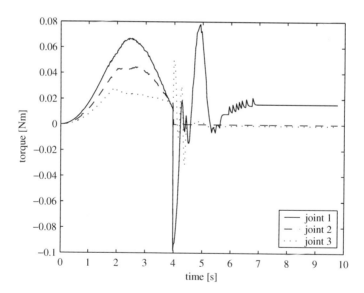

Fig. 8.8. Applied torque on each joint

8.4 Examples of Linear Filtering with $\theta(k)$ Unknown

In this section we present some numerical examples with the stationary and robust LMMSE filters presented in Sections 5.4 and 5.5. For the sake of comparison, we also present simulations of the IMM filter [32], which is a time-variant suboptimal filter for MJLS.

8.4.1 Stationary LMMSE Filter

In order to implement the stationary LMMSE filter, we solve the system of linear equations (5.58) with unique solution $Q_i, i \in \mathbb{N}$, we plug $Q = (Q_1, \ldots, Q_N)$ into (5.60) and solve the corresponding ARE to obtain P. The stationary LMMSE estimator $\hat{x}(k|k)$ is given by (5.43), where $\hat{z}(k|k)$ satisfies the time-invariant recursive equations (5.45) replacing $\widetilde{Z}(k|k-1)$ by P. Notice that all these calculations can be performed off-line. Consider a scalar MJLS described by the following equations,

$$x(k+1) = a_{\theta(k)}x(k) + g_{\theta(k)}w(k)$$
$$y(k) = l_{\theta(k)}x(k) + h_{\theta(k)}w(k) \tag{8.21}$$

with $x(0)$ Gaussian with mean 10 and variance 10. $\theta(k) \in \{1, 2\}$, and $\{w(k)\}$ is an independent noise sequence, and $\pi_1(0) = \pi_2(0) = 0.5$. Table 8.13 presents 6 cases with different parameters a_i, g_i, l_i, h_i and p_{ij}. For each case, 4000 Monte Carlo simulations were performed and both filters (stationary LMMSE and IMM) were compared under the same conditions. The results obtained for

Table 8.13. Simulation parameters

Cases	p_{11}	p_{22}	a_1	a_2	g_1	g_2	l_1	l_2	h_1	h_2
1	0.975	0.95	0.995	0.99	$[0.1\ 0]$	$[0.1\ 0]$	1.0	1.0	$[0\ 5.0]$	$[0\ 5.0]$
2	0.995	0.99	0.995	0.995	$[0.5\ 0]$	$[0.5\ 0]$	1.0	0.8	$[0\ 0.8]$	$[0\ 0.8]$
3	0.975	0.95	0.995	0.995	$[0.1\ 0]$	$[5.0\ 0]$	1.0	1.0	$[0\ 1.0]$	$[0\ 1.0]$
4	0.975	0.95	0.995	0.25	$[1.0\ 0]$	$[1.0\ 0]$	1.0	1.0	$[0\ 1.0]$	$[0\ 1.0]$
5	0.975	0.95	0.995	0.25	$[0.1\ 0]$	$[0.1\ 0]$	1.0	1.0	$[0\ 5.0]$	$[0\ 5.0]$
6	0.975	0.95	0.995	0.25	$[0.1\ 0]$	$[5.0\ 0]$	1.0	1.0	$[0\ 5.0]$	$[0\ 5.0]$

the above configuration are in Figure 8.9, showing the square root of the mean square error (rms) for each of the 6 cases studied with three types of noise distribution: normal, uniform and exponential. Both IMM and stationary LMMSE filter results are presented in the figure.

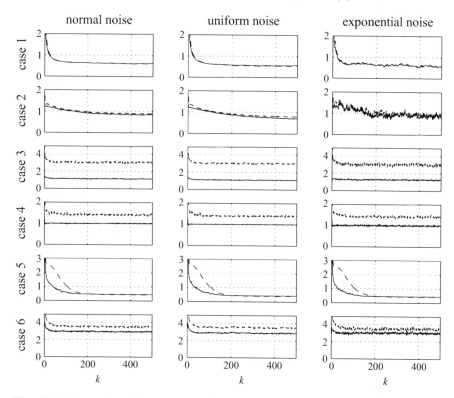

Fig. 8.9. Comparison between IMM (solid line) and stationary LMMSE (dashed line) filters

8.4.2 Robust LMMSE Filter

We now investigate the performance of the robust LMMSE filter for the nominal system. As in the previous subsection, we included the IMM filter performance under the same conditions for comparison. The robust LMMSE filter parameters for the case with uncertainties are as in (5.82) of Theorem 5.16, using the optimal solution of the LMI optimization problem posed in Theorem 5.19 with restrictions (5.86), (5.87) and (5.88). For the scalar MJLS (8.21), we consider that matrices A, G, L, H can be written as a convex combination of the vertices A^j, G^j, L^j, H^j respectively with $j = 1, \ldots, \rho$ for ρ vertices. Table 8.14 shows nominal parameters used to design the IMM filters for 4 cases. Uncertainties are considered only on matrix A, defined in (5.52). Table 8.15 presents parameters a_i^j, $i \in \mathbb{N}, j = 1, \cdots, \rho$, associated to the corresponding extreme points of the convex set, that is, the values a_1^1, a_1^2, a_2^1, a_2^2 for the robust LMMSE filter. Parameters $\bar{a}_1$ and $\bar{a}_2$ are the effective values of a_1 and a_2. We have $\rho = 4$ in (5.88), with A^j taking the following values:

$$A^1 = \begin{bmatrix} p_{11}a_1^1 & p_{12}a_2^1 \\ p_{21}a_1^1 & p_{22}a_2^1 \end{bmatrix} \qquad A^2 = \begin{bmatrix} p_{11}a_1^1 & p_{12}a_2^2 \\ p_{21}a_1^1 & p_{22}a_2^2 \end{bmatrix}$$

$$A^3 = \begin{bmatrix} p_{11}a_1^2 & p_{12}a_2^1 \\ p_{21}a_1^2 & p_{22}a_2^1 \end{bmatrix} \qquad A^4 = \begin{bmatrix} p_{11}a_1^2 & p_{12}a_2^2 \\ p_{21}a_1^2 & p_{22}a_2^2 \end{bmatrix}.$$

Table 8.14. Simulation parameters

Cases	p_{11}	p_{22}	a_1	a_2	g_1	g_2	l_1	l_2	h_1	h_2
1	0.975	0.95	0.995	0.99	$[0.1\ 0]$	$[0.1\ 0]$	1.0	1.0	$[0\ 5.0]$	$[0\ 5.0]$
2	0.995	0.99	0.8	0.75	$[0.5\ 0]$	$[0.5\ 0]$	1.0	0.8	$[0\ 0.8]$	$[0\ 0.8]$
3	0.975	0.95	0.3	0.3	$[0.1\ 0]$	$[5.0\ 0]$	1.0	1.0	$[0\ 1.0]$	$[0\ 1.0]$
4	0.975	0.95	0.5	0.199	$[1.0\ 0]$	$[1.0\ 0]$	1.0	1.0	$[0\ 1.0]$	$[0\ 1.0]$

Table 8.15. Extreme values for the parameters a_i

Cases	$\bar{a}_1$	$\bar{a}_2$	a_1^1	a_1^2	a_2^1	a_2^2
1	-0.8	-0.796	0.995	-0.995	0.99	-0.99
2	-0.4	-0.398	0.8	-0.995	0.75	-0.99
3	-0.95	-0.95	0.3	-0.95	0.3	-0.95
4	-0.5	-0.199	0.5	-0.5	0.199	-0.199

Figure 8.10 shows the mean square root errors for 4000 Monte Carlo simulations.

There are some important differences between the IMM filter and the robust LMMSE filter. The LMI-based design method of the latter may produce conservative filters, with poorer performance than the IMM filter. On the other hand the robust LMMSE filter is a time-invariant mean square stable filter, therefore not requiring on-line calculations for its implementation, while the IMM filter is time-variant and provides no stability guarantees.

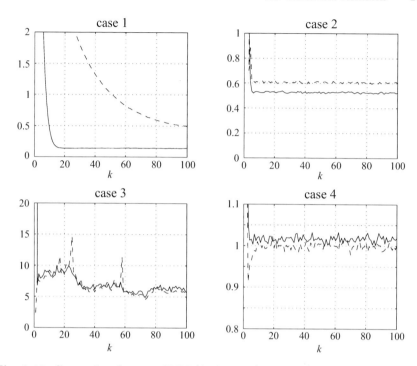

Fig. 8.10. Comparison between IMM (dashed line) and robust LMMSE (solid line) filters

8.5 Historical Remarks

Application oriented works involving MJLS modeling appeared as early as 1975, with the work of W.P. Blair and D.D. Sworder ([27] and [28]). A historical perspective on the subject shows that the applications reported evolved apace with the theory. The early works dealt basically with linear quadratic regulators, while more recent ones employ the newest advances (for example, [200] and [202]).

The works of M.A. Rami and L. El Ghaoui ([191] and later [192]) started a new and prolific trend in the area: the use of LMIs and related techniques. Authors such as E.K. Boukas, P. Shi and many others successfully applied the LMI framework to problems involving robustness, time delays, constrained control and other issues ([36], [38], and [199] are a small sample of such works).

A

Coupled Algebraic Riccati Equations

The results in this appendix are concerned about coupled Riccati difference equations and the associated coupled algebraic Riccati equations, which are used throughout this book. We deal, essentially, with the existence of solutions and asymptotic convergence. Regarding the existence of solutions, we are particularly interested in maximal and stabilizing solutions. The appeal of the maximal solution has to do with the fact that it can be obtained numerically via a certain LMI optimization problem. Although in control and filtering applications the interest lies essentially in the stabilizing solution, it is shown that the two concepts of solution coincide whenever the stabilizing solution exists.

A.1 Duality Between the Control and Filtering CARE

In this appendix we study the asymptotic behavior of a set of coupled Riccati difference equations, and some properties of the corresponding stationary solution, which satisfies a set of coupled algebraic Riccati equations (CARE). We concentrate on the control CARE but, as will be seen, the filtering CARE can be regarded as dual of the control CARE, provided that duality is properly defined. In order to do this, we have to generalize the definition of the control CARE, and show that it can be written in two equivalent forms. The second form will be useful to establish the duality with the filtering equations.

We consider a set of positive numbers $s = \{s_{ij}\}$, $0 \leq s_{ij} \leq 1$, $i, j \in \mathbb{N}$ and define, for $V = (V_1, \ldots, V_N) \in \mathbb{H}^{m,n}$, $V^* \in \mathbb{H}^{n,m}$ as $V^* = (V_1^*, \ldots, V_N^*)$, and $s^* = \{s_{ij}^*\}$ as $s_{ij}^* = s_{ji}$. We redefine the operators $\mathcal{E}(.) = (\mathcal{E}_1(.), \ldots, \mathcal{E}_N(.))$, $\mathcal{L}(.) = (\mathcal{L}_1(.), \ldots, \mathcal{L}_N(.))$, $\mathcal{T}(.) = (\mathcal{T}_1(.), \ldots, \mathcal{T}_N(.))$, $\mathcal{J}(.) = (\mathcal{J}_1(.), \ldots, \mathcal{J}_N(.))$, $\mathcal{V}(.) = (\mathcal{V}_1(.), \ldots, \mathcal{V}_N(.))$ in (3.6), (3.7), (3.8), (3.9), (3.10) respectively replacing p_{ij} by s_{ij}. Thus, the only difference with respect to (3.6)–(3.10) is that the sum over j of s_{ij} is not necessarily one. This extra generality will be useful to establish the duality between the control and filtering coupled Riccati equations. Notice that, from Remark 3.3, $r_\sigma(\mathcal{L}) = r_\sigma(\mathcal{T}) = r_\sigma(\mathcal{J}) = r_\sigma(\mathcal{V})$.

We recall now the definitions of (mean square) stabilizability and detectability (see Definitions 3.40 and 3.41 respectively), conveniently modified for our purposes.

Definition A.1. Let $A = (A_1, \ldots, A_N) \in \mathbb{H}^n$, $B = (B_1, \ldots, B_N) \in \mathbb{H}^{m,n}$. We say that (A, B, s) is stabilizable if there is $F = (F_1, \ldots, F_N) \in \mathbb{H}^{n,m}$ such that $r_\sigma(\mathcal{T}) < 1$ when $\Gamma_i = A_i + B_i F_i$ in (3.7) for $i \in \mathbb{N}$. In this case, F is said to stabilize (A, B, s).

Definition A.2. Let $L = (L_1, \ldots, L_N) \in \mathbb{H}^{n,p}$. We say that (s, L, A) is detectable if there is $M = (M_1, \ldots, M_N) \in \mathbb{H}^{p,n}$ such that $r_\sigma(\mathcal{T}) < 1$ when $\Gamma_i = A_i + M_i L_i$ in (3.7) for $i \in \mathbb{N}$. In this case, M is said to stabilize (s, L, A).

Therefore (A, B, s) stabilizability is equivalent, when $s_{ij} = p_{ij}$, to mean square stabilizability according to Definition 3.40. Similarly, (s, L, A) detectability is equivalent, when $s_{ij} = p_{ij}$, to mean square detectability according to Definition 3.41. The next proposition shows the duality between these two concepts.

Proposition A.3. (A, B, s) is stabilizable if and only if (s^*, B^*, A^*) is detectable.

Proof. From Definition A.1, there exists $F = (F_1, \ldots, F_N) \in \mathbb{H}^{n,m}$ such that $r_\sigma(\mathcal{T}) < 1$ when $\Gamma_i = A_i + B_i F_i$ in (3.7) for each $i \in \mathbb{N}$. From (3.9), the operator $\mathcal{V}$ is defined, for $V = (V_1, \ldots, V_N) \in \mathbb{H}^n$, as

$$\mathcal{V}_i(V) = (A_i^* + F_i^* B_i^*)^* \Big(\sum_{j=1}^N s_{ij}^* V_j\Big)(A_i^* + F_i^* B_i^*)$$

and since $r_\sigma(\mathcal{V}) = r_\sigma(\mathcal{T}) < 1$, we conclude from Definition A.2 that F^* stabilizes (s^*, B^*, A^*). $\qquad\square$

In what follows, we need to consider $\varsigma(k) = (\varsigma_1(k), \ldots, \varsigma_N(k))$ and $\varsigma = (\varsigma_1, \ldots, \varsigma_N)$, with $\varsigma_i(k) > 0$ and $\varsigma_i > 0$ for $i \in \mathbb{N}$ and $k = 0, 1, \ldots$, and such that

$$\lim_{k \to \infty} \varsigma_i(k) = \varsigma_i \tag{A.1}$$

exponentially fast for $i \in \mathbb{N}$. Consider also

$$A = (A_1, \ldots, A_N) \in \mathbb{H}^n,$$
$$B = (B_1, \ldots, B_N) \in \mathbb{H}^{m,n},$$
$$C = (C_1, \ldots, C_N) \in \mathbb{H}^{n,q},$$
$$D = (D_1, \ldots, D_N) \in \mathbb{H}^{m,q},$$

with (4.26) and (4.27) satisfied. We set

$$\mathbb{W} = \{X \in \mathbb{H}^{n*}, \alpha \in \mathbb{R}^N; \alpha_i \geq 0, i \in \mathbb{N}, \text{ and}$$
$$B_i^* \mathcal{E}_i(X) B_i + \alpha_i D_i^* D_i \text{ is non singular for each } i \in \mathbb{N}\}$$

We need also the following definitions:

Definition A.4. *We define*

$$\mathcal{F}(.,.) : \mathbb{W} \to \mathbb{H}^m,$$
$$\mathcal{X}(.,.) : \mathbb{W} \to \mathbb{H}^n,$$
$$\mathcal{Y}(.,.) : \mathbb{W} \to \mathbb{H}^n,$$

in the following way; for $(X, \alpha) \in \mathbb{W}$, $\alpha = (\alpha_1, \ldots, \alpha_N)$, *with* $\alpha_i > 0$ *for each* $i \in \mathbb{N}$, $X = (X_1, \ldots, X_N)$, *and* $F = (F_1, \ldots, F_N) \in \mathbb{H}^{m,n}$, $\mathcal{F}(X, \alpha) = (\mathcal{F}_1(X, \alpha), \ldots, \mathcal{F}_N(X, \alpha))$, $\mathcal{X}(X, \alpha) = (\mathcal{X}_1(X, \alpha), \ldots, \mathcal{X}_N(X, \alpha))$ *and* $\mathcal{Y}(X, \alpha) = (\mathcal{Y}_1(X, \alpha), \ldots, \mathcal{Y}_N(X, \alpha))$ *are defined as*

$$\mathcal{F}_i(X, \alpha) \triangleq - (B_i^* \mathcal{E}_i(X) B_i + \alpha_i D_i^* D_i)^{-1} B_i^* \mathcal{E}_i(X) A_i,$$
$$\mathcal{X}_i(X, \alpha) \triangleq A_i^* \mathcal{E}_i(X) A_i + \alpha_i C_i^* C_i$$
$$- A_i^* \mathcal{E}_i(X) B_i \Big[\alpha_i D_i^* D_i + B_i^* \mathcal{E}_i(X) B_i \Big]^{-1} B_i^* \mathcal{E}_i(X) A_i,$$

and

$$\mathcal{Y}_i(X, \alpha) \triangleq \sum_{j=1}^{N} s_{ij} \{ A_j^* X_j A_j + \alpha_j C_j^* C_j$$
$$- A_j^* X_j B_j \Big[\alpha_j D_j^* D_j + B_j^* X_j B_j \Big]^{-1} B_j^* X_j A_j \}$$

for each $i \in \mathbb{N}$. *In particular we set* $\mathcal{F}(X) = \mathcal{F}(X, \varsigma), \mathcal{X}(X) = \mathcal{X}(X, \varsigma), \mathcal{Y}(X) = \mathcal{Y}(X, \varsigma)$.

In addition, for any $F = (F_1, \ldots, F_N) \in \mathbb{H}^{m,n}$, $(V, \alpha) \in \mathbb{W}$, $V = (V_1, \ldots, V_N) \in \mathbb{H}^n$, and $i \in \mathbb{N}$, the following useful identity can be easily established:

$$(A_i + B_i F_i)^* V_i (A_i + B_i F_i) + \alpha_i F_i^* D_i^* D_i F_i$$
$$= (A_i + B_i \mathcal{F}_i(V, \alpha))^* V_i (A_i + B_i \mathcal{F}_i(V, \alpha))$$
$$+ \alpha_i \mathcal{F}_i(V, \alpha)^* D_i^* D_i \mathcal{F}_i(V, \alpha)$$
$$+ (F_i - \mathcal{F}_i(V, \alpha))^* (B_i^* V_i B_i + \alpha_i D_i^* D_i)(F_i - \mathcal{F}_i(V, \alpha)). \tag{A.2}$$

We want to study the asymptotic behavior of the set of coupled Riccati difference equations defined recursively for $k = 0, 1, \ldots$ as

$$X(k+1) = \mathcal{X}(X(k), \varsigma(k)), \tag{A.3}$$

to a $X \in \mathbb{H}^{n+}$ satisfying

$$X = \mathcal{X}(X). \tag{A.4}$$

where $X(0) \in \mathbb{H}^{n+}$. Notice that from the identity (A.2) it is easy to check by induction on k that $X(k) \in \mathbb{H}^{n+}$ for all $k = 0, 1, \ldots$.

Associated to the recursive equation (A.3), we have another set of coupled Riccati difference equations defined recursively for $k = 0, 1, \ldots$ as

$$Y(k+1) = \mathcal{Y}(Y(k), \varsigma(k)), \tag{A.5}$$

with $Y(0) \in \mathbb{H}^{n+}$. As before, it is easy to check from the identity (A.2) that $Y(k) \in \mathbb{H}^{n+}$ for all $k = 0, 1, \ldots$. We will also be interested in studying the asymptotic behavior of $Y(k)$ to a $Y \in \mathbb{H}^{n+}$ satisfying

$$Y = \mathcal{Y}(Y). \tag{A.6}$$

In what follows set, for $V \in \mathbb{H}^{n+}$, $K(V) = (K_1(V), \ldots, K_N(V))$ as

$$K_i(V) = -(B_i^* V_i B_i + \alpha_i D_i^* D_i)^{-1} B_i^* V_i A_i.$$

We have then the following proposition.

Proposition A.5. *The following assertions are equivalent:*

1. *For any $X(0) \in \mathbb{H}^{n+}$, $X(k)$ defined by (A.3) converges to a $X \in \mathbb{H}^{n+}$ which satisfies $X = \mathcal{X}(X)$.*
2. *For any $Y(0) \in \mathbb{H}^{n+}$, $Y(k)$ defined by (A.5) converges to a $Y \in \mathbb{H}^{n+}$ which satisfies $Y = \mathcal{Y}(Y)$.*

Moreover there exists $X \in \mathbb{H}^{n+}$ which satisfies $X = \mathcal{X}(X)$ if and only if there exists $Y \in \mathbb{H}^{n+}$ which satisfies $Y = \mathcal{Y}(Y)$. Furthermore, in this case, $\mathcal{F}(X)$ stabilizes (A, B, s) if and only if $K(Y)$ stabilizes (A, B, s).

Proof. Suppose that 1) holds. For any $Y(0) \in \mathbb{H}^{n+}$, set $X(0) = (X_1(0), \ldots, X_N(0)) \in \mathbb{H}^{n+}$ as

$$X_i(0) = A_i^* Y_i(0) A_i + \varsigma_i(0) C_i^* C_i$$
$$- A_i^* Y_i(0) B_i \left[\varsigma_i(0) D_i^* D_i + B_i^* Y_i(0) B_i \right]^{-1} B_i^* Y_i(0) A_i.$$

According to 1), $\lim_{k \to \infty} X(k) = X \in \mathbb{H}^{n+}$ with X satisfying $X = \mathcal{X}(X)$. Then it is easy to verify that

$$X_i(k) = A_i^* Y_i(k) A_i + \varsigma_i(k) C_i^* C_i$$
$$- A_i^* Y_i(k) B_i \left[\varsigma_i(k) D_i^* D_i + B_i^* Y_i(k) B_i \right]^{-1} B_i^* Y_i(k) A_i,$$
$$Y_i(k+1) = \mathcal{E}_i(X(k))$$

and taking the limit as $k \to \infty$ we conclude that $\lim_{k \to \infty} Y(k) = Y = \mathcal{E}(X) \in \mathbb{H}^{n+}$, with $Y = \mathcal{Y}(Y)$.

Suppose now that 2) holds. For any $X(0) \in \mathbb{H}^{n+}$, set $Y(0) = \mathcal{E}(X(0)) \in \mathbb{H}^{n+}$. Then according to 1), $\lim_{k \to \infty} Y(k) = Y \in \mathbb{H}^{n+}$ with Y satisfying $Y = \mathcal{Y}(Y)$. It is easy to verify that

$$X_i(k+1) = A_i^* Y_i(k) A_i + \varsigma_i(k) C_i^* C_i$$
$$- A_i^* Y_i(k) B_i \left[\varsigma_i(k) D_i^* D_i + B_i^* Y_i(k) B_i \right]^{-1} B_i^* Y_i(k) A_i,$$
$$Y_i(k) = \mathcal{E}_i(X(k))$$

and taking the limit as $k \to \infty$ we conclude that $\lim_{k \to \infty} X(k) = X \in \mathbb{H}^{n+}$, with $X = \mathcal{X}(X)$.

Finally, suppose that $X \in \mathbb{H}^{n+}$ satisfies $X = \mathcal{X}(X)$. Then by defining $Y = \mathcal{E}(X) \in \mathbb{H}^{n+}$ we get that $Y = \mathcal{Y}(Y)$ and if $\mathcal{F}(X)$ stabilizes (A, B, s) then clearly $\mathrm{K}(Y)$ stabilizes (A, B, s). On the other hand, if $Y \in \mathbb{H}^{n+}$ satisfies $Y = \mathcal{Y}(Y)$ then, by defining $X \in \mathbb{H}^{n+}$ as

$$X_i = A_i^* Y_i A_i + \varsigma_i C_i^* C_i - A_i^* Y_i B_i \left[\varsigma_i D_i^* D_i + B_i^* Y_i B_i \right]^{-1} B_i^* Y_i A_i$$
$$= (A_i + B_i \mathcal{F}_i(Y))^* Y_i (A_i + B_i \mathcal{F}_i(Y)) + \varsigma_i C_i^* C_i + \varsigma_i \mathcal{F}_i(Y)^* D_i^* D_i \mathcal{F}_i(Y)$$

we conclude that $Y = \mathcal{E}(X)$ and thus $X = \mathcal{X}(X)$. Clearly if $\mathrm{K}(Y)$ stabilizes (A, B, s) then $\mathcal{F}(X)$ stabilizes (A, B, s). $\qquad \square$

We can see now how (A.3), (A.4), (A.5) and (A.6) are related to the control and filtering Riccati equations (4.14), (4.35), (5.13) and (5.31). For the control equations (4.14) and (4.35) we just have to make $s_{ij} = p_{ij}$, $\varsigma_i(k) = 1$, $\varsigma_i = 1$ in (A.3) and (A.4) respectively. We have that the filtering equations (5.13) (with $\mathbb{J}(k) = \mathbb{N}$ and p_{ij}, A, L, G, H time-invariant) and (5.31) can be seen as dual of (A.5) and (A.6) respectively when $s_{ij} = p_{ij}$, $\varsigma_i(k) = \pi_i(k)$, $\varsigma_i = \pi_i$, after we make the following correspondence between the two problems:

$$s \leftrightarrow s^*$$
$$A_i \leftrightarrow A_i^*$$
$$B_i \leftrightarrow L_i^*$$
$$C_i \leftrightarrow G_i^*$$
$$D_i \leftrightarrow H_i^*. \tag{A.7}$$

This means that if we take the coupled Riccati equations (A.5) and (A.6) with $s_{ij} = p_{ij}$, $\varsigma_i(k) = \pi_i(k)$, $\varsigma_i = \pi_i$, and relabel s_{ij}, A_i, B_i, C_i and D_i by respectively s_{ji}, A_i^*, L_i^*, G_i^*, and H_i^*, we get precisely (5.13) (with $\mathbb{J}(k) = \mathbb{N}$ and p_{ij}, A, L, G, H time-invariant) and (5.31). The same relabelling applied to $\mathrm{K}_i(Y)$ produces $M_i(Y)^*$ in (5.32). Thus the coupled Riccati equations (A.5) and (A.6), and (5.13) and (5.31), are the same in all but notation. From Proposition A.5 we have that the facts deduced from (A.3) and (A.4), can also be applied to the filtering coupled Riccati equations (5.13) and (5.31).

Notice that to have the assumption $\varsigma_i(k) > 0$ satisfied we could iterate (5.13) up to a time k_0 such that $\pi_i(k) > 0$ for all $k > k_0$ (such k_0 exists since by the ergodic assumption (Assumption 3.31 in Chapter 3) made in Chapters 4 and 6 for the infinite horizon problems, $\pi_i(k) \to \pi_i > 0$ exponentially fast for each $i \in \mathbb{N}$), and consider $\varsigma_i(k) = \pi_i(k + k_0)$, and $Y(k_0)$ from (5.13) as the initial condition for (A.5).

In the next sections we study the control coupled Riccati equations (A.3) and (A.4). As mentioned before, from Proposition A.5, the results also hold for the coupled Riccati equations (A.5) and (A.6). As seen in (A.7), the dual of (A.5) and (A.6) lead to the filtering equations (5.13) and (5.31), and therefore the results derived for the control coupled Riccati equations can also be used for the filtering coupled Riccati equations.

A.2 Maximal Solution for the CARE

This section discusses mainly the class of solutions to the CARE named the *maximal solutions*. While in control and filtering applications we are more interested in another class of solutions, the so-called *stabilizing solutions*, the former class is also of interest, as there is a very intimate relation between them. Besides, it can be obtained numerically via a certain LMI optimization problem, as will be seen in Problem A.11. We need the following definitions:

Definition A.6. *For* $(X, \alpha) \in \mathbb{W}$, *with* $\alpha = (\alpha_1, \ldots, \alpha_N)$, $\alpha_i > 0$ *for each* $i \in \mathbb{N}$, $X = (X_1, \ldots, X_N)$, *and* $F = (F_1, \ldots, F_N) \in \mathbb{H}^{m,n}$, *we define*

$$\mathcal{O}(.,.) : \mathbb{H}^{m,n} \times \mathbb{R}^{N+} \to \mathbb{H}^n,$$
$$\mathcal{R}(.,.) : \mathbb{W} \to \mathbb{H}^m,$$

with

$$\mathcal{O}(F, \alpha) = (\mathcal{O}_1(F, \alpha), \ldots, \mathcal{O}_N(F, \alpha)),$$
$$\mathcal{R}(X, \alpha) = (\mathcal{R}_1(X, \alpha), \ldots, \mathcal{R}_N(X, \alpha)),$$

in the following way;

$$\mathcal{O}_i(F, \alpha) \triangleq \alpha_i(C_i^* C_i + F_i^* D_i^* D_i F_i),$$
$$\mathcal{R}_i(X, \alpha) \triangleq (B_i^* \mathcal{E}_i(X) B_i + \alpha_i D_i^* D_i).$$

We make $\mathcal{O}_i(F) = \mathcal{O}_i(F, \varsigma)$, $\mathcal{R}_i(X) = \mathcal{R}_i(X, \varsigma)$, and set

$$\mathbb{L} = \{X = (X_1, \ldots, X_N) \in \mathbb{H}^{n*}; \mathcal{R}_i(X) \text{ is non-singular for each } i \in \mathbb{N}\}$$
$$\mathbb{M} = \{X = (X_1, \ldots, X_N) \in \mathbb{H}^{n*}; \mathcal{R}(X) > 0 \text{ and } -X + \mathcal{X}(X) \geq 0\}.$$

The following two lemmas present auxiliary results which will be useful in the proof of Theorem A.10. The proof of Lemma A.7 is straightforward, but otherwise very long, and therefore will be omitted; [184] and [193] present similar results for the deterministic case, with a complete proof in [184].

Lemma A.7. *Suppose that $X \in \mathbb{L}$ and for some $\widehat{F} = (\widehat{F}_1, \ldots, \widehat{F}_N) \in \mathbb{H}^{n,m}$,*

1. $\widehat{X} = (\widehat{X}_1, \ldots, \widehat{X}_N) \in \mathbb{H}^{n*}$ satisfies for $i \in \mathbb{N}$

$$\widehat{X}_i - (A_i + B_i\widehat{F}_i)^*\mathcal{E}_i(\widehat{X})(A_i + B_i\widehat{F}_i) = \mathcal{O}_i(\widehat{F}).$$

Then, for $i \in \mathbb{N}$,

$$(\widehat{X}_i - X_i) - (A_i + B_i\widehat{F}_i)^*\mathcal{E}_i(\widehat{X} - X)(A_i + B_i\widehat{F}_i)$$
$$= \mathcal{X}_i(X) - X_i + (\widehat{F}_i - \mathcal{F}_i(X))^*\mathcal{R}_i(X)(\widehat{F}_i - \mathcal{F}_i(X)). \quad \text{(A.8a)}$$

2. Moreover, if $\widehat{X} \in \mathbb{L}$ then, for $i \in \mathbb{N}$,

$$(\widehat{X}_i - X_i) - (A_i + B_i\mathcal{F}_i(\widehat{X}))^*\mathcal{E}_i(\widehat{X} - X)(A_i + B_i\mathcal{F}_i(\widehat{X}))$$
$$= \mathcal{X}_i(X) - X_i + (\mathcal{F}_i(\widehat{X}) - \mathcal{F}_i(X))^*\mathcal{R}_i(X)(\mathcal{F}_i(\widehat{X}) - \mathcal{F}_i(X))$$
$$+ (\widehat{F}_i - \mathcal{F}_i(\widehat{X}))^*\mathcal{R}_i(\widehat{X})(\widehat{F}_i - \mathcal{F}_i(\widehat{X})). \quad \text{(A.8b)}$$

3. Furthermore, if $\widehat{Y} = (\widehat{Y}_1, \ldots, \widehat{Y}_N) \in \mathbb{H}^{n*}$ and satisfies, for $i \in \mathbb{N}$

$$\widehat{Y}_i - (A_i + B_i\mathcal{F}_i(\widehat{X}))^*\mathcal{E}_i(\widehat{Y})(A_i + B_i\mathcal{F}_i(\widehat{X})) = \mathcal{O}_i(\mathcal{F}(\widehat{X})), \quad \text{(A.8c)}$$

then for $i \in \mathbb{N}$,

$$(\widehat{X}_i - \widehat{Y}_i) - (A_i + B_i\mathcal{F}_i(\widehat{X}))^*\mathcal{E}_i(\widehat{X} - \widehat{Y})(A_i + B_i\mathcal{F}_i(\widehat{X}))$$
$$= (\widehat{F}_i - \mathcal{F}_i(\widehat{X}))^*\mathcal{R}_i(\widehat{X})(\widehat{F}_i - \mathcal{F}_i(\widehat{X})). \quad \text{(A.8d)}$$

For the next lemma, set $\Gamma_i = A_i + B_iF_i$, $i \in \mathbb{N}$, for some $F = (F_1, \ldots, F_N) \in \mathbb{H}^{n,m}$. Consider also the operator $\bar{\mathcal{L}}(.) = (\bar{\mathcal{L}}_1(.), \ldots, \bar{\mathcal{L}}_N(.)) \in \mathbb{B}(\mathbb{H}^n)$ defined as

$$\bar{\mathcal{L}}_i(.) \triangleq \Lambda_i^*\mathcal{E}_i(.)\Lambda_i, \quad i \in \mathbb{N} \quad \text{(A.9)}$$

where $\Lambda_i = A_i + B_iK_i$, $i \in \mathbb{N}$, for some $K = (K_1, \ldots, K_N) \in \mathbb{H}^{n,m}$.

The next lemma provides a crucial result for the development of this appendix.

Lemma A.8. *Let $\mathcal{L}$ and $\bar{\mathcal{L}}$ be as defined in (3.8) and (A.9), with $\Gamma_i = A_i + B_iF_i$ and $\Lambda_i = A_i + B_iK_i$, $i \in \mathbb{N}$. Suppose that $r_\sigma(\mathcal{L}) < 1$ and for some $X = (X_1, \ldots, X_N) \geq 0$ and $\delta > 0$,*

$$X_i - \Lambda_i^*\mathcal{E}_i(X)\Lambda_i \geq \delta(K_i - F_i)^*(K_i - F_i), i \in \mathbb{N}. \quad \text{(A.10)}$$

Then $r_\sigma(\bar{\mathcal{L}}) < 1$.

Proof. Set $\bar{\mathcal{T}} = \bar{\mathcal{L}}^*$, so that for any $V = (V_1, \ldots, V_N) \in \mathbb{H}^n$,

$$\bar{\mathcal{T}}_j(V) \triangleq \sum_{i=1}^N s_{ij}\Lambda_i V_i \Lambda_i^*, j \in \mathbb{N}. \quad \text{(A.11)}$$

Note that for arbitrary $\epsilon > 0$ and $V = (V_1, \ldots, V_N) \geq 0$,

$$0 \leq \left[\epsilon(A_i + B_i F_i) - \frac{1}{\epsilon}B_i(K_i - F_i)\right] V_i \left[\epsilon(A_i + B_i F_i) - \frac{1}{\epsilon}B_i(K_i - F_i)\right]^*$$

$$= \epsilon^2(A_i + B_i F_i)V_i(A_i + B_i F_i)^* + \frac{1}{\epsilon^2}B_i(K_i - F_i)V_i(K_i - F_i)^*B_i^*$$

$$- (A_i + B_i F_i)V_i(K_i - F_i)^*B_i^* - B_i(K_i - F_i)V_i(A_i + B_i F_i)^*$$

so that

$$(A_i + B_i F_i)V_i(K_i - F_i)^*B_i^* + B_i(K_i - F_i)V_i(A_i + B_i F_i)^*$$

$$\leq \epsilon^2(A_i + B_i F_i)V_i(A_i + B_i F_i)^* + \frac{1}{\epsilon^2}B_i(K_i - F_i)V_i(K_i - F_i)^*B^*. \quad (A.12)$$

Combining (A.11) with (A.12) leads to

$$0 \leq \bar{\mathcal{T}}_j(V) = \sum_{i=1}^{N} s_{ij} \Lambda_i V_i \Lambda_i^*$$

$$= \sum_{i=1}^{N} s_{ij}(A_i + B_i F_i + B_i(K_i - F_i))V_i(A_i + B_i F_i + B_i(K_i - F_i))^*$$

$$= \sum_{i=1}^{N} s_{ij}[(A_i + B_i F_i)V_i(A_i + B_i F_i)^* + B_i(K_i - F_i)V_i(A_i + B_i F_i)^*$$

$$+ (A_i + B_i F_i)V_i(K_i - F_i)^*B_i^* + B_i(K_i - F_i)V_i(K_i - F_i)^*B_i^*]$$

$$\leq (1 + \epsilon^2) \sum_{i=1}^{N} s_{ij}(A_i + B_i F_i)V_i(A_i + B_i F_i)^*$$

$$+ (1 + \frac{1}{\epsilon^2}) \sum_{i=1}^{N} s_{ij}B_i(K_i - F_i)V_i(K_i - F_i)^*B_i^*$$

$$= (1 + \epsilon^2)\mathcal{T}_j(V) + (1 + \frac{1}{\epsilon^2})\mathcal{Q}_j(V) \quad (A.13)$$

where $\mathcal{Q}(V) = (\mathcal{Q}_1(V), \ldots, \mathcal{Q}_N(V))$ is defined as

$$\mathcal{Q}_j(V) \triangleq \sum_{i=1}^{N} s_{ij}B_i(K_i - F_i)V_i(K_i - F_i)^*B_i^*$$

and $\mathcal{T} = \mathcal{L}^*$. Since $r_\sigma(\mathcal{T}) = r_\sigma(\mathcal{L}) < 1$ by hypothesis, we can choose $\epsilon > 0$ such that $r_\sigma(\tilde{\mathcal{T}}) < 1$, where $\tilde{\mathcal{T}}(.) = (1 + \epsilon^2)\mathcal{T}(.)$. Define for $t = 0, 1, \ldots$ the sequences

$$Y(t+1) \triangleq \tilde{\mathcal{T}}(X(t)), \qquad\qquad Y(0) \in \mathbb{H}^{n+}$$
$$Z(t+1) \triangleq \tilde{\mathcal{T}}(Z(t)) + \tilde{\mathcal{Q}}(Y(t)), \ Z(0) = Y(0) \quad (A.14)$$

with

$$\widetilde{\mathcal{Q}}(.) = (1 + \frac{1}{\epsilon^2})\mathcal{Q}(.).$$

Then for $t = 0, 1, 2, \ldots$

$$Z(t) \geq Y(t) \geq 0. \tag{A.15}$$

Indeed, (A.15) is immediate from (A.14) for $t = 0$. Suppose by induction that (A.15) holds for t. Then from (A.13)

$$Z(t+1) = \widetilde{\mathcal{T}}(Z(t)) + \widetilde{\mathcal{Q}}(Y(t))$$

$$= (1 + \epsilon^2)\mathcal{T}(Z(t)) + (1 + \frac{1}{\epsilon^2})\mathcal{Q}(Y(t))$$

$$\geq (1 + \epsilon^2)\mathcal{T}(Y(t)) + (1 + \frac{1}{\epsilon^2})\mathcal{Q}(Y(t))$$

$$\geq \widetilde{\mathcal{T}}(Y(t)) = Y(t+1)$$

showing the result for $t + 1$. From (A.14) it follows that

$$Z(t) = \widetilde{\mathcal{T}}^t(Y(0)) + \sum_{s=0}^{t-1} \widetilde{\mathcal{T}}^{t-1-s}\left(\widetilde{\mathcal{Q}}(Y(s))\right)$$

and taking the 1-norm of the above expression, we have

$$\|Z(t)\|_1 \leq \left\|\widetilde{\mathcal{T}}^t\right\| \|Y(0)\|_1 + \sum_{s=0}^{t-1} \left\|\widetilde{\mathcal{T}}^{t-1-s}\right\| \left\|\widetilde{\mathcal{Q}}(Y(s))\right\|_1.$$

Since $r_\sigma(\widetilde{\mathcal{T}}) < 1$, it is possible to find (see Proposition 2.5) $\beta \geq 1$, $0 < \zeta < 1$, such that

$$\left\|\widetilde{\mathcal{T}}^s\right\| \leq \beta\zeta^s, \ s = 0, 1, \ldots$$

and thus,

$$\|Z(t)\|_1 \leq \beta\zeta^t \|Y(0)\|_1 + \beta \sum_{s=0}^{t-1} \zeta^{t-1-s} \left\|\widetilde{\mathcal{Q}}(Y(s))\right\|_1.$$

Suppose for the moment that $\sum_{s=0}^{\infty} \left\|\widetilde{\mathcal{Q}}(Y(s))\right\|_1 < \infty$. Then

$$\sum_{t=0}^{\infty} \|Z(t)\|_1 \leq \frac{\beta}{1-\zeta} \|Y(0)\|_1 + \beta \sum_{t=0}^{\infty} \sum_{s=0}^{t-1} \zeta^{t-1-s} \left\|\widetilde{\mathcal{Q}}(Y(s))\right\|_1$$

$$= \frac{\beta}{1-\zeta} \|Y(0)\|_1 + \frac{\beta}{1-\zeta} \sum_{s=0}^{\infty} \left\|\widetilde{\mathcal{Q}}(Y(s))\right\|_1 < \infty$$

and therefore from (A.14) and (A.15), for any $Y(0) = (Y_1(0), \ldots, Y_N(0)) \geq 0$,

$$0 \le \sum_{t=0}^{\infty} \left\| \bar{\mathcal{T}}^t(Y(0)) \right\|_1 = \sum_{t=0}^{\infty} \| Y(t) \|_1 \le \sum_{t=0}^{\infty} \| Z(t) \|_1 < \infty.$$

From Proposition 2.5, $r_\sigma(\bar{\mathcal{T}}) = r_\sigma(\bar{\mathcal{L}}) < 1$. It remains to prove that

$$\sum_{s=0}^{\infty} \left\| \tilde{\mathcal{Q}}(X(s)) \right\|_1 < \infty.$$

Indeed, setting

$$c_0 = \left[(1 + \frac{1}{\epsilon^2}) \| B \|_{\max}^2 \right] \frac{N}{\delta}$$

we obtain from (A.10) that

$$\left\| \tilde{\mathcal{Q}}(Y(s)) \right\|_1 = \sum_{j=1}^{N} \left\| \tilde{\mathcal{Q}}_j(Y(s)) \right\|$$

$$\le (1 + \frac{1}{\epsilon^2}) \sum_{i=1}^{N} \sum_{j=1}^{N} s_{ij} \left\| B_i(K_i - F_i)Y_i(s)(K_i - F_i)^* B_i^* \right\|$$

$$\le (1 + \frac{1}{\epsilon^2}) \sum_{i=1}^{N} \sum_{j-1}^{N} s_{ij} \| B_i \|^2 \operatorname{tr}((K_i - F_i)Y_i(s)(K_i - F_i)^*)$$

$$\le \left[(1 + \frac{1}{\epsilon^2}) \| B_i \|_{\max}^2 \right] N$$

$$\times \sum_{i=1}^{N} \operatorname{tr} \left(\{Y_i(s)\}^{1/2} (K_i - F_i)^* (K_i - F_i) \{Y_i(s)\}^{1/2} \right)$$

$$\le c_0 \sum_{i=1}^{N} \operatorname{tr} \left(Y_i(s) (X_i - \mathcal{L}_i(X)) \right)$$

$$= c_0 \langle Y(s); X - \mathcal{L}(X) \rangle$$

$$= c_0 \{ \langle Y(s); X \rangle - \langle Y(s); \mathcal{L}(X) \rangle \}$$

$$= c_0 \{ \langle Y(s); X \rangle - \langle \mathcal{T}(Y(s)); X \rangle \}$$

$$= c_0 \{ \langle Y(s); X \rangle - \langle Y(s+1); X \rangle \}.$$

Taking the sum from $s = 0$ to τ, we get

$$\sum_{s=0}^{\tau} \left\| \tilde{\mathcal{Q}}(Y(s)) \right\|_1 \le c_0 \{ \langle Y(0); X \rangle - \langle Y(\tau + 1); X \rangle \} \le c_0 \langle Y(0); X \rangle$$

since $X = (X_1, \ldots, X_N) \ge 0$ and thus $\langle Y(r+1); X \rangle \ge 0$. Taking the limit as $\tau \to \infty$, we obtain the desired result.   □

The following theorem presents the main result of this section, regarding the existence of a maximal solution for (A.4) in $\mathbb{M}$. Let us first consider the following definition.

Definition A.9. *We say that $\widehat{X} \in \mathbb{M}$ is the maximal solution for (A.4) if it satisfies (A.4), and for any $X \in \mathbb{M}$, $\widehat{X} \geq X$.*

Theorem A.10. *Suppose that (A, B, s) is stabilizable. Then for $l = 0, 1, 2, \ldots,$ there exists $X^l = (X_1^l, \ldots, X_N^l)$ which satisfies the following properties.*

1.
$$X^l \in \mathbb{H}^{n+}, \tag{A.16a}$$

2.
$$X^0 \geq X^1 \geq \cdots \geq X^l \geq X \text{ for any arbitrary } X \in \mathbb{M}; \tag{A.16b}$$

3.
$$r_\sigma(\mathcal{L}^l) < 1 \tag{A.16c}$$

where $\mathcal{L}^l(.) = (\mathcal{L}_1^l(.), \ldots, \mathcal{L}_N^l(.))$ and for $i \in \mathbb{N}$,

$$\mathcal{L}_i^l(X^l) \triangleq A_i^{l*} \mathcal{E}_i(X^l) A_i^l,$$
$$A_i^l \triangleq A_i + B_i F_i^l,$$
$$F_i^l \triangleq \mathcal{F}_i(X^{l-1}) \text{ for } l = 1, 2, \ldots$$

4.
$$X_i^l - A_i^{l*} \mathcal{E}_i(X^l) A_i^l = \mathcal{O}_i(F^l), i \in \mathbb{N}. \tag{A.16d}$$

Moreover there exists $X^+ = (X_1^+, \ldots, X_N^+) \in \mathbb{H}^{n+}$ such that $X^+ = \mathcal{X}(X^+)$, $X^+ \geq X$ for any $X \in \mathbb{M}$ and $X^l \to X^+$, as $l \to \infty$. Furthermore $r_\sigma(\mathcal{L}^+) \leq 1$, where $\mathcal{L}^+(.) = (\mathcal{L}_1^+(.), \ldots, \mathcal{L}_N^+(.))$ is defined as $\mathcal{L}_i^+(.) \triangleq A_i^{+*} \mathcal{E}_i(.) A_i^+$, for $i \in \mathbb{N}$, and

$$F_i^+ = \mathcal{F}_i(X^+)$$
$$A_i^+ = A_i + B_i F_i^+.$$

Proof. Let us apply induction on l to show that (A.16) are satisfied. Consider an arbitrary $X \in \mathbb{M}$. By the hypothesis that (A, B, s) is stabilizable we can find F^0 such that $r_\sigma(\mathcal{L}^0) < 1$, where $\mathcal{L}^0(.) = (\mathcal{L}_1^0(.), \ldots, \mathcal{L}_N^0(.))$ and $\mathcal{L}_i^0(.) = A_i^{0*} \mathcal{E}_i(.) A_i^0$ with $A_i^0 = A_i + B_i F_i^0$. Thus, from Proposition 3.20, there exists a unique $X^0 = (X_1^0, \ldots, X_N^0) \in \mathbb{H}^n$ solution of

$$X_i^0 - A_i^{0*} \mathcal{E}_i(X^0) A_i^0 = \mathcal{O}_i(F^0), \ i \in \mathbb{N}.$$

Moreover, since $\mathcal{O}(F^0) \in \mathbb{H}^{n+}$, we have that $X^0 \in \mathbb{H}^{n+}$. Setting for $i \in \mathbb{N}$,

$$R_i \triangleq \mathcal{R}_i(X)$$
$$F_i \triangleq \mathcal{F}_i(X).$$

we have from Lemma A.7, (A.8a), that for $i \in \mathbb{N}$,

$$(X_i^0 - X_i) - A_i^{0*} \mathcal{E}_i (X^0 - X) A_i^0 = \mathcal{X}_i(X) - X_i + (F_i^0 - F_i)^* R_i (F_i^0 - F_i)$$

and since $\mathcal{X}_i(X) - X_i + (F_i^0 - F_i)^* R_i (F_i^0 - F_i) \geq 0$, $i \in \mathbb{N}$ and $r_\sigma(\mathcal{L}^0) < 1$, we have from Proposition 3.20 again that $X^0 - X \geq 0$, and thus (A.16) hold for $l = 0$. Suppose now that we already have a sequence $\{X^l\}_{l=0}^{k-1}$ satisfying (A.16). Set

$$R_i^{k-1} \triangleq R_i(X^{k-1})$$
$$F_i^k \triangleq \mathcal{F}_i(X^{k-1})$$
$$A_i^k \triangleq A_i + B_i F_i^k.$$

From (A.8b) in Lemma A.7, we get that

$$(X_i^{k-1} - X_i) - A_i^{k*} \mathcal{E}_i (X^{k-1} - X) A_i^k$$
$$= \mathcal{X}_i(X) - X_i + (F_i^k - F_i)^* R_i (F_i^k - F_i) + (F_i^k - F_i^{k-1})^* R_i^{k-1} (F_i^k - F_i^{k-1})$$
$$\geq (F_i^k - F_i^{k-1})^* R_i^{k-1} (F_i^k - F_i^{k-1})$$

and since $R_i^{k-1} > 0$ for $i \in \mathbb{N}$, we can find $\delta^{k-1} > 0$ such that $R_i^{k-1} > \delta^{k-1} I$. Thus, for $i \in \mathbb{N}$,

$$(X_i^{k-1} - X_i) - A_i^{k*} \mathcal{E}_i (X^{k-1} - X) A_i^k \geq \delta^{k-1} (F_i^k - F_i^{k-1})^* (F_i^k - F_i^{k-1})$$

and from Lemma A.8, $r_\sigma(\mathcal{L}^k) < 1$. Let $X^k \in \mathbb{H}^{n+}$ be the unique solution of (see Proposition 3.20 and recall that $\mathcal{O}(F^l) \in \mathbb{H}^{n+}$)

$$X_i^k - A_i^{k*} \mathcal{E}_i (X^k) A_i^k = \mathcal{O}_i(F^l), \ i \in \mathbb{N}.$$

Equation (A.8a) in Lemma A.7 yields, for $i \in \mathbb{N}$,

$$(X_i^k - X_i) - A_i^{k*} \mathcal{E}_i (X^k - X) A_i^k = \mathcal{X}_i(X) - X_i + (F_i^k - F_i)^* D_i (F_i^k - F_i)$$

and since $r_\sigma(\mathcal{L}^k) < 1$, we get from Proposition 3.20 that $X^k \geq X$. Equation (A.8d) in Lemma A.7 yields

$$(X_i^{k-1} - X_i^k) - A_i^{k*} \mathcal{E}_i (X^{k-1} - X^k) A_i^k = (F_i^k - F_i^{k-1})^* R_i^{k-1} (F_i^k - F_i^{k-1})$$

for $i \in \mathbb{N}$, which shows, from the fact that $r_\sigma(\mathcal{L}^k) < 1$, $(F_i^k - F_i^{k-1})^* D_i^{k-1} (F_i^k - F_i^{k-1})$ is positive semi-definite for each $i \in \mathbb{N}$, and Proposition 3.20, that $X^{k-1} \geq X^k \geq X$. This completes the induction argument for (A.16). Since $\{X^l\}_{l=0}^\infty$ is a decreasing sequence with $X^l \geq X$, for all $l = 0, 1, \ldots$, we get that there exists $X^+ \in \mathbb{H}^{n+}$ such that (see [216], p. 79) $X^l \downarrow X^+$ as $l \to \infty$. Clearly, $X^+ \geq X$. Moreover, substituting $F_i^l = \mathcal{F}_i(X^{l-1})$ in (A.16d) and taking the limit as $l \to \infty$, we get that for $i \in \mathbb{N}$,

$$0 = X_i^+ - (A_i + \mathcal{F}_i(X^+))^* \mathcal{E}_i(X^+)(A_i + \mathcal{F}_i(X^+)) - \mathcal{O}_i(\mathcal{F}(X^+)).$$

Rearranging the terms, we obtain that $\mathcal{X}(X^+) - X^+ = 0$, showing the desired result. Since X is arbitrary in $\mathbb{M}$, it follows that $X^+ \geq X$ for all $X \in \mathbb{M}$. Finally notice that since $r_\sigma(\mathcal{L}^k) < 1$, we get that (see [201], p. 328, for continuity of the eigenvalues on finite dimensional linear operator entries) $r_\sigma(\mathcal{L}^+) \leq 1$, where $\mathcal{L}_i^+(.) = A_i^{+*}\mathcal{E}_i(.)A_i^+$, $A_i^+ = A_i + B_i F_i^+$ and $F_i^+ = \mathcal{F}_i(X^+)$. $\square$

Theorem A.10 provides conditions for the existence of a maximal solution to the CARE (A.4), based on the concept of stabilizability of (A, B, s). The maximal solution, as will be shown in detail in the next section, coincides with the stabilizing solution to the CARE (these are of real interest) when the latter exists. As will be seen in the following, maximal solutions are also relatively easy to obtain.

The next result establishes a link between a LMI optimization problem and the maximal solution X^+ in $\mathbb{M}$, a result originally presented in [191]. Suppose without loss of generality that all matrices involved below are real. Consider the following convex programming problem.

Problem A.11.

$$\max \operatorname{tr}\left(\sum_{i=1}^{N} X_i\right) \tag{A.17a}$$

subject to

$$\begin{bmatrix} -X_i + A_i^*\mathcal{E}_i(X)A_i + \varsigma_i C_i^* C_i & A_i^*\mathcal{E}_i(X)B_i \\ B_i^*\mathcal{E}_i(X)A_i & B_i^*\mathcal{E}_i(X)B_i + \varsigma_i D_i^* D_i \end{bmatrix} \geq 0 \tag{A.17b}$$

$$B_i^*\mathcal{E}_i(X)B_i + \varsigma_i D_i^* D_i > 0 \tag{A.17c}$$

and $X = (X_1, \ldots, X_N) \in \mathbb{H}^{n*}$, with $i \in \mathbb{N}$.

Theorem A.12. *Suppose that (A, B, s) is stabilizable. Then there exists $X^+ \in \mathbb{M}$ satisfying (A.4) such that $X^+ \geq X$ for all $X \in \mathbb{M}$ if and only if there exists a solution $\widehat{X}$ for the above convex programming problem. Moreover, $\widehat{X} = X^+$.*

Proof. First of all notice that, from the Schur complement (see Lemma 2.23), $X = (X_1, \ldots, X_N)$ satisfies (A.17) if and only if

$$- X_i + A_i^*\mathcal{E}_i(X)A_i$$
$$+ \varsigma_i C_i^* C_i - A_i^*\mathcal{E}_i(X)B_i(B_i\mathcal{E}_i(X)B_i + \varsigma_i D_i^* D_i)^{-1}B_i^*\mathcal{E}_i(X)A_i \geq 0$$

and $\mathcal{R}_i(X) > 0$, for $i \in \mathbb{N}$, that is, if and only if $X \in \mathbb{M}$. Thus if $X^+ \in \mathbb{M}$ is such that $X^+ \geq X$ for all $X \in \mathbb{M}$, clearly $\operatorname{tr}(X_1^+ + \ldots + X_N^+) \geq \operatorname{tr}(X_1 + \cdots + X_N)$ for all $X \in \mathbb{M}$ and it follows that X^+ is a solution of the Convex Programming Problem A.11. On the other hand, suppose that $\widehat{X}$ is a solution of the convex programming problem. Thus $\widehat{X} \in \mathbb{M}$ and from Theorem A.10, there exists $X^+ \in \mathbb{M}$ satisfying (A.4) such that $X^+ \geq \widehat{X}$. From the optimality of $\widehat{X}$ and the fact that $X^+ \in \mathbb{M}$,

$$\operatorname{tr}(X_1^+ - \widehat{X}_1) + \cdots + \operatorname{tr}(X_N^+ - \widehat{X}_N) \le 0.$$

This, with the fact that $X_1^+ - \widehat{X}_1 \ge 0, \ldots, X_N^+ - \widehat{X}_N \ge 0$ can only hold if $X_1^+ = \widehat{X}_1, \ldots, X_N^+ = \widehat{X}_N$. □

Theorem A.12 presents a connection between the solution of Problem A.11 and the maximal solution of the CARE (A.4), which is: they are the same. Thus Problem A.11 provides a viable numerical way for obtaining the maximal solution of the CARE (A.4).

A.3 Stabilizing Solution for the CARE

This section presents results concerning the existence of a stabilizing solution for the CARE and its relation with the maximal solution. As we are going to see in this section, the maximal solution coincides with the stabilizing solution, whenever the latter exists. Also in this section, a set of both sufficient and necessary and sufficient conditions for the existence of the stabilizing solution is presented.

A.3.1 Connection Between Maximal and Stabilizing Solutions

We start by defining the meaning of stabilizing solutions for the CARE, along the same lines as in Definitions 4.4 and 5.7.

Definition A.13. *We say that $X = (X_1, \ldots, X_N) \in \mathbb{L}$ is a stabilizing solution for the CARE if it satisfies (A.4) and $\mathcal{F}(X)$ stabilizes (A, B, s).*

The next lemma establishes the connection between maximal and stabilizing solutions for the CARE, and it is crucial for the computational solution of the problem of finding stabilizing solutions (it also justifies the use of "the mean square stabilizing solution" in Definitions 4.4 and 5.7).

Lemma A.14. *There exists at most one stabilizing solution for the CARE (A.4), which will coincide with the maximal solution in $\mathbb{M}$.*

Proof. Suppose that $\widehat{X} = (\widehat{X}_1, \ldots, \widehat{X}_N)$ is a stabilizing solution for the CARE (A.4), so that (A, B, s) is stabilizable. From Theorem A.10, there is a maximal solution $X^+ \in \mathbb{M}$ satisfying (A.4). We have that

$$\widehat{X}_i - (A_i + B_i\mathcal{F}_i(\widehat{X}))^*\mathcal{E}_i(\widehat{X})(A_i + B_i\mathcal{F}_i(\widehat{X})) = \mathcal{O}_i(\mathcal{F}_i(\widehat{X}))$$

for $i \in \mathbb{N}$, and from Lemma A.7, (A.8b),

$$(\widehat{X}_i - X_i^+) - (A_i + B_i\mathcal{F}_i(\widehat{X}))^*\mathcal{E}_i(\widehat{X} - X^+)(A_i + B_i\mathcal{F}_i(\widehat{X}))$$
$$= (\mathcal{F}_i(\widehat{X}) - \mathcal{F}_i(X^+))^*\mathcal{R}_i(X^+)(\mathcal{F}_i(\widehat{X}) - \mathcal{F}_i(X^+))$$

for $i \in \mathbb{N}$. Since $\mathcal{R}(X^+) > 0$ we get that

$$(\mathcal{F}_i(\widehat{X}) - \mathcal{F}_i(X^+))^* \mathcal{R}_i(X^+)(\mathcal{F}_i(\widehat{X}) - \mathcal{F}_i(X^+)) \geq 0$$

for $i \in \mathbb{N}$. Combining the last two equations and recalling that $\widehat{X}$ is a stabilizing solution, we have from Proposition 3.20 that $\widehat{X} - X^+ \geq 0$. But this also implies that $\mathcal{R}(\widehat{X}) \geq \mathcal{R}(X^+) > 0$ and thus $\widehat{X} \in \mathbb{M}$. From Theorem A.10, $\widehat{X} - X^+ \leq 0$, showing the desired result. □

The lemma above suggests the following approach for obtaining the stabilizing solution for the CARE. First obtain the maximal solution, solving Problem A.11. The maximal solution is not necessarily a stabilizing one, but Lemma A.14 states that if a stabilizing solution exists, then it coincides with the maximal one. The second step is to verify if the maximal solution stabilizes the system, which is easily done. If it does, then the attained solution is the stabilizing solution. Otherwise there is no stabilizing solution.

A.3.2 Conditions for the Existence of a Stabilizing Solution

We start by presenting a necessary and sufficient condition for the existence of the stabilizing solution for the CARE. Afterwards we present some sufficient conditions. In what follows, we write

$$\mathcal{Z}_i(X) \triangleq \mathcal{X}_i(X) - X_i, \ i \in \mathbb{N}$$
$$\mathcal{Z}(X)^{1/2} \triangleq (\mathcal{Z}_1(X)^{1/2}, \ldots, \mathcal{Z}_N(X)^{1/2}),$$
$$A + B\mathcal{F}(X) \triangleq (A_1 + B_1\mathcal{F}_1(X), \ldots, A_N + B_N\mathcal{F}_N(X)).$$

Theorem A.15. *The following assertions are equivalent:*

1. *(A, B, s) is stabilizable and $(s, \mathcal{Z}(X)^{1/2}, A + B\mathcal{F}(X))$ is detectable for some $X \in \mathbb{M}$.*
2. *There exists the stabilizing solution to the CARE (A.4).*

Moreover, if 1) is satisfied for some $X \in \mathbb{M}$ then there exists a unique solution for (A.4) over the set $\{\bar{X} \in \mathbb{M}; \bar{X} \geq X\}$.

Proof. Let us show first that 1) implies 2) and moreover, if 1) is satisfied for some $X \in \mathbb{M}$ then there exists a unique solution for (A.4) over the set $\{\bar{X} \in \mathbb{M}; \bar{X} \geq X\}$. Take $X \in \mathbb{M}$ such that $(s, \mathcal{Z}(X)^{1/2}, A + B\mathcal{F}(X))$ is detectable, and $\bar{X} = (\bar{X}_1, \ldots, \bar{X}_N) \in \mathbb{M}$ such that $\bar{X} \geq X$ and satisfies (A.4). From the hypothesis that (A, B, s) is stabilizable we get, from Theorem A.10, that at least the maximal solution $X^+ = (X_1^+, \ldots, X_N^+) \in \mathbb{M}$ would satisfy these requirements, since it satisfies (A.4) and $X^+ \geq X$. Setting $\bar{F}_i = \mathcal{F}_i(\bar{X})$, $\bar{A}_i = A_i + B_i\bar{F}_i$, $R_i = \mathcal{R}_i(X)$, $F_i = \mathcal{F}_i(X)$, $i \in \mathbb{N}$, and recalling that

$$\bar{X}_i - \bar{A}_i^* \mathcal{E}_i(\bar{X})\bar{A}_i = \mathcal{O}_i(\bar{F}_i)$$

for $i \in \mathbb{N}$, we get from Lemma A.7, (A.8b), that

$$(\bar{X}_i - X_i) - \bar{A}_i^* \mathcal{E}_i(\bar{X} - X)\bar{A}_i = \mathcal{Z}_i(X) + (\bar{F}_i - F_i)^* R_i(\bar{F}_i - F_i) \qquad \text{(A.18)}$$

for $i \in \mathbb{N}$. From the fact that $\mathcal{Z}_i(X) \geq 0$ and $R_i > 0$, $i \in \mathbb{N}$, we can find $\delta > 0$ such that for $i \in \mathbb{N}$,

$$\mathcal{Z}_i(X) + (\bar{F}_i - F_i)^* R_i(\bar{F}_i - F_i) \geq \delta \left(\mathcal{Z}_i(X) + (\bar{F}_i - F_i)^*(\bar{F}_i - F_i) \right). \qquad \text{(A.19)}$$

From Definition A.2 and the detectability hypothesis, we can find $H = (H_1, \ldots, H_N) \in \mathbb{H}^n$ such that $r_\sigma(\mathcal{L}) < 1$, where $\mathcal{L}(.) = (\mathcal{L}_1(.), \ldots, \mathcal{L}_N(.))$ is as in (3.8) with $\Gamma_i = A_i + B_i F_i + H_i \mathcal{Z}_i(X)^{1/2}$, $i \in \mathbb{N}$. Define $\bar{\Phi} = (\bar{\Phi}_1, \ldots, \bar{\Phi}_N) \in \mathbb{H}^{n,n+m}$, $\Phi = (\Phi_1, \ldots, \Phi_N) \in \mathbb{H}^{n,n+m}$, $\Xi = (\Xi_1, \ldots, \Xi_N) \in \mathbb{H}^{n+m,n}$ as

$$\bar{\Phi}_i \triangleq \begin{bmatrix} 0 \\ \bar{F}_i \end{bmatrix}, \quad \Phi_i \triangleq \begin{bmatrix} \mathcal{Z}_i(X)^{1/2} \\ F_i \end{bmatrix}, \quad \Xi_i \triangleq \begin{bmatrix} H_i & B_i \end{bmatrix}.$$

Then it is easy to verify that for $i \in \mathbb{N}$

$$A_i + \Xi_i \Phi_i = A_i + B_i F_i + H_i \mathcal{Z}_i(X)^{1/2} = \Gamma_i$$

$$A_i + \Xi_i \bar{\Phi}_i = A_i + B_i \bar{F}_i = \bar{A}_i$$

and from (A.18) and (A.19) we get

$$
\begin{aligned}
\delta(\bar{\Phi}_i - \Phi_i)^*(\bar{\Phi}_i - \Phi_i) &= \delta(\mathcal{Z}_i(X) + (\bar{F}_i - F_i)^*(\bar{F}_i - F_i)) \\
&\leq (\bar{X}_i - X_i) - \bar{A}_i^* \mathcal{E}_i(\bar{X} - X)\bar{A}_i \\
&= (\bar{X}_i - X_i) - (A_i + \Xi_i \bar{\Phi}_i)^* \mathcal{E}_i(\bar{X} - X)(A_i + \Xi_i \bar{\Phi}_i).
\end{aligned}
$$

Setting $\bar{\mathcal{L}}(.) = (\bar{\mathcal{L}}_1(.), \ldots, \bar{\mathcal{L}}_N(.))$, $\bar{\mathcal{L}}_i(.) = \bar{A}_i^* \mathcal{E}_i(.)\bar{A}_i$, for $i \in \mathbb{N}$, and recalling that $\bar{X} - X \geq 0$, we get from Lemma A.8 that $r_\sigma(\bar{\mathcal{L}}) < 1$. Moreover from the uniqueness of the stabilizing solution proved in Lemma A.14 we have that $\bar{X}$ is the unique solution for the CARE (A.4) in $\{\bar{X} \in \mathbb{M}; \bar{X} \geq X\}$.

Let us show now that 2) implies 1). Suppose that $X^+ = (X_1^+, \ldots, X_N^+) \in \mathbb{M}$ is the stabilizing solution for the CARE (A.4). Then clearly (A, B, s) will be stabilizable and $(s, \mathcal{Z}(X^+)^{1/2}, A + B\mathcal{F}(X^+)) = (s, 0, A + B\mathcal{F}(X^+))$ will be detectable. $\qquad\qquad \square$

We proceed now with some sufficient conditions for the existence of a stabilizing solution. The first one is immediate from Theorem A.15 and (2.5), and resembles the detectability condition in Theorem 2.21.

Corollary A.16. *If (A, B, s) is stabilizable and (s, C, A) is detectable then the stabilizing solution to the CARE (A.4) exists.*

Proof. If (s, C, A) is detectable then we can find $M = (M_1, \ldots, M_N) \in \mathbb{H}^{q,n}$ such that $r_\sigma(\mathcal{L}) < 1$ when $\Gamma_i = A_i + M_i C_i$ in (3.8) for $i \in \mathbb{N}$. From Remark 2.3 and (2.5), we can find $U_i \in \mathbb{B}(\mathbb{C}^n), i \in \mathbb{N}$ such that $C_i = U_i |C_i| = U_i(C_i^* C_i)^{1/2}$. Defining $\bar{M}_i = \frac{1}{(\varsigma_i)^{1/2}} M_i U_i$, $\bar{M} = (\bar{M}_1, \ldots, \bar{M}_N)$ and recalling that $\mathcal{Z}_i(0) = \varsigma_i C_i^* C_i$, we have that $\bar{M}$ stabilizes $(s, \mathcal{Z}(0)^{1/2}, A)$ and the result follows after taking $X = 0$ in Theorem A.15 and noticing that $\mathcal{F}(0) = 0$. $\qquad \square$

In the next results we will replace the detectability condition of Theorem A.15 by conditions based on the observable and non-observable modes of each pair $(\mathcal{Z}_i(X)^{1/2}, A_i + B_i\mathcal{F}_i(X))$, $i \in \mathbb{N}$. For the remainder of this section we will assume that all matrices involved in the CARE (A.4) are real. First we present the following definition.

Definition A.17. *Consider* $\Gamma = (\Gamma_1, \ldots, \Gamma_N) \in \mathbb{H}^n$, *with* $r_\sigma(s_{ii}^{1/2}\Gamma_i) < 1$, $i \in \mathbb{N}$. *We define* $\mathcal{A} \in \mathbb{B}(\mathbb{H}^n)$ *in the following way: for* $V = (V_1, \ldots, V_N)$, *we set* $\mathcal{A}(V) = (\mathcal{A}_1(V), \ldots, \mathcal{A}_N(V))$ *as*

$$\mathcal{A}_i(V) \triangleq \sum_{k=0}^{\infty} (s_{ii}^{1/2}\Gamma_i^*)^k \left(\Gamma_i^* \left(\sum_{j=1,j\neq i}^{N} s_{ij}V_j \right) \Gamma_i \right) (s_{ii}^{1/2}\Gamma_i)^k. \tag{A.20}$$

It is clear that since $r_\sigma(s_{ii}^{1/2}\Gamma_i) < 1$ for each $i \in \mathbb{N}$ then $\mathcal{A} \in \mathbb{B}(\mathbb{H}^n)$. Moreover $\mathcal{A}$ maps $\mathbb{H}^{n+}$ into $\mathbb{H}^{n+}$.

In what follows it will be convenient to consider the following norm $\|.\|_{1t}$ in $\mathbb{H}^n$ (see [216], p. 173). For $V = (V_1, \ldots, V_N) \in \mathbb{H}^n$,

$$\|V\|_{1t} = \sum_{i=1}^{N} \mathrm{tr}((V_i^*V_i)^{1/2}) = \sum_{i=1}^{N} \mathrm{tr}(|V_i|).$$

An important property is as follows: If $V(k) = (V_1(k), \ldots, V_N(k)) \in \mathbb{H}^{n+}$ for each $k = 1, \ldots, t$ then, writing $S(t) = (S_1(t), \ldots, S_N(t)) = V(1) + \ldots + V(t)$, it follows that $S_i(t) = V_i(1) + \ldots + V_i(t) \geq 0$ and thus $S(t) \in \mathbb{H}^{n+}$. Moreover $(V_i(t)^*V_i(t))^{1/2} = V_i(t)$, $(S_i(t)^*S_i(t))^{1/2} = S_i(t)$, so that

$$\left\| \sum_{k=1}^{t} V(k) \right\|_{1t} = \|V(1) + \ldots + V(t)\|_{1t} = \|S(t)\|_{1t}$$

$$= \sum_{i=1}^{N} \mathrm{tr}(S_i(t)) = \sum_{i=1}^{N} \mathrm{tr}\left(\sum_{k=1}^{t} V_i(k) \right)$$

$$= \sum_{k=1}^{t} \sum_{i=1}^{N} \mathrm{tr}(V_i(k)) = \sum_{k=1}^{t} \|V(k)\|_{1t}. \tag{A.21}$$

We have the following result, proved in [60]:

Lemma A.18. *Consider* $\mathcal{L}$ *and* $\mathcal{A}$ *as in (3.8) and (A.20) respectively. The following assertions are equivalent:*

1. $r_\sigma(\mathcal{L}) < 1$.

2. $r_\sigma(s_{ii}^{1/2}\Gamma_i) < 1$, *for* $i \in \mathbb{N}$, *and* $r_\sigma(\mathcal{A}) < 1$.

3. $r_\sigma(s_{ii}^{1/2}\Gamma_i) < 1$, *for* $i \in \mathbb{N}$, *and* $\sum_{k=0}^{\infty} \mathcal{A}^k(L) < \infty$, *for some* $L > 0$ *in* $\mathbb{H}^{n+}$.

Proof. Let us show first that 2) and 3) are equivalent. Indeed, if 2) holds then from Proposition 2.5 we have that 3) also holds. On the other hand, suppose that 3) holds. From continuity of the norm operator, and recalling that, since $\mathcal{A}$ maps $\mathbb{H}^{n+}$ into $\mathbb{H}^{n+}$, so that $\mathcal{A}_i^k(L) \in \mathbb{H}^{n+}$ for all $k = 0, 1, \ldots$, it follows from (A.21) that

$$\left\| \lim_{t\to\infty} \sum_{k=0}^{t} \mathcal{A}^k(L) \right\|_{1t} = \lim_{t\to\infty} \left\| \sum_{k=0}^{t} \mathcal{A}^k(L) \right\|_{1t}$$

$$= \lim_{t\to\infty} \sum_{k=0}^{t} \left\| \mathcal{A}^k(L) \right\|_{1t}.$$

Since $L > 0$, we can find, for each $H \in \mathbb{H}^{n+}$, some $\rho > 0$ such that $0 \le H \le \rho L$. Linearity of the operator $\mathcal{A}$ yields for each $k = 0, 1, \ldots$,

$$0 \le \mathcal{A}^k(H) \le \rho \mathcal{A}^k(L),$$

and thus

$$\lim_{t\to\infty} \sum_{k=0}^{t} \left\| \mathcal{A}^k(H) \right\|_{1t} \le \rho \lim_{t\to\infty} \sum_{k=3}^{t} \left\| \mathcal{A}^k(L) \right\|_{1t} < \infty$$

and from Proposition 2.5, $r_\sigma(\mathcal{A}) < 1$. Let us show now that 1) and 2) are equivalent. Suppose that 2) holds and consider $V = (V_1, \ldots, V_N) \in \mathbb{H}^{n+}$, $V_i > 0$ for each $i \in \mathbb{N}$. Define $\mathcal{S}(H) = (\mathcal{S}_1(H), \ldots, \mathcal{S}_N(H))$ for $H = (H_1, \ldots, H_N) \in \mathbb{H}^n$ as $\mathcal{S}(H) = L + \mathcal{A}(H)$ where $L = (L_1, \ldots, L_N) \in \mathcal{H}^{n+}$ and

$$L_i = \sum_{k=0}^{\infty} (s_{ii}^{1/2} \Gamma_i^*)^k V_i (s_{ii}^{1/2} \Gamma_i)^k \ge V_i > 0, \quad i \in \mathbb{N}.$$

By induction, it follows that $\mathcal{S}^t(H) = \mathcal{A}^t(H) + \sum_{k=0}^{t-1} \mathcal{A}^{t-1-k}(L)$, and from the hypothesis that $r_\sigma(\mathcal{A}) < 1$ we have that (cf. Proposition 2.6) there exists a unique $Q \in \mathbb{H}^n$ such that $\mathcal{S}(Q) = Q = \sum_{k=0}^{\infty} \mathcal{A}^k(L) = \lim_{t\to\infty} \mathcal{S}^t(H)$, and hence $Q \ge L > 0$. Thus for each $i \in \mathbb{N}$,

$$Q_i = \mathcal{S}_i(Q) = \sum_{k=0}^{\infty} (s_{ii}^{1/2} \Gamma_i^*)^k \left\{ V_i + \Gamma_i^* \Big(\sum_{j\neq i} s_{ij} Q_j \Big) \Gamma_i \right\} (s_{ii}^{1/2} \Gamma_i)^k$$

and therefore $Q_i - \Gamma_i^*(\mathcal{E}_i(Q))\Gamma_i = V_i$, $i \in \mathbb{N}$, that is, $Q - \mathcal{L}(Q) = V$. From Theorem 3.9 in Chapter 3 we have that $r_\sigma(\mathcal{L}) < 1$. Let us show now that 1) implies 3) (and consequently 2)). Indeed we have from Theorem 3.9 that for any $V = (V_1, \ldots, V_N) > 0$ in $\mathbb{H}^{n+}$ there exists $Q = (Q_1, \ldots, Q_N) \in \mathcal{H}^{n+}$, $Q > 0$, such that $Q - \mathcal{L}(Q) = V$. Thus for each $i \in \mathbb{N}$,

$$Q_i - s_{ii} \Gamma_i^* Q_i \Gamma_i = V_i + \Gamma_i^* \Big(\sum_{j\neq i} s_{ij} Q_j \Big) \Gamma_i = V_i > 0.$$

From the above Lyapunov equation and standard results in linear systems (see Theorem 2.14) it follows that $r_\sigma(s_{ii}^{1/2}\Gamma_i) < 1$ for each $i \in \mathbb{N}$. Moreover, for $i \in \mathbb{N}$,

$$Q_i = \sum_{k=0}^\infty (s_{ii}^{1/2}\Gamma_i^*)^k \left\{ V_i + \Gamma_i^* \left(\sum_{j\neq i} s_{ij}Q_j \right) \Gamma_i \right\} (s_{ii}^{1/2}\Gamma_i)^k = L_i + \mathcal{A}_i(Q)$$

where

$$L_i = \sum_{k=0}^\infty (s_{ii}^{1/2}\Gamma_i^*)^k V_i (s_{ii}^{1/2}\Gamma_i)^k \geq V_i > 0, \quad i \in \mathbb{N}.$$

Writing $L = (L_1, \ldots, L_N) > 0$, we have $Q = L + \mathcal{A}(Q)$. Iterating this equation yields

$$Q = \mathcal{A}^{t+1}(Q) + \sum_{k=0}^t \mathcal{A}^k(L) \geq \sum_{k=0}^t \mathcal{A}^k(L) > 0$$

which shows that $\lim_{t\to\infty} \sum_{k=0}^t \mathcal{A}^k(L)$ exists for some $L > 0$ and therefore 3) holds. □

We also need the following result.

Lemma A.19. *Suppose that (A, B, s) is stabilizable, and for some $X \in \mathbb{M}$, $(\mathcal{Z}_i(X)^{1/2}, A_i + B_i\mathcal{F}_i(X))$ has no unobservable modes over the unitary complex disk for each $i \in \mathbb{N}$. Then $r_\sigma(s_{ii}^{1/2}A_i^+) < 1$ for each $i \in \mathbb{N}$, where $A_i^+ = A_i + B_i\mathcal{F}_i(X^+)$ and X^+ is the maximal solution of the CARE (A.4) in $\mathbb{M}$.*

Proof. From the hypothesis made and Theorem A.10, it is clear that there exists the maximal solution X^+ for the CARE (A.4) in $\mathbb{M}$. As seen in Theorem A.10, $r_\sigma(\mathcal{L}^+) \leq 1$ and from this it is easy to verify that $r_\sigma(s_{ii}^{1/2}A_i^+) \leq 1$ for each $i \in \mathbb{N}$. Suppose by contradiction that, for some $i \in \mathbb{N}$ and some $\lambda \in \mathbb{C}$, with $\|\lambda\| = 1$, and $x \neq 0$ in $\mathbb{C}^n$,

$$s_{ii}^{1/2} A_i^+ x = \lambda x.$$

From (A.18) and setting $F^+ = \mathcal{F}(X^+)$, $F = \mathcal{F}(X)$, we get that

$$0 = x^*((X_i^+ - X_i) - s_{ii}A_i^{+*}(X_i^+ - X_i)A_i^+)x$$

$$= x^* A_i^{+*} \left(\sum_{j=1,j\neq i}^N s_{ij}(X_j^+ - X_j) \right) A_i^+ x$$

$$+ x^*(\mathcal{Z}_i(X) + (F_i^+ - F_i)^*\mathcal{R}_i(X)(F_i^+ - F_i))x$$

and since $X^+ - X \geq 0$, $\mathcal{R}(X) > 0$, $\mathcal{Z}(X) \geq 0$, we conclude that

$$x^* A_i^{+*} \left(\sum_{j=1,j\neq i}^N s_{ij}(X_j^+ - X_j) \right) A_i^+ x = 0,$$

$$x^* \mathcal{Z}_i(X)x = 0,$$

$$x^*(F_i^+ - F_i)^* \mathcal{R}_i(X)(F_i^+ - F_i)x = 0,$$

which implies that $\mathcal{Z}_i^{1/2}(X)x = 0, (F_i^+ - F_i)x = 0$. Thus

$$s_{ii}^{1/2}(A_i + B_i F_i^+)x = s_{ii}^{1/2}(A_i + B_i F_i)x = \lambda x,$$

which is equivalent to saying that λ is an unobservable mode of $(\mathcal{Z}_i(X)^{1/2}, A_i + B_i F_i)$ over the unitary complex disk, in contradiction to the hypothesis made. $\qquad \square$

In what follows, we define

$$s_{ij}^T \triangleq \sum_{k=1}^N s_{ik}^{T-1} s_{kj}, \quad T = 2, 3, \dots, \quad s_{ij}^1 = s_{ij}$$

and set $\min\{\emptyset\} \triangleq \infty$.

From Lemmas A.18 and A.19, we have the following result.

Theorem A.20. *Suppose that (A, B, s) is stabilizable. Suppose also that for some $X \in \mathbb{M}$ and each $i \in \mathbb{N}$, one of the conditions below is satisfied:*

1. *$(\mathcal{Z}_i(X)^{1/2}, s_{ii}^{1/2}(A_i + B_i \mathcal{F}_i(X)))$ has no unobservable modes inside the closed unitary complex disk, or*
2. *$(\mathcal{Z}_i(X)^{1/2}, s_{ii}^{1/2}(A_i + B_i \mathcal{F}_i(X)))$ has no unobservable modes over the unitary complex disk, 0 is not an unobservable mode of $(\mathcal{Z}_i(X)^{1/2}, A_i + B_i \mathcal{F}_i(X))$ and*

$$\zeta(i) \triangleq \min\{T \geq 1; s_{ij}^T > 0 \text{ for some } j \in \Upsilon\} < \infty$$

where $\Upsilon \triangleq \{\kappa \in \mathbb{N}; \kappa \text{ satisfies condition 1)}\}$.

Then X^+, the maximal solution of the CARE (A.4) in $\mathbb{M}$, is also the mean square stabilizing solution. Moreover, $X^+ - X > 0$.

Proof. From the hypothesis made, Theorem A.10 and Lemma A.19, it is clear that there is a maximal solution $X^+ \in \mathbb{M}$ and that $r_\sigma(s_{ii}^{1/2} A_i^+) < 1$, for each $i \in \mathbb{N}$. Notice now that

$$(X_i^+ - X_i) - s_{ii} A_i^{+*}(X_i^+ - X_i)A_i^+ = A_i^{+*}\left(\sum_{j=1, j \neq i}^N s_{ij}(X_j^+ - X_j)\right)A_i^+ + Z_i^* Z_i$$

where $Z_i^* = [\mathcal{Z}_i(X)^{1/2} (F_i^+ - F_i)^* \mathcal{R}_i(X)^{1/2}]$. This implies that

$$(X^+ - X) = H + \mathcal{A}(X^+ - X) \geq H \qquad (A.22)$$

where $\mathcal{A}(.) = (\mathcal{A}(.), \ldots, \mathcal{A}(.))$ is defined as in (A.20) with $\Gamma = A^+$, $F^+ = \mathcal{F}(X^+)$, $F = \mathcal{F}(X)$, and $H = (H_1, \ldots, H_N)$ is defined as

$$H_i = \sum_{k=0}^{\infty} (s_{ii}^{1/2} A_i^{+*})^k Z_i^* Z_i (s_{ii}^{1/2} A_i^+)^k \geq 0. \tag{A.23}$$

Iterating (A.22) we have

$$X^+ - X = \mathcal{A}^{t-1}(X^+ - X) + \sum_{k=0}^{t} \mathcal{A}^k(H) \geq \sum_{k=0}^{t} \mathcal{A}^k(H) \geq 0 \tag{A.24}$$

which shows that $\sum_{k=0}^{\infty} \mathcal{A}^k(H)$ exists. If we can show that for some integer ζ,

$$\sum_{k=0}^{\zeta} \mathcal{A}^k(H) > 0, \tag{A.25}$$

then

$$0 \leq \sum_{k=0}^{\infty} \mathcal{A}^k \left(\sum_{s=0}^{\zeta} \mathcal{A}^s(H) \right) = \sum_{k=s}^{\infty} \mathcal{A}^k(H) \leq \sum_{k=0}^{\infty} \mathcal{A}^k(H)$$

and from Lemma A.18 we get $r_\sigma(\mathcal{L}^+) < 1$. It remains to prove that (A.25) holds for some integer ζ. We shall show first that for any $i \in \Upsilon$, $H_i > 0$. From (A.23), it is enough to show that $(Z_i, s_{ii}^{1/2} A_i^+)$ is observable (see Theorem 2.19). By contradiction, suppose that $\lambda \in \mathbb{C}$ is an unobservable mode of $(Z_i, s_{ii}^{1/2} A_i^+)$, that is, for some $x \neq 0$ in $\mathbb{C}^n$,

$$s_{ii}^{1/2} A_i^+ x = \lambda x,$$

$$Z_i x = \begin{bmatrix} \mathcal{Z}_i(X)^{1/2} x \\ \mathcal{R}_i(X)^{1/2}(F_i^+ - F_i)x \end{bmatrix} = 0,$$

which implies that $F_i x = F_i^+ x$, and $\mathcal{Z}_i(X)^{1/2} x = 0$. Thus we get that

$$s_{ii}^{1/2} A_i^+ x = s_{ii}^{1/2}(A_i + B_i F_i^+)x = s_{ii}^{1/2}(A_i + B_i F_i)x = \lambda x,$$

$$\mathcal{Z}_i(X)^{1/2} x = 0,$$

that is, λ is an unobservable mode of $(\mathcal{Z}_i(X)^{1/2}, s_{ii}^{1/2}(A_i + B_i F_i))$. Since $r_\sigma(s_{ii}^{1/2}(A_i + B_i F_i^+)) < 1$, we must have that $\|\lambda\| < 1$, which is a contradiction with 1).

Suppose now that i satisfies condition 2). Set $T = \zeta(i)$. Since $s_{ij}^T > 0$ for some $j \in \Upsilon$ and finite T, and T is the minimal integer with this property, we can find a sequence of distinct elements $\{i_0, i_1, \ldots, i_{T-1}, i_T\}$, $i_0 = i, i_T = j$, such that $s_{ii_1} s_{i_1 i_2} \ldots s_{i_{T-1}j} > 0$ and each $i_k, k = 0, \ldots, T-1$, satisfies condition 2) (otherwise T would not be the minimum). Let us show by induction that

$$H_j > 0$$

$$H_{i_{T-1}} + \mathcal{A}_{i_{T-1}}(H) > 0$$

$$\vdots$$

$$H_i + \mathcal{A}_i(H) + \ldots + \mathcal{A}_i^T(H) > 0.$$

As seen in 1), $H_j > 0$. Suppose that $H_{i_k} + \mathcal{A}_{i_k}(H) + \ldots + \mathcal{A}_{i_k}^{T-k}(H) > 0$. Let us show that

$$H_{i_{k-1}} + \mathcal{A}_{i_{k-1}}(H) + \ldots + \mathcal{A}_{i_{k-1}}^{T-k+1}(H) > 0.$$

Suppose by contradiction that for some $x \neq 0$ in $\mathbb{C}^n$, $(H_{i_{k-1}} + \mathcal{A}_{i_{k-1}}(H) + \ldots + \mathcal{A}_{i_{k-1}}^{T-k+1}(H))x = 0$. Then we must have $H_{i_{k-1}}x = 0, \ldots, \mathcal{A}_{i_{k-1}}^{T-k+1}(H)x = 0$, and thus

$$
\begin{aligned}
0 &= x^* \big(\mathcal{A}_{i_{k-1}}(H) + \ldots + \mathcal{A}_{i_{k-1}}^{T-k+1}(H) \big) x \\
&= x^* \mathcal{A}_{i_{k-1}} \big(H + \mathcal{A}(H) + \ldots + \mathcal{A}^{T-k}(H) \big) x \\
&= x^* \bigg(\sum_{s=0}^{\infty} (s_{i_{k-1}i_{k-1}}^{1/2} A_{i_{k-1}}^{+*})^s A_{i_{k-1}}^{+*} \bigg(\sum_{l=1, l \neq i_{k-1}}^{N} s_{i_{k-1}l}(H_l + \ldots \\
&\quad + \mathcal{A}_l^{T-k}(H)) \bigg) (s_{i_{k-1}i_{k-1}}^{1/2} A_{i_{k-1}}^{+})^s A_{i_{k-1}}^{+} \bigg) x \\
&\geq x^* \bigg(A_{i_{k-1}}^{+*} \bigg(\sum_{l=1, l \neq i_{k-1}}^{N} s_{i_{k-1}l}(H_l + \ldots + \mathcal{A}_l^{T-k}(H)) \bigg) A_{i_{k-1}}^{+} \bigg) x \geq 0
\end{aligned}
$$

and since $s_{i_{k-1}i_k} > 0$, $i_{k-1} \neq i_k$, $H_{i_k} + \ldots + \mathcal{A}_{i_k}^{T-k}(H) > 0$, we conclude that $A_{i_{k-1}}^{+}x = 0$. Notice now that $H_{i_{k-1}}x = 0$ implies (see (A.23)) that $Z_{i_{k-1}}x = 0$, and thus $\mathcal{Z}_{i_{k-1}}(X)x = 0$ and $F_{i_{k-1}}^{+}x = F_{i_{k-1}}x$. Therefore,

$$A_{i_{k-1}}^{+}x = (A_{i_{k-1}} + B_{i_{k-1}}F_{i_{k-1}}^{+})x = (A_{i_{k-1}} + B_{i_{k-1}}F_{i_{k-1}})x = 0$$

$$\mathcal{Z}_{i_{k-1}}(X)^{1/2}x = 0$$

which implies that 0 is an unobservable mode of $(\mathcal{Z}_{i_{k-1}}(X)^{1/2}, A_{i_{k-1}} + B_{i_{k-1}}F_{i_{k-1}})$, contradicting hypothesis 2) of the theorem. Therefore (A.25) is satisfied for $\zeta = \max\{\zeta(i); i \notin \Upsilon\}$. Finally, notice from (A.24) and (A.25) that

$$X^+ - X \geq \sum_{s=0}^{\zeta} \mathcal{A}^s(H) > 0.$$

$\square$

Corollary A.21. *Suppose that (A, B, s) is stabilizable. Suppose also that for some $X \in \mathbb{M}$ and each $i \in \mathbb{N}$, one of the conditions below is satisfied:*

1. $(\mathcal{Z}_i(X)^{1/2}, s_{ii}^{1/2}(A_i + B_i\mathcal{F}_i(X)))$ is observable, or
2. $(\mathcal{Z}_i(X)^{1/2}, s_{ii}^{1/2}(A_i + B_i\mathcal{F}_i(X)))$ is detectable, 0 is not an unobservable mode of $(\mathcal{Z}_i(X)^{1/2}, A_i + B_i\mathcal{F}_i(X))$, and $\zeta(i) < \infty$.

Then X^+, the maximal solution of the CARE (A.4) in $\mathbb{M}$, is the unique solution such that $X^+ - X \geq 0$. Moreover, $X^+ - X > 0$ and X^+ is the stabilizing solution of the CARE.

Proof. As seen in Theorem A.12, X^+ is the unique mean square stabilizing solution of the CARE (A.4) and $X^+ - X > 0$. Suppose that $\bar{X}$ is another solution of the CARE and that $\bar{X} - X \geq 0$ (thus $\bar{X} \in \mathbb{M}$). If we show that $r_\sigma(A_i + B_i\mathcal{F}_i(\bar{X})) < 1$, $i = 1, \ldots, N$, then by repeating the same arguments of the proof of Theorem A.12, we get that $\bar{X}$ is a stabilizing solution, and thus $\bar{X} = X^+$. Suppose by contradiction that for some $i \in \mathbb{N}$ and some $\lambda \in \mathbb{C}$, with $\|\lambda\| \geq 1$, and $x \neq 0$ in $\mathbb{C}^n$,

$$s_{ii}^{1/2}(A_i + B_i\mathcal{F}_i(\bar{X}))x = \lambda x.$$

Then from Lemma A.7, (A.8b) and setting $\bar{F}_i = \mathcal{F}_i(\bar{X})$, we get that

$$
\begin{aligned}
0 &= x^*((\bar{X}_i - X_i) - s_{ii}(A_i + B_i\bar{F}_i)^*(\bar{X}_i - X_i)(A_i + B_i\bar{F}_i))x \\
&= (1 - \|\lambda\|)x^*(\bar{X}_i - X_i)x \\
&= x^*\Bigg((A_i + B_i\bar{F}_i)^*\Bigg(\sum_{j=1,j\neq i}^N s_{ij}(\bar{X}_j - X_j)\Bigg)(A_i + B_i\bar{F}_i) \\
&\qquad + \mathcal{Z}_i(X) + (\bar{F}_i - F_i)^*\mathcal{R}_i(X)(\bar{F}_i - F_i)\Bigg)x
\end{aligned}
$$

and since $\|\lambda\| \geq 1$ and $\bar{X} - X \geq 0, \mathcal{R}(X) > 0, \mathcal{Z}(X) \geq 0$, we can conclude that

$$x^*\Bigg((A_i + B_i\bar{F}_i)^*\Bigg(\sum_{j=1,j\neq i}^N s_{ij}(\bar{X}_j - X_j)\Bigg)(A_i + B_i\bar{F}_i)\Bigg)x = 0,$$

$$x^*\mathcal{Z}_i(X)x = 0,$$

$$x^*(\bar{F}_i - F_i)^*\mathcal{R}_i(X)(\bar{F}_i - F_i)x = 0$$

and therefore

$$s_{ii}^{1/2}(A_i + B_i\bar{F}_i)x = s_{ii}^{1/2}(A_i + B_iF_i)x = \lambda x,$$

$$\mathcal{Z}_i(X)^{1/2}x = 0,$$

which shows that λ is an unobservable mode of $(\mathcal{Z}_i(X)^{1/2}, s_{ii}^{1/2}(A_i + B_iF_i))$, contradicting (1) or (2). □

Remark A.22. As in Corollary A.16, by taking $X = 0$ and using (2.5), the above results could be written in terms of the non-observable modes of each pair (C_i, A_i), $i \in \mathbb{N}$.

A.4 Asymptotic Convergence

In this section we will assume that the stabilizing solution, named $\widehat{X}$, for (A.4) exists. We will establish the asymptotic convergence of the sequence $X(k)$, defined in (A.3), to $\widehat{X}$. For this end we need to define the following sequences $\alpha(k) = (\alpha_1(k), \ldots, \alpha_N(k))$ and $P(k) = (P_1(k), \ldots, P_N(k))$:

$$\alpha_i(k) = \inf_{l \geq k} \varsigma_i(l),$$

$$P_i(k+1) = (A_i + B_i \mathcal{F}_i(\widehat{X}))^* \mathcal{E}_i(P(k))(A_i + B_i \mathcal{F}_i(\widehat{X})) + \mathcal{O}_i(F(\widehat{X}), \varsigma(k))$$

$$P(0) = X(0).$$

It is immediate to check that $\varsigma_i(k) \geq \alpha_i(k)$, $\alpha_i(k+1) \geq \alpha_i(k) > 0$, and from (A.1) that

$$\lim_{k \to \infty} \alpha_i(k) = \varsigma_i$$

exponentially fast. We have the following result, showing the desired convergence.

Proposition A.23. *Suppose that the stabilizing solution $\widehat{X} \in \mathbb{H}^{n+}$ for (A.4) exists, and it is the unique solution for (A.4) over $\mathbb{H}^{n+}$. Then $X(k)$ defined in (A.3) converges to $\widehat{X}$ as k goes to infinity whenever $X(0) \in \mathbb{H}^{n+}$.*

Proof. Define $V(k) = (V_1(k), \ldots, V_N(k))$ as

$$V(k+1) = \mathcal{X}(V(k), \alpha(k)),$$

$$V(0) = 0.$$

Let us show by induction on k that

$$0 \leq V(k) \leq V(k+1), \tag{A.26}$$

$$V(k) \leq X(k) \leq P(k). \tag{A.27}$$

For (A.26), the result is clearly true for $k = 1$. Suppose it holds for k, that is, $0 \leq V(k-1) \leq V(k)$. Then

$$\mathcal{R}_i(V(k), \alpha(k)) > 0$$

and from (A.2),

$$\begin{aligned}
V_i(k+1) =& (A_i + B_i \mathcal{F}_i(V(k), \alpha(k)))^* \mathcal{E}_i(V(k))(A_i + B_i \mathcal{F}_i(V(k), \alpha(k))) \\
&+ \mathcal{O}_i(\mathcal{F}_i(V(k), \alpha(k)), \alpha(k)), \\
V_i(k) =& (A_i + B_i \mathcal{F}_i(V(k), \alpha(k)))^* \mathcal{E}_i(V(k-1))(A_i + B_i \mathcal{F}_i(V(k), \alpha(k))) \\
&+ \mathcal{O}_i(\mathcal{F}_i(V(k), \alpha(k)), \alpha(k-1)) \\
&- (\mathcal{F}_i(V(k), \alpha(k)) - \mathcal{F}_i(V(k-1), \alpha(k-1)))^* \mathcal{R}_i(V(k-1), \alpha(k-1)) \\
&\times (\mathcal{F}_i(V(k), \alpha(k)) - \mathcal{F}_i(V(k-1), \alpha(k-1))).
\end{aligned}$$

By taking the difference and recalling that $\alpha(k) \geq \alpha(k-1)$, so that

$$\mathcal{O}_i(\mathcal{F}_i(V(k), \alpha(k)), \alpha(k)) \geq \mathcal{O}_i(\mathcal{F}_i(V(k), \alpha(k)), \alpha(k-1))$$

we get that

$$
\begin{aligned}
V_i(k+1) - V_i(k) \geq &(A_i + B_i \mathcal{F}_i(V(k), \alpha(k)))^* \\
&\times \mathcal{E}_i(V(k) - V(k-1))(A_i + B_i \mathcal{F}_i(V(k), \alpha(k))) \geq 0
\end{aligned}
$$

showing (A.26). For (A.27) with $k = 0$ we have by definition that

$$0 = V(0) \leq X(0) = P(0).$$

Suppose that (A.27) holds for k. Then from (A.2) again,

$$
\begin{aligned}
X_i(k+1) =& (A_i + B_i \mathcal{F}_i(X(k), \varsigma(k)))^* \mathcal{E}_i(X(k))(A_i + B_i \mathcal{F}_i(X(k), \varsigma(k))) \\
&+ \mathcal{O}_i(\mathcal{F}_i(X(k), \varsigma(k)), \varsigma(k)), \\
V_i(k+1) =& (A_i + B_i \mathcal{F}_i(X(k), \varsigma(k)))^* \mathcal{E}_i(V(k))(A_i + B_i \mathcal{F}_i(X(k), \varsigma(k))) \\
&+ \mathcal{O}_i(\mathcal{F}_i(X(k), \varsigma(k)), \alpha(k)) \\
&- (\mathcal{F}_i(X(k), \varsigma(k)) - \mathcal{F}_i(V(k), \alpha(k)))^* \mathcal{R}_i(V(k-1), \alpha(k-1)) \\
&\times (\mathcal{F}_i(X(k), \varsigma(k)) - \mathcal{F}_i(V(k), \alpha(k))).
\end{aligned}
$$

By taking the difference and recalling that $\varsigma(k) \geq \alpha(k)$, so that

$$\mathcal{O}_i(\mathcal{F}_i(X(k), \varsigma(k)), \varsigma(k)) \geq \mathcal{O}_i(\mathcal{F}_i(X(k), \varsigma(k)), \alpha(k)),$$

we get that

$$
\begin{aligned}
X_i(k+1) - V_i(k+1) \geq &(A_i + B_i \mathcal{F}_i(X(k), \varsigma(k)))^* \\
&\times \mathcal{E}_i(X(k) - V(k))(A_i + B_i \mathcal{F}_i(X(k), \varsigma(k))) \\
&\geq 0.
\end{aligned}
$$

Once more, from (A.2), we have that

$$
\begin{aligned}
X_i(k+1) =& (A_i + B_i \mathcal{F}_i(\widehat{X}))^* \mathcal{E}_i(X(k))(A_i + B_i \mathcal{F}_i(\widehat{X})) + \mathcal{O}_i(\widehat{\mathcal{F}}_i(X), \varsigma(k)) \\
&- (\mathcal{F}_i(X(k), \varsigma(k)) - \mathcal{F}_i(\widehat{X}))^* \mathcal{R}_i(X(k), \varsigma(k))(\mathcal{F}_i(X(k), \varsigma(k)) - \mathcal{F}_i(\widehat{X}))
\end{aligned}
$$

so that

$$
\begin{aligned}
P_i(k+1) - X_i(k+1) \geq &(A_i + B_i \mathcal{F}_i(\widehat{X}))^* \mathcal{E}_i(P(k) - X(k))(A_i + B_i \mathcal{F}_i(\widehat{X})) \\
&\geq 0
\end{aligned}
$$

completing the proof of (A.27). From Proposition 3.36 in Chapter 3 and the fact that $\mathcal{F}(X)$ stabilizes (A, B, s), we have that $P(k)$ converges to a $P \in \mathbb{H}^{n+}$, which is the unique solution of the set of matricial equations in $S = (S_1, \ldots, S_N) \in \mathbb{H}^{n+}$

$$S_i = (A_i + B_i \mathcal{F}_i(\widehat{X}))^* \mathcal{E}_i(S)(A_i + B_i \mathcal{F}_i(\widehat{X})) + \mathcal{O}_i(\mathcal{F}_i(\widehat{X}), \varsigma). \qquad (A.28)$$

Since $\widehat{X}$ also satisfies the above equation, we have from the uniqueness of (A.28) that $\widehat{X} = P$. From (A.26) the sequence $V(k)$ is monotone increasing, and bounded above by $\widehat{X}$. Thus there exists $V = (V_1, \ldots, V_N) \in \mathbb{H}^{n+}$ such that $V(k)$ converges to V, and moreover V satisfies $V = \mathcal{X}(V)$. From uniqueness of the solution of (A.4) over $\mathbb{H}^{n+}$, it follows that $V = \widehat{X}$. This and (A.27) shows that $X(k)$ also converges to $\widehat{X}$.   □

Remark A.24. As in Corollary A.16, by taking $X = 0$ in Theorem A.15 and using (2.5), the above result could be written in terms of the detectability of (s, C, A) and stabilizability of (A, B, s), leading to a result similar to Theorem 2.21.

B

Auxiliary Results for the Linear Filtering Problem with $\theta(k)$ Unknown

In this appendix we present some proofs of the results of Sections 5.4 and 5.5.

B.1 Optimal Linear Filter

B.1.1 Proof of Theorem 5.9 and Lemma 5.11

We present in this subsection the proof of Theorem 5.9 and Lemma 5.11.

Proof of Theorem 5.9.

Proof. Since

$$y(k) = L(k)z(k) + H_{\theta(k)}(k)w(k), \tag{B.1}$$
$$E(y(k)) = L(k)q(k) \tag{B.2}$$

we get that

$$y^c(k) = L(k)z^c(k) + H_{\theta(k)}(k)w(k). \tag{B.3}$$

From the independence of $w(k)$ and $\{\theta(k), (y^c)^{k-1}\}$, we have that

$$
\begin{aligned}
\langle \gamma^* H_{\theta(k)}(k)w(k); \alpha^*(y^c)^{k-1} \rangle &= E(\gamma^* H_{\theta(k)}(k)w(k)\alpha^*(y^c)^{k-1}) \\
&= E(w(k)^*)E(H_{\theta(k)}^* \gamma \alpha^*(y^c)^{k-1}) \\
&= 0 \tag{B.4}
\end{aligned}
$$

showing that $H(k)w(k)$ is orthogonal to $\mathfrak{L}((y^c)^{k-1})$. Similar reasoning shows the orthogonality between $H(k)w(k)$ and $\tilde{z}(k|k-1)$. Thus from (B.1) and (B.3) and recalling from the results seen in Subsection 5.4.1 that $\hat{z}^c(k|k-1) \in \mathfrak{L}((y^c)^{k-1})$ and $\tilde{z}(k|k-1)$ is orthogonal to $\mathfrak{L}((y^c)^{k-1})$ we obtain that (note that $H(k)H(k)^* > 0$ from (5.3))

$$\widehat{y}^c(k|k-1) = \mathsf{L}(k)\widehat{z}^c(k|k-1), \tag{B.5}$$

$$\widetilde{y}(k|k-1) = \mathsf{L}(k)\widetilde{z}(k|k-1) + H_{\theta(k)}w(k), \tag{B.6}$$

$$E(\widetilde{y}(k|k-1)\widetilde{y}(k|k-1)^*) = \mathsf{L}(k)\widetilde{Z}(k|k-1)\mathsf{L}(k)^* + \sum_{j=1}^N \pi_{ij}(k)H_i(k)H_i(k)^*$$

$$= \mathsf{L}(k)\widetilde{Z}(k|k-1)\mathsf{L}(k)^* + \mathsf{H}(k)\mathsf{H}(k)^* > 0, \tag{B.7}$$

$$E(z^c(k)\widetilde{y}(k|k-1)^*) = E((\widetilde{z}(k|k-1) + \widehat{z}^c(k|k-1)\widetilde{y}(k|k-1)^*)$$

$$= \widetilde{Z}(k|k-1)\mathsf{H}(k)^*. \tag{B.8}$$

Recalling that $E(y^c(k)) = 0$ we have from (5.36) that

$$\widehat{z}^c_j(k|k-1) = E(z_j(k) - E(z_j(k)))((y^c)^{k-1})^*) \operatorname{cov}((y^c)^{k-1})^{-1}(y^c)^{k-1}$$

$$= E(E(z_j(k)((y^c)^{k-1})^* | \mathfrak{F}_{k-1}) \operatorname{cov}((y^c)^{k-1})^{-1}(y^c)^{k-1}$$

$$= E(E(\sum_{i=1}^N A_i(k-1)z_i(k-1)\mathbf{1}_{\{\theta(k-1)=i\}}\mathbf{1}_{\{\theta(k)=j\}}$$

$$\times ((y^c)^{k-1})^* | \mathfrak{F}_{k-1})) \operatorname{cov}((y^c)^{k-1})^{-1}(y^c)^{k-1}$$

$$= \sum_{i=1}^N \mathcal{P}(\theta(k) = j|\theta(k-1) = i)E((A_i(k-1)z_i(k-1)$$

$$\times ((y^c)^{k-1})^*)) \operatorname{cov}((y^c)^{k-1})^{-1}(y^c)^{k-1}$$

$$= \sum_{i=1}^N p_{ij}(k-1)A_i(k-1)E(z_i(k-1) - E(z_i(k-1)))$$

$$\times ((y^c)^{k-1})^* \operatorname{cov}((y^c)^{k-1})^{-1}(y^c)^{k-1}$$

$$= \sum_{i=1}^N p_{ij}(k-1)A_i(k-1)\widehat{z}^c_i(k-1|k-1) \tag{B.9}$$

where it can be easily shown that (5.3) implies that $\operatorname{cov}((y^c)^{k-1}) > 0$ for all $k \geq 1$, and in the second equality above we used the fact that $G_{\theta(k-1)}w(k-1)$ and $\mathfrak{L}((y^c)^{k-1})$ are orthogonal (same reasoning as in (B.4) above). From (5.37),

$$\widehat{z}^c(k|k) = \widehat{z}^c(k|k-1)$$

$$+ E(z^c(k)\widetilde{y}(k|k-1)^*)E(\widetilde{y}(k|k-1)\widetilde{y}(k|k-1)^*)^{-1}\widetilde{y}(k|k-1) \tag{B.10}$$

and (B.5)–(B.10) lead to

$$\widehat{z}^c(k|k) = \widehat{z}^c(k|k-1)$$

$$+ \widetilde{Z}(k|k-1)\mathsf{L}(k)^*(\mathsf{L}(k)\widetilde{Z}(k|k-1)\mathsf{L}(k)^* + \mathsf{H}(k)\mathsf{H}(k)^*)^{-1}$$

$$\times (y^c(k) - \mathsf{H}(k)\widehat{z}^c(k|k-1)) \tag{B.11}$$

and

$$\widehat{z}^c(k|k-1) = \mathsf{A}(k)\widehat{z}^c(k-1|k-1), \ \ \widehat{z}^c(0|-1) = 0. \tag{B.12}$$

Equation (5.45) now follows from (B.10) and (B.11) after noting from (B.1) that

$$y^c(k) - \mathsf{L}(k)\widehat{z}^c(k|k-1) = y(k) - \mathsf{L}(k)q(k) - \mathsf{L}(k)(\widehat{z}(k|k-1) - q(k))$$
$$= y(k) - \mathsf{L}(k)\widehat{z}(k|k-1)$$

and that

$$\widehat{z}(k|k-1) = \widehat{z}^c(k|k-1) + q(k),$$

with

$$E(z_j(k+1)) = E(E(z_j(k+1)|\mathfrak{F}_k)) = \sum_{i=1}^{N} p_{ij}(k)A_i(k)E(z_i(k))$$

or, in other words,

$$q(k+1) = \mathsf{A}(k)q(k).$$

Equation (5.46) easily follows from $\widetilde{z}(k|t) = z^c(k) - \widehat{z}^c(k|t)$ and the fact that $\widetilde{z}(k|t)$ is orthogonal to $\mathfrak{L}((y^c)^t)$ and therefore orthogonal to $\widehat{z}^c(k|t)$. Equation (5.47) is readily derived, and (5.48) follows from (5.44) after noting that

$$y(k) - \mathsf{L}(k)\widehat{z}(k|k-1) = \mathsf{L}(k)\widetilde{z}(k|k-1) + H_{\theta(k)}(k)w(k)$$

and recalling that $\widehat{z}^c(k|k-1)$, $\widetilde{z}(k|k-1)$ and $H_{\theta(k)}w(k)$ are all orthogonal among themselves (same reasoning as in (B.4) above). Finally (5.49) follows immediately from (5.45) and (5.43) from the identity $x(k) = \sum_{i=1}^{N} z_i(k)$. □

Proof of Lemma 5.11

Proof. First we define

$$\mathsf{T}(k) \triangleq -\mathsf{A}(k)\widetilde{Z}(k|k-1)\mathsf{L}(k)^*(\mathsf{L}(k)\widetilde{Z}(k|k-1)\mathsf{L}(k)^* + \mathsf{H}(k)\mathsf{H}(k)^*)^{-1} \tag{B.13}$$

(recall from Remark 5.10 that the inverse in (B.13) is well defined). From standard results on Riccati difference equations (see [80] or the identity (A.2)) the recursive equation (5.51) is equivalent to

$$\widetilde{Z}(k+1|k) = (\mathsf{A}(k) + \mathsf{T}(k)\mathsf{L}(k))\widetilde{Z}(k|k-1)(\mathsf{A}(k) + \mathsf{T}(k)\mathsf{L}(k))^* + \mathfrak{V}(Q(k),k)$$
$$+ \mathsf{G}(k)\mathsf{G}(k)^* + \mathsf{T}(k)\mathsf{H}(k)\mathsf{H}(k)^*\mathsf{T}(k)^*. \tag{B.14}$$

Writing (5.38) in terms of $z(k)$ we obtain that

$$z(k+1) = A(k)z(k) + M(k+1)z(k) + \vartheta(k) \tag{B.15}$$

where

$$M(k+1,j) = \begin{bmatrix} m_1(k+1,j) & \dots & m_N(k+1,j) \end{bmatrix},$$
$$m_i(k+1,j) = (\mathbf{1}_{\{\theta(k+1)=j\}} - p_{ij}(k))A_i(k)\mathbf{1}_{\{\theta(k)=i\}},$$
$$M(k+1) = \begin{bmatrix} M(k+1,1) \\ \vdots \\ M(k+1,N) \end{bmatrix},$$
$$\vartheta(k) = \begin{bmatrix} \mathbf{1}_{\{\theta(k+1)=1\}}G_{\theta(k)}(k)w(k) \\ \vdots \\ \mathbf{1}_{\{\theta(k+1)=N\}}G_{\theta(k)}(k)w(k) \end{bmatrix}.$$

From (5.45) and (B.13), we obtain that

$$\hat{z}(k+1|k) = A(k)\hat{z}(k|k-1) + T(k)L(k)\tilde{z}(k|k-1) + T(k)H_{\theta(k)}(k)w(k) \tag{B.16}$$

and thus, from (B.15) and (B.16) we get that:

$$\tilde{z}(k+1|k) = (A(k) + T(k)L(k))\tilde{z}(k|k-1)$$
$$+ M(k+1)z(k) + T(k)H_{\theta(k)}(k)w(k) + \vartheta(k). \tag{B.17}$$

Therefore from (B.17) the recursive equation for $\tilde{Z}(k|k-1)$ is given by

$$\tilde{Z}(k+1|k) = (A(k) + T(k)L(k))\tilde{Z}(k|k-1)(A(k) + T(k)L(k))^*$$
$$+ E(M(k+1)z(k)z(k)^*M(k+1)^*) + E(\vartheta(k)\vartheta(k)^*)$$
$$+ T(k)E(H_{\theta(k)}(k)w(k)w(k)^*H_{\theta(k)}(k)^*)T(k)^* \tag{B.18}$$

and after some algebraic manipulations we obtain that

$$E(M(k+1)z(k)z(k)^*M(k+1)^*) = \mathfrak{V}(Q(k),k), \tag{B.19}$$
$$E(\vartheta(k)\vartheta(k)^*) = G(k)G(k)^*, \tag{B.20}$$
$$E(H_{\theta(k)}(k)w(k)w(k)^*H_{\theta(k)}(k)) = H(k)H(k)^*. \tag{B.21}$$

Replacing (B.19), (B.20) and (B.21) into (B.18), we get (B.14).   □

B.1.2 Stationary Filter

In order to establish the convergence of $\tilde{Z}(k|k-1)$ in Theorem 5.12, Section 5.4, we shall need two auxiliary results. Let κ be such that $\inf\limits_{\ell \geq \kappa} \pi_i(\ell) > 0$ for all $i \in \mathbb{N}$ (since $\pi_i(k) \overset{k \to \infty}{\to} \pi_i > 0$ we have that this number exists). Define

$$\alpha_i(k) \triangleq \inf_{\ell \geq k} \pi_i(\ell + \kappa).$$

Obviously

$$\pi_i(k + \kappa) \geq \alpha_i(k) \geq \alpha_i(k - 1), \quad k = 1, 2, \ldots, \quad i \in \mathbb{N} \qquad \text{(B.22)}$$

and $\alpha_i(k) \overset{k \to \infty}{\to} \pi_i$ exponentially fast. Define now $\bar{Q}(k) \triangleq (\bar{Q}_1(k), \ldots, \bar{Q}_N(k)) \in \mathbb{H}^{n+}$ with $\bar{Q}_j(0) = 0, j \in \mathbb{N}$ and

$$\bar{Q}_j(k + 1) = \sum_{i=1}^{N} p_{ij}(A_i \bar{Q}_i(k) A_i^* + \alpha_i(k) G_i G_i^*).$$

In the next lemma recall that $Q = (Q_1, \ldots, Q_N) \in \mathbb{H}^{n+}$ is the unique solution that satisfies (5.58).

Lemma B.1. $\bar{Q}(k) \overset{k \to \infty}{\to} Q$ and for each $k = 0, 1, 2, \ldots$,

$$Q(k + \kappa) \geq \bar{Q}(k) \geq \bar{Q}(k - 1). \qquad \text{(B.23)}$$

Proof. From Proposition 3.36 in Chapter 3, we get that $\bar{Q}(k) \overset{k \to \infty}{\to} Q$. Let us now show (B.23) by induction on k. For $k = 0$ the result is immediate, since $Q(\kappa) \geq 0 = \bar{Q}(0)$ and $\bar{Q}(1) \geq 0 = \bar{Q}(0)$. Suppose that (B.23) holds for k. Then from (B.22) and (B.23) we have that

$$Q_j(k + 1 + \kappa) = \sum_{i=1}^{N} p_{ij}(A_i Q_i(k + \kappa) A_i^* + \pi_i(k + \kappa) G_i G_i^*)$$

$$\geq \sum_{i=1}^{N} p_{ij}(A_i \bar{Q}_i(k) A_i^* + \alpha_i(k) G_i G_i^*)$$

$$= \bar{Q}_j(k + 1) \geq \sum_{i=1}^{N} p_{ij}(A_i \bar{Q}_i(k - 1) A_i^* + \alpha_i(k - 1) G_i G_i^*)$$

$$= \bar{Q}_j(k)$$

completing the induction argument in (B.23). □

Define now

$$R(k + 1) = AR(k)A^* + \mathfrak{V}(\bar{Q}(k)) + \text{diag}\left[\sum_{i=1}^{N} \alpha_i(k) p_{ij} G_i G_i^*\right]$$
$$+ AR(k)L^*(LR(k)L^* + \bar{H}(k)\bar{H}(k)^*)^{-1} LR(k)A^*$$

where $R(0) = 0$ and $\bar{H}(k) = [H_1 \alpha_1(k)^{1/2} \cdots H_N \alpha_N(k)^{1/2}]$. Notice that from the definition of κ and condition (5.3) we have that the inverse of $LR(k)L^* + \bar{H}(k)\bar{H}(k)^*$ is well defined.

Lemma B.2. *For each* $k = 0, 1, \ldots,$

$$0 \le R(k) \le R(k+1) \le \widetilde{Z}(k+1+\kappa|k+\kappa). \qquad (B.24)$$

Proof. Let us show (B.24) by induction on k. Setting

$$S(k) = -AR(k)L^*(LR(k)L^* + \bar{H}(k)\bar{H}(k)^*)^{-1}$$

it follows that, if $R(k) \le \widetilde{Z}(k+\kappa|k-1+\kappa)$, then from (B.22) and (B.23):

$$R(k+1) = (A + T(k+\kappa)L)R(k)(A + T(k+\kappa)L)^* + \mathfrak{V}(\bar{Q}(k))$$

$$+ \operatorname{diag}\left[\sum_{i=1}^{N} \alpha_i(k)p_{ij}G_iG_i^*\right] + T(k+\kappa)\bar{H}(k)\bar{H}(k)^*T(k+\kappa)^*$$

$$- (T(k+\kappa) - S(k))(LR(k)L^* + \bar{H}(k)\bar{H}(k)^*)(T(k+\kappa) - S(k))^*$$

$$\le (A + T(k+\kappa)L)\widetilde{Z}(k+\kappa|k-1+\kappa)(A + T(k+\kappa)L)^*$$

$$+ \mathfrak{V}(Q(k+\kappa)) + \operatorname{diag}\left[\sum_{i=1}^{N} \pi_i(k+\kappa)p_{ij}G_iG_i^*\right]$$

$$+ T(k+\kappa)H(k+\kappa)H(k+\kappa)^*T(k+\kappa)^*$$

$$= \widetilde{Z}(k+1+\kappa|k+\kappa).$$

Obviously $R(0) = 0 \le \widetilde{Z}(\kappa|\kappa - 1)$, showing that $R(k) \le \widetilde{Z}(k+\kappa|k-1+\kappa)$ for all $k = 0, 1, 2, \ldots$. Similarly if $R(k) \ge R(k-1)$, then from (B.22) and (B.23):

$$R(k) = (A + S(k)L)R(k-1)(A + S(k)L)^* + \mathfrak{V}(\bar{Q}(k-1))$$

$$+ \operatorname{diag}\left[\sum_{i=1}^{N} \alpha_i(k-1)p_{ij}G_iG_i^*\right] + S(k)\bar{H}(k-1)\bar{H}(k-1)^*S(k)^*$$

$$- (S(k) - S(k-1))(LR(k-1)L^*$$

$$+ \bar{H}(k-1)\bar{H}(k-1)^*)(S(k) - S(k-1))^*$$

$$\le (A + S(k)L)R(k)(A + S(k)L)^* + \mathfrak{V}(Q(k))$$

$$+ \operatorname{diag}\left[\sum_{i=1}^{N} \alpha_i(k)p_{ij}G_iG_i^*\right] + S(k)\bar{H}(k)\bar{H}(k)^*S(k)^*$$

$$= R(K+1)$$

and since $R(0) = 0 \le R(1)$ the induction argument is completed for (B.24). $\qquad \square$

Proof of Theorem 5.12.

Proof. From MSS of (5.38), we have from Proposition 3.6 in Chapter 3 that $r_\sigma(A) < 1$ and thus according to standard results for algebraic Riccati equations there exists a unique positive semi-definite solution $P \in \mathbb{B}(\mathbb{R}^{Nn})$ to (5.60) and moreover $r_\sigma(A + T(P)L) < 1$ (see [48]). Furthermore P satisfies

$$P = (A + T(P)L)P(A + T(P)L)^* + \mathfrak{V}(Q)$$
$$+ GG^* + T(P)HH^*T(P)^*. \tag{B.25}$$

Define $P(0) = \widetilde{Z}(0| - 1)$ and

$$P(k+1) = (A + T(P)L)P(k)(A + T(P)L)^* + \mathfrak{V}(Q(k))$$
$$+ G(k)G(k)^* + T(P)H(k)H(k)^*T(P)^*. \tag{B.26}$$

Let us show by induction on k that $P(k) \geq \widetilde{Z}(k|k-1)$. Since

$$\widetilde{Z}(k+1|k) = \mathfrak{V}(Q(k)) + G(k)G(k)^*$$
$$+ (A + T(P)L)\widetilde{Z}(k|k-1)(A + T(P)L)^* + T(P)H(k)H(k)^*T(P)^*$$
$$- (T(k) - T(P))(L\widetilde{Z}(k|k-1)L^* + H(k)H(k)^*)(T(k) - T(P))^* \tag{B.27}$$

we have from (B.26) and (B.27) that

$$(P(k+1) - \widetilde{Z}(k+1|k)) = (A + T(P)L)(P(k) - \widetilde{Z}(k|k-1))(A + T(P)L)^*$$
$$+ (T(k) - T(P))(L\widetilde{Z}(k|k-1)L^* + H(k)H(k)^*)(T(k) - T(P))^*. \tag{B.28}$$

By definition $P(0) = \widetilde{Z}(0| - 1)$. Suppose that $P(k) \geq \widetilde{Z}(k|k - 1)$. From (B.28) we have that $P(k + 1) \geq \widetilde{Z}(k + 1|k)$. Therefore we have shown by induction that $P(k) \geq \widetilde{Z}(k|k - 1)$ for all $k = 0, 1, 2, \ldots$. From MSS and ergodicity of the Markov chain we have that $Q(k) \overset{k\to\infty}{\to} Q$, $H(k) \overset{k\to\infty}{\to} H$ and $G(k) \overset{k\to\infty}{\to} G$ exponentially fast. From $r_\sigma(A + T(P)L) < 1$ and the same reasoning as in the proof of Proposition 3.36 in Chapter 3 we get that $P(k) \overset{k\to\infty}{\to} \bar{P}$, where $\bar{P}$ satisfies

$$\bar{P} = (A + T(P)L)\bar{P}(A + T(P)L)^*$$
$$+ \mathfrak{V}(Q) + GG^* + T(P)HH^*T(P)^*. \tag{B.29}$$

and $\bar{P}$ is the unique solution to (B.29). Recalling that P satisfies (B.25), we have that P is also a solution to (B.29) and from uniqueness, $\bar{P} = P$. Therefore

$$\widetilde{Z}(k|k - 1) \leq P(k) \tag{B.30}$$

and $P(k) \overset{k\to\infty}{\to} P$. From (B.30) and (B.24) it follows that $0 \leq R(k) \leq R(k + 1) \leq P(k + 1 + \kappa)$ and thus we can conclude that $R(k) \uparrow R$ whenever $k \to \infty$ for some $R \geq 0$. Moreover, from the fact that $\alpha_i(k) \overset{k\to\infty}{\to} \pi_i$ and $\bar{Q}(k) \overset{k\to\infty}{\to} Q$ we have that R satisfies (5.60). From uniqueness of the positive semi-definite solution to (5.60) we can conclude that $R = P$. From (B.30) and (B.24), $R(k) \leq \widetilde{Z}(k + \kappa|k - 1 + \kappa) \leq P(k + \kappa)$ and since $R(k) \uparrow P$, and $P(k) \to P$ as $k \to \infty$, we get that $\widetilde{Z}(k|k - 1) \overset{k\to\infty}{\to} P$. $\square$

B.2 Robust Filter

We present in this section some proofs of the results of Section 5.5.

Proof of Proposition 5.13.

Proof. For any vector $v^* = \begin{bmatrix} v_1 & \cdots & v_N \end{bmatrix}$, we have that

$$v^* \Delta_i v = \sum_{j=1}^N p_{ij} v_j^2 - \left(\sum_{j=1}^N p_{ij} v_j \right)^2 = E_i\left(V(\theta(1))^2 \right) - E_i\left(V(\theta(1)) \right)^2 \geq 0$$

where $V(j) = v_j$ and $E_i(\cdot)$ denotes the expected value when $\theta(0) = i$. This shows the desired result. □

Proof of Proposition 5.14.

Proof. By straightforward calculations from (5.63) and noting that $B_f L_{\theta(k)} x(k) = B_f L z(k)$, we have that:

$$\begin{aligned}
\widehat{Z}(k+1) =& E\left(\widehat{z}(k+1)\widehat{z}(k+1)^* \right) \\
=& E\big((B_f L z(k) + A_f \widehat{z}(k) + B_f H_{\theta(k)} w(k)) \\
& \times (B_f L z(k) + A_f \widehat{z}(k) + B_f H_{\theta(k)} w(k))^* \big) \\
=& B_f L E(z(k)z(k)^*)L^* B_f^* + A_f E(\widehat{z}(k)\widehat{z}(k)^*)A_f^* \\
& + B_f L E(z(k)\widehat{z}(k)^*)A_f^* + A_f E(\widehat{z}(k)z(k)^*)L^* B_f^* \\
& + B_f \left(\sum_{i=1}^N \pi_i(k) H_i H_i^* \right) B_f^* \\
=& \begin{bmatrix} B_f L & A_f \end{bmatrix} \begin{bmatrix} Z(k) & U(k) \\ U(k)^* & \widehat{Z}(k) \end{bmatrix} \begin{bmatrix} L^* B_f^* \\ A_f^* \end{bmatrix} + B_f H(k)H(k)^* B_f^*. \quad \text{(B.31)}
\end{aligned}$$

Similarly,

$$\begin{aligned}
U_j(k+1) =& E\left(z(k+1,j)\widehat{z}(k+1)^* \right) \\
=& E\big(1_{\{\theta(k)=j\}} \left(A_{\theta(k)} x(k) + G_{\theta(k)} w(k) \right) \\
& \times \left(B_f L z(k) + A_f \widehat{z}(k) + B_f H_{\theta(k)} w(k) \right)^* \big) \\
=& \sum_{i=1}^N p_{ij} \left[A_i Q_i(k) L_i^* B_f^* + A_i U_i(k) A_f^* + \pi_i(k) G_i H_i^* B_f \right]
\end{aligned}$$

and recalling that $G_i H_i^* = 0$, we have that:

$$U_j(k+1) = \sum_{i=1}^N p_{ij} A_i Q_i(k) L_i^* B_f^* + \sum_{i=1}^N p_{ij} A_i U_i(k) A_f^* \quad \text{(B.32)}$$

or, in other words,

$$U_j(k+1) = \begin{bmatrix} p_{1j}A_1 & \cdots & p_{Nj}A_N \end{bmatrix} \begin{bmatrix} Q_1(k) & \cdots & 0 \\ \vdots & \ddots & \vdots \\ 0 & \cdots & Q_N(k) \end{bmatrix} \mathsf{L}^* B_f^*$$

$$+ \begin{bmatrix} p_{1j}A_1 & \cdots & p_{Nj}A_N \end{bmatrix} \begin{bmatrix} U_1(k) \\ \vdots \\ U_N(k) \end{bmatrix} A_f^*$$

so that

$$U(k+1) = \begin{bmatrix} \mathsf{A} & 0 \end{bmatrix} \begin{bmatrix} Z(k) & U(k) \\ U(k)^* & \widehat{Z}(k) \end{bmatrix} \begin{bmatrix} \mathsf{L}^* B_f^* \\ A_f^* \end{bmatrix}. \tag{B.33}$$

From (5.67), (B.31) and (B.33) we obtain (5.69).   □

The next results present necessary and sufficient conditions for System (5.63) to be MSS.

Proposition B.3. *System (5.63) is MSS if and only if $r_\sigma(\mathcal{T}) < 1$ and $r_\sigma(A_f) < 1$.*

Proof. If System (5.63) is MSS then for any initial condition $x_e(0), \theta(0)$,

$$E\big(\|x_e(k)\|^2\big) = E\big(\|x(k)\|^2\big) + E\big(\|\widehat{z}(k)\|^2\big) \overset{k\to\infty}{\longrightarrow} 0,$$

that is, $E\big(\|x_e(k)\|^2\big) \overset{k\to\infty}{\longrightarrow} 0$ and $E\big(\|\widehat{z}(k)\|^2\big) \overset{k\to\infty}{\longrightarrow} 0$. From Theorem 3.9 in Chapter 3, $r_\sigma(\mathcal{T}) < 1$. Consider now an initial condition $x_e(0) = \big(0^* \ \widehat{z}(0)^* \big)^*$. Then clearly $\widehat{z}(k) = A_f^k \widehat{z}(0)$ and since $\widehat{z}(k) \overset{k\to\infty}{\longrightarrow} 0$ for any initial condition $\widehat{z}(0)$, it follows that $r_\sigma(A_f) < 1$.

Suppose now that $r_\sigma(\mathcal{T}) < 1$ and $r_\sigma(A_f) < 1$. From Theorem 3.9 in Chap. 3, $r_\sigma(\mathcal{T}) < 1$ implies that $E\big(\|x(k)\|^2\big) \leq ab^k$ for some $a > 0$, $0 < b < 1$. From (5.63),

$$\widehat{z}(k) = A_f \widehat{z}(0) + \sum_{t=0}^{k-1} A_f^{k-1-t} B_f L_{\theta(t)} x(t)$$

and from the triangular inequality,

$$E\big(\|\widehat{z}(k)\|^2\big)^{1/2} \leq E\big(\|A_f^k \widehat{z}(0)\|^2\big)^{1/2} + \sum_{t=0}^{k-1} E\big(\|A_f^{k-1-t} B_f L_{\theta(t)} x(t)\|^2\big)^{1/2}$$

$$\leq \|A_f^k\| F\big(\|\widehat{z}(0)\|^2\big)^{1/2}$$

$$+ \|B_f\| \|L\|_{\max} \sum_{t=0}^{k-1} \|A_f^{k-1-t}\| E\big(\|x(t)\|^2\big)^{1/2}$$

where we recall that (see Chapter 2) $\|L\|_{\max} = \max\{\|L_i\|, i \in \mathbb{N}\}$. Since $r_\sigma(A_f) < 1$, we can find $a' > 0$, $0 < b' < 1$, such that $\|A_f^k\| \le a'(b')^k$. Then, for some $\bar{a} > 0$, $0 < \bar{b} < 1$,

$$E\big(\|\widehat{z}(k)\|^2\big)^{1/2} \le \bar{a}\big((\bar{b})^k + \sum_{t=0}^{k-1}(\bar{b})^{k-1-t}(\bar{b})^t\big)$$

$$= \bar{a}\big((\bar{b})^k + \sum_{t=0}^{k-1}(\bar{b})^{k-1}\big) = \bar{a}(\bar{b})^k\big(1 + \frac{k}{\bar{b}}\big) \overset{k\to\infty}{\longrightarrow} 0$$

showing that

$$E\big(\|x_e(k)\|^2\big) = E\big(\|x(k)\|^2\big) + E\big(\|\widehat{z}(k)\|^2\big) \overset{k\to\infty}{\longrightarrow} 0.$$

$\square$

Proposition B.4. *System (5.63) is MSS if and only if there exists* $\mathsf{S} > 0$, *with*

$$\mathsf{S} = \begin{bmatrix} \begin{bmatrix} Q_1 & \cdots & 0 \\ \vdots & \ddots & \vdots \\ 0 & \cdots & Q_N \end{bmatrix} & U \\ U^* & \widehat{\widehat{Z}} \end{bmatrix}$$

such that

$$\mathsf{S} - \left\{ \begin{bmatrix} A & 0 \\ B_f L & A_f \end{bmatrix} \mathsf{S} \begin{bmatrix} A & 0 \\ B_f L & A_f \end{bmatrix}^* + \sum_{i=1}^{N} \begin{bmatrix} D_i \\ 0 \end{bmatrix} \mathrm{dg}[Q_i] \begin{bmatrix} D_i^* & 0 \end{bmatrix} \right\} > 0. \qquad (B.34)$$

Proof. Consider the operator $\widetilde{\mathcal{T}} \in \mathbb{B}(\mathbb{H}^{(N+1)n})$ as follows: for $\widetilde{V} = (\widetilde{V}_1, \ldots, \widetilde{V}_N) \in \mathbb{H}^{(N+1)n}$, $\widetilde{\mathcal{T}}(\widetilde{V}) \triangleq (\widetilde{\mathcal{T}}_1(\widetilde{V}), \ldots, \widetilde{\mathcal{T}}_N(\widetilde{V}))$ is given by:

$$\widetilde{\mathcal{T}}_j(\widetilde{V}) \triangleq \sum_{i=1}^{N} p_{ij} \widetilde{A}_i \widetilde{V}_i \widetilde{A}_i^*, \qquad j \in \mathbb{N} \qquad (B.35)$$

where

$$\widetilde{A}_i \triangleq \begin{bmatrix} A_i & 0 \\ B_f L_i & A_f \end{bmatrix}.$$

Consider model (5.63) with $w(k) = 0$, for $k \ge 0$,

$$\widetilde{\mathsf{S}}_i(k) \triangleq \begin{bmatrix} E(z_i(k)z_i(k)^*) & E(z_i(k)\widehat{z}(k)^*) \\ E(\widehat{z}(k)z_i(k)^*) & E(\widehat{z}(k)\widehat{z}(k)^* 1_{\theta(k)=i)}) \end{bmatrix}, \qquad (B.36)$$

and $\widetilde{S}(k) \triangleq (\widetilde{S}_1(k), \ldots, \widetilde{S}_N(k)) \in \mathbb{H}^{(N+1)n}$. From Proposition 3.1 in Chapter 3, $\widetilde{S}(k+1) = \widetilde{\mathcal{T}}(\widetilde{S}(k))$, and System (5.63) is MSS if and only if there exists $\widetilde{S} = (\widetilde{S}_1, \ldots, \widetilde{S}_N) \in \mathbb{H}^{(N+1)n}$, $\widetilde{S}_i > 0$, $i \in \mathbb{N}$, such that (see Theorem 3.9 in Chapter 3)

$$\widetilde{S}_j - \widetilde{\mathcal{T}}_j(\widetilde{S}) > 0, \qquad j \in \mathbb{N}. \tag{B.37}$$

For each $j \in \mathbb{N}$, partitionate $\widetilde{S}_j$ as follows

$$\widetilde{S}_j = \begin{bmatrix} Q_j & U_j \\ U_j^* & \widehat{Z}_j \end{bmatrix},$$

where $Q_j \in \mathbb{B}(\mathbb{R}^n)$, and $\widehat{Z}_j \in \mathbb{B}(\mathbb{R}^{Nn})$, and define

$$\widehat{Z} \triangleq \sum_{j=1}^N \widehat{Z}_j, \quad Z \triangleq \mathrm{diag}[Q_i],$$

$$U \triangleq \begin{bmatrix} U_1 \\ \vdots \\ U_N \end{bmatrix}, \quad S \triangleq \begin{bmatrix} Z & U \\ U^* & \widehat{Z} \end{bmatrix}.$$

Notice that $S > 0$. Indeed since for each $j \in \mathbb{N}$, $\widetilde{S}_j > 0$, it follows from the Schur complement (Lemma 2.23) that $\widehat{Z}_j > U_j^* Z_j^{-1} U_j$, for $j \in \mathbb{N}$, so that

$$\widehat{Z} = \sum_{j=1}^N \widehat{Z}_j > \sum_{j=1}^N U_j^* Q_j^{-1} U_j = U^* Z^{-1} U.$$

From the Schur complement again it follows that

$$\begin{bmatrix} Z & U \\ U^* & \widehat{Z} \end{bmatrix} = S > 0.$$

Reorganizing (B.37), we obtain that (B.34) holds.

To prove necessity, we have from (B.34) that $Q_j - \mathcal{T}_j(Q) > 0$ for each $j \in \mathbb{N}$ where $Q = (Q_1, \ldots, Q_N) \in \mathbb{H}^n$. From Theorem 3.9 in Chapter 3, $r_\sigma(\mathcal{T}) < 1$. From the Lyapunov equation (B.34), it is easy to see that $r_\sigma \left(\begin{bmatrix} A & 0 \\ B_f L & A_f \end{bmatrix} \right) < 1$ and thus $r_\sigma(A_f) < 1$. Since $r_\sigma(\mathcal{T}) < 1$, $r_\sigma(A_f) < 1$ it follows from Proposition B.3 that System (5.63) is MSS. □

Proof of Proposition 5.15.

Proof. Consider the operator $\widetilde{\mathcal{T}}(\cdot)$ as in (B.35), $\widetilde{S}_i(k)$ as in (B.36). As shown in Proposition 3.1 and Proposition 3.36 of Chap. 3,

$$\tilde{S}_j(k+1) = \tilde{T}_j(\tilde{S}(k)) + \sum_{i=1}^{N} p_{ij}\pi_i(k) \begin{bmatrix} G_i G_i^* & 0 \\ 0 & B_f H_i H_i^* B_f^* \end{bmatrix}$$

and $\tilde{S}_j(k) \stackrel{k\to\infty}{\longrightarrow} \tilde{S}_j \geq 0$, with $\tilde{S} = (\tilde{S}_1, \ldots, \tilde{S}_N)$ satisfying

$$\tilde{S}_j = \tilde{T}_j(\tilde{S}) + \sum_{i=1}^{N} p_{ij}\pi_i \begin{bmatrix} G_i G_i^* & 0 \\ 0 & B_f H_i H_i^* B_f^* \end{bmatrix}. \tag{B.38}$$

Note that

$$\tilde{S}_i(k) = \begin{bmatrix} Q_i(k) & U_i(k) \\ U_i^*(k) & \hat{Q}_i(k) \end{bmatrix} \stackrel{k\to\infty}{\longrightarrow} \begin{bmatrix} Q_i & U_i \\ U_i^* & \hat{Q}_i \end{bmatrix}$$

where $\hat{Q}_i(k) = E(\hat{z}(k)\hat{z}(k)^* \mathbf{1}_{\{\theta(k)=i\}})$. Moreover

$$\hat{Z}(k) = \sum_{i=1}^{N} \hat{Q}_i(k) \stackrel{k\to\infty}{\longrightarrow} \sum_{i=1}^{N} \hat{Q}_i = \hat{Z}.$$

Defining

$$S = \begin{bmatrix} Z & U \\ U^* & \hat{Z} \end{bmatrix}, \qquad Z = \begin{bmatrix} Q_1 & \cdots & 0 \\ \vdots & \ddots & \vdots \\ 0 & \cdots & Q_N \end{bmatrix}, \qquad U = \begin{bmatrix} U_1 \\ \vdots \\ U_N \end{bmatrix},$$

it follows that $S(k) \stackrel{k\to\infty}{\longrightarrow} S$. Furthermore from (B.38) we have that S satisfies (5.70) and (5.71). Suppose that V also satisfies (5.70), (5.71). Then

$$X_j = \sum_{i=1}^{N} p_{ij} A_i X_i A_i^* + \sum_{i=1}^{N} p_{ij}\pi_i G_i G_i^*$$

$$Q_j = \sum_{i=1}^{N} p_{ij} A_i Q_i A_i^* + \sum_{i=1}^{N} p_{ij}\pi_i G_i G_i^* \tag{B.39}$$

and since $r_\sigma(\mathcal{T}) < 1$, it follows from Proposition 3.20 in Chapter 3 that $Q_j = X_j$, $j \in \mathbb{N}$. Then,

$$(V - S) = \begin{bmatrix} A & 0 \\ B_f L & A_f \end{bmatrix} (V - S) \begin{bmatrix} A & 0 \\ B_f L & A_f \end{bmatrix}^*. \tag{B.40}$$

From Proposition 3.6 in Chap. 3, $r_\sigma(\mathcal{T}) < 1$ implies that $r_\sigma(A) < 1$ and thus

$$r_\sigma\left(\begin{bmatrix} A & 0 \\ B_f L & A_f \end{bmatrix} \right) < 1.$$

From (B.40) it follows that $V - S = 0$. Finally suppose that V is such that (5.71) and (5.72) are satisfied. Then

$$X_j \geq \sum_{i=1}^{N} p_{ij} A_i X_i A_i^* + \sum_{i=1}^{N} p_{ij} \pi_i G_i G_i^* \tag{B.41}$$

and subtracting (B.39) from (B.41) it follows that $(X_j - Q_j) \geq \sum_{i=1}^{N} p_{ij} A_i (X_i - Q_i) A_i^*$, that is, $(X - Q) \geq \mathcal{T}(X - Q)$ where $X = (X_1, \ldots, X_N)$. This implies that $X \geq Q$. Using this fact, we conclude that

$$(V - S) \geq \begin{bmatrix} A & 0 \\ B_f L & A_f \end{bmatrix} (V - S) \begin{bmatrix} A & 0 \\ B_f L & A_f \end{bmatrix}^*$$

and since $r_\sigma \left(\begin{bmatrix} A & 0 \\ B_f L & A_f \end{bmatrix} \right) < 1$ it follows that $V - S \geq 0$.   □

Proof of Theorem 5.16.

Proof. For (A_f, B_f, J_f) fixed, consider S, W satisfying (5.75)-(5.77). Without loss of generality, suppose further that U is non-singular (if not, redefine U as $U + \epsilon I$ so that it is non-singular). As in [129] define

$$S^{-1} = \begin{bmatrix} Y & V \\ V^* & \widehat{Y} \end{bmatrix} > 0$$

where $Y > 0$ and $\widehat{Y} > 0$ are $Nn \times Nn$. We have that

$$ZY + UV^* = I \tag{B.42}$$
$$U^*Y + \widehat{Z}V^* = 0, \tag{B.43}$$

and from (B.42) and (B.43), $Y^{-1} = Z + UV^*Y^{-1} = Z - U\widehat{Z}^{-1}U^* < Z$, $V^* = U^{-1} - U^{-1}ZY = U^{-1}(Y^{-1} - Z)Y$ implying that V is non-singular. Define the non-singular $2Nn \times 2Nn$ matrix

$$T = \begin{bmatrix} Z^{-1} & Y \\ 0 & V^* \end{bmatrix}$$

and the non-singular $(2Nn+p) \times (2Nn+p)$ and $(5Nn+Nr(N+1)) \times (5Nn+Nr(N+1))$ matrices T_1, T_2 as follows:

$$T_1 = \begin{bmatrix} T & 0 \\ 0 & I \end{bmatrix}, \quad T_2 = \begin{bmatrix} T & 0 & 0 & 0 & 0 & 0 & 0 \\ 0 & \mathrm{dg}[Q_1^{-1}] & 0 & 0 & 0 & 0 & 0 \\ 0 & 0 & \ddots & 0 & 0 & 0 & 0 \\ 0 & 0 & 0 & \mathrm{dg}[Q_N^{-1}] & 0 & 0 & 0 \\ 0 & 0 & 0 & 0 & I & 0 & 0 \\ 0 & 0 & 0 & 0 & 0 & I & 0 \\ 0 & 0 & 0 & 0 & 0 & 0 & T \end{bmatrix}.$$

We apply the following transformations of similarity to (5.76) and (5.77). We pre and post multiply (5.76) by T_1^* and T_1 respectively, and (5.77) by T_2^* and T_2 respectively. From (B.42) and (B.43), we obtain that

$$T^*\mathsf{S} = \begin{bmatrix} I & Z^{-1}U \\ I & 0 \end{bmatrix}, \quad T^*\mathsf{ST} = \begin{bmatrix} Z^{-1} & Z^{-1} \\ Z^{-1} & \mathsf{Y} \end{bmatrix} = \begin{bmatrix} \mathsf{X} & \mathsf{X} \\ \mathsf{X} & \mathsf{Y} \end{bmatrix}$$

where

$$\mathsf{X} = Z^{-1} = \mathrm{diag}[\mathsf{X}_i], \quad \mathsf{X}_i = Q_i^{-1}, \quad i \in \mathbb{N},$$

and

$$T^*\mathsf{S}\begin{bmatrix} \begin{bmatrix} J^* \\ \vdots \\ J^* \end{bmatrix} \\ J^* \\ \vdots \\ -J_f^* \end{bmatrix} = \begin{bmatrix} \begin{bmatrix} J^* \\ \vdots \\ J^* \end{bmatrix} - Z^{-1}UJ_f^* \\ \begin{bmatrix} J^* \\ \vdots \\ J^* \end{bmatrix} \end{bmatrix} = \begin{bmatrix} \begin{bmatrix} J^* \\ \vdots \\ J^* \end{bmatrix} - J_{aux}^* \\ \begin{bmatrix} J^* \\ \vdots \\ J^* \end{bmatrix} \end{bmatrix}$$

where $J_{aux} = J_f U^* Z^{-1} = J_f U^* \mathsf{X}$, and also

$$T^*\begin{bmatrix} \mathsf{D}_i \\ 0 \end{bmatrix} = \begin{bmatrix} Z^{-1}\mathsf{D}_i \\ \mathsf{YD}_i \end{bmatrix} = \begin{bmatrix} \mathsf{X} \\ \mathsf{Y} \end{bmatrix} \mathsf{D}_i,$$

$$T^*\begin{bmatrix} \mathsf{G} \\ 0 \end{bmatrix} = \begin{bmatrix} Z^{-1}\mathsf{G} \\ \mathsf{YG} \end{bmatrix} = \begin{bmatrix} \mathsf{X} \\ \mathsf{Y} \end{bmatrix} \mathsf{G},$$

$$T^*\begin{bmatrix} 0 \\ B_f \mathsf{H} \end{bmatrix} = \begin{bmatrix} 0 \\ VB_f \mathsf{H} \end{bmatrix} = \begin{bmatrix} 0 \\ \mathsf{FH} \end{bmatrix},$$

where $F = VB_f$, and

$$\begin{aligned}
T^*\begin{bmatrix} A & 0 \\ B_f \mathsf{L} & A_f \end{bmatrix}\mathsf{ST} &= \begin{bmatrix} Z^{-1}A & 0 \\ \mathsf{YA} + VB_f \mathsf{L} & VA_f \end{bmatrix}\begin{bmatrix} I & I \\ U^*Z^{-1} & 0 \end{bmatrix} \\
&= \begin{bmatrix} Z^{-1}A & Z^{-1}A \\ \mathsf{YA} + VB_f \mathsf{L} + VA_f U^*Z^{-1} & \mathsf{YA} + VB_f \mathsf{L} \end{bmatrix} \\
&= \begin{bmatrix} \mathsf{XA} & \mathsf{XA} \\ \mathsf{YA} + \mathsf{FL} + R & \mathsf{YA} + \mathsf{FL} \end{bmatrix},
\end{aligned}$$

where $R = VA_f U^* Z^{-1} = VA_f U^* \mathsf{X}$. With these transformations of similarity, (5.76) becomes (5.80) and (5.77) becomes (5.81). Similarly by applying the inverse transformations of similarity we have that (5.80) becomes (5.76) and (5.81) becomes (5.77).   □

Proof of Proposition 5.17.

Proof. Consider any A_f, B_f, J_f such that $r_\sigma(A_f) < 1$ and write

$$\widehat{z}(k+1) = A_f\widehat{z}(k) + B_f y(k)$$
$$\widehat{z}(k+1|k) = (\mathsf{A} + \mathsf{T}(k)\mathsf{L})\widehat{z}(k|k-1) + \mathsf{T}(k)y(k)$$

where

$$\widetilde{Z}(k+1|k) = \mathsf{A}\widetilde{Z}(k|k-1)\mathsf{A}^* + \mathsf{G}\mathsf{G}^*$$
$$- \mathsf{A}\widetilde{Z}(k|k-1)\mathsf{L}^*(\mathsf{L}\widetilde{Z}(k|k-1)\mathsf{L}^* + \mathsf{H}(k)\mathsf{H}(k)^*)^{-1}\mathsf{L}\widetilde{Z}(k|k-1)\mathsf{A}^*$$
$$\mathsf{T}(k) = -\mathsf{A}\widetilde{Z}(k|k-1)\mathsf{L}^*(\mathsf{L}\widetilde{Z}(k|k-1)\mathsf{L}^* + \mathsf{H}(k)\mathsf{H}(k)^*)^{-1}$$

and

$$\widetilde{Z}(k|k-1) = E\big(\widetilde{z}(k|k-1)\widetilde{z}(k|k-1)^*\big)$$
$$\widetilde{z}(k|k-1) = z(k) - \widehat{z}(k|k-1).$$

From the orthogonality between $\widetilde{z}(k|k-1)$ and $\begin{bmatrix} J \cdots J \end{bmatrix}\widehat{z}(k|k-1) - J_f\widehat{z}(k)$ (see Subsect. 5.4.2), we have that

$$E\big(\|e(k)\|^2\big) = \operatorname{tr}\Bigg(E\bigg(\Big(\begin{bmatrix} J \cdots J \end{bmatrix}\widetilde{z}(k|k-1) + \begin{bmatrix} J \cdots J \end{bmatrix}\widehat{z}(k|k-1) - J_f\widehat{z}(k)\Big)$$
$$\times \Big(\begin{bmatrix} J \cdots J \end{bmatrix}\widetilde{z}(k|k-1) + \begin{bmatrix} J \cdots J \end{bmatrix}\widehat{z}(k|k-1) - J_f\widehat{z}(k)\Big)^*\bigg)\Bigg)$$

$$= \operatorname{tr}\left(\begin{bmatrix} J \cdots J \end{bmatrix}\widetilde{Z}(k|k-1)\begin{bmatrix} J^* \\ \vdots \\ J^* \end{bmatrix} \right)$$
$$+ E\left(\| \begin{bmatrix} J \cdots J \end{bmatrix}\widehat{z}(k|k-1) - J_f\widehat{z}(k)\|^2\right)$$

$$\geq \operatorname{tr}\left(\begin{bmatrix} J \cdots J \end{bmatrix}\widetilde{Z}(k|k-1)\begin{bmatrix} J^* \\ \vdots \\ J^* \end{bmatrix} \right)$$

$$\overset{k\to\infty}{\to} \operatorname{tr}\left(\begin{bmatrix} J \cdots J \end{bmatrix} P \begin{bmatrix} J^* \\ \vdots \\ J^* \end{bmatrix} \right)$$

since, as shown in Theorem 5.12, $\widetilde{Z}(k|k-1) \overset{k\to\infty}{\to} P$. This shows that

$$\operatorname{tr}\left(\begin{bmatrix} J \cdots J & J_f \end{bmatrix} \mathsf{S} \begin{bmatrix} J^* \\ \vdots \\ J^* \\ -J_f \end{bmatrix} \right) \geq \operatorname{tr}\left(\begin{bmatrix} J \cdots J \end{bmatrix} P \begin{bmatrix} J^* \\ \vdots \\ J^* \end{bmatrix} \right).$$

On the other hand, consider

$$\widehat{z}_{op}(k+1) = (A + T(P)L)\widehat{z}_{op}(k) + T(P)y(k)$$
$$\widehat{v}_{op}(k) = \begin{bmatrix} J & \cdots & J \end{bmatrix} \widehat{z}_{op}(k),$$

where

$$T(P) = -APL^*(LPL^* + HH^*)^{-1}.$$

Then

$$E(\|e_{op}(k)\|^2) = \mathrm{tr}\left(\begin{bmatrix} J & \cdots & J \end{bmatrix} \widetilde{Z}(k) \begin{bmatrix} J \\ \vdots \\ J \end{bmatrix} \right)$$

where

$$\widetilde{Z}(k) = E((z(k) - \widehat{z}_{op}(k))(z(k) - \widehat{z}_{op}(k))^*)$$

which satisfies (see Theorem 5.12)

$$\widetilde{Z}(k+1) = (A + T(P)L)\widetilde{Z}(k)(A + T(P)L)^*$$
$$+ \mathfrak{V}(Q(k)) + G(k)G(k)^* + T(P)H(k)H(k)^*T(P)^*.$$

As shown in Theorem 5.12, $\widetilde{Z}(k) \overset{k\to\infty}{\to} P$ and thus

$$\lim_{k\to\infty} E(\|e_{op}(k)\|^2) = \mathrm{tr}\left(\begin{bmatrix} J & \cdots & J \end{bmatrix} P \begin{bmatrix} J \\ \vdots \\ J \end{bmatrix} \right)$$

showing the desired result. □

Proof of Proposition 5.18.

Proof. Notice that rearranging the terms of (5.80), we obtain

$$\begin{bmatrix} X & \begin{bmatrix} J^* \\ \vdots \\ J^* \end{bmatrix} - J^*_{aux} & X \\ \begin{bmatrix} J & \cdots & J \end{bmatrix} - J_{aux} & W & \begin{bmatrix} J & \cdots & J \end{bmatrix} \\ X & \begin{bmatrix} J^* \\ \vdots \\ J^* \end{bmatrix} & Y \end{bmatrix} > 0$$

and from the Schur complement (see Lemma 2.23), we have that (5.80) is equivalent to

$$\begin{bmatrix} X \\ \begin{bmatrix} J & \cdots & J \end{bmatrix} - J_{aux} \end{bmatrix} \frac{\begin{bmatrix} J^* & \cdots & J^* \end{bmatrix}^* - J^*_{aux}}{W} > \begin{bmatrix} X \\ \begin{bmatrix} J & \cdots & J \end{bmatrix} \end{bmatrix} Y^{-1} \begin{bmatrix} X & \begin{bmatrix} J^* & \cdots & J^* \end{bmatrix}^* \end{bmatrix}$$

that is

$$
\begin{bmatrix}
X - XY^{-1}X & \begin{bmatrix} J^* \\ \vdots \\ J^* \end{bmatrix} - J_{aux}^* - XY^{-1} \begin{bmatrix} J^* \\ \vdots \\ J^* \end{bmatrix} \\
\begin{bmatrix} J \cdots J \end{bmatrix} - J_{aux} - \begin{bmatrix} J \cdots J \end{bmatrix} Y^{-1}X & W - \begin{bmatrix} J \cdots J \end{bmatrix} Y^{-1} \begin{bmatrix} J^* \\ \vdots \\ J^* \end{bmatrix}
\end{bmatrix} > 0
$$

and choosing

$$
J_{aux} = [J \quad \cdots \quad J](I - Y^{-1}X), \tag{B.44}
$$

(5.80) becomes

$$
\begin{bmatrix}
Y - XY^{-1}X & 0 \\
0 & W - \begin{bmatrix} J \cdots J \end{bmatrix} Y^{-1} \begin{bmatrix} J^* \cdots J^* \end{bmatrix}^*
\end{bmatrix} > 0.
$$

that is, $X - XY^{-1}X > 0$ and

$$
W - \begin{bmatrix} J \cdots J \end{bmatrix} Y^{-1} \begin{bmatrix} J^* \\ \vdots \\ J^* \end{bmatrix} > 0.
$$

Applying the Schur complement, we have that

$$
\begin{bmatrix} X & X \\ X & Y \end{bmatrix} > 0
$$

and

$$
\begin{bmatrix} Y & \begin{bmatrix} J^* \cdots J^* \end{bmatrix}^* \\ \begin{bmatrix} J \cdots J \end{bmatrix} & W \end{bmatrix} > 0.
$$

The first inequality follows from (5.81). Therefore (5.80), with the choice $J_{aux} = [J \quad \cdots \quad J](I - Y^{-1}X)$, can be rewritten as

$$
\begin{bmatrix} Y & \begin{bmatrix} J^* \\ \vdots \\ J^* \end{bmatrix} \\ \begin{bmatrix} J \cdots J \end{bmatrix} & W \end{bmatrix} > 0.
$$

With the choice $U = U^* = (X^{-1} - Y^{-1}) > 0$, we obtain that (5.82) becomes

$$
V = Y(Y^{-1} - X^{-1})(X^{-1} - Y^{-1})^{-1} = -Y,
$$
$$
A_f = (-Y)^{-1}R((X^{-1} - Y^{-1})X)^{-1} = -Y^{-1}R(I - Y^{-1}X)^{-1},
$$
$$
B_f = -Y^{-1}F,
$$
$$
J_f = J_{aux}((X^{-1} - Y^{-1})X)^{-1} = J_{aux}(I - Y^{-1}X)^{-1} = [J \cdots J].
$$

We apply now the following transformation of similarity to (5.81). Multiplying on the left and on the right by T_3^* and T_3 respectively, where

$$T_3 = \begin{bmatrix} \begin{bmatrix} \mathsf{X}^{-1} & \mathsf{X}^{-1} \\ 0 & -\mathsf{Y}^{-1} \end{bmatrix} & 0 & 0 & 0 \\ 0 & \begin{bmatrix} \mathrm{dg}[X_1^{-1}] & & \\ & \cdots & \\ & & \mathrm{dg}[X_N^{-1}] \end{bmatrix} & 0 & 0 \\ 0 & 0 & I & 0 \\ 0 & 0 & 0 & \begin{bmatrix} \mathsf{X}^{-1} & -\mathsf{X}^{-1} \\ 0 & -\mathsf{Y}^{-1} \end{bmatrix} \end{bmatrix}$$

we have, choosing $R = -(\mathsf{Y}\mathsf{A} + \mathsf{F}\mathsf{L})(I - \mathsf{Y}^{-1}\mathsf{X})$, that (5.81) becomes

$$\begin{bmatrix} \Upsilon_{11} & \Upsilon_{12}^* \\ \Upsilon_{12} & \Upsilon_{22} \end{bmatrix}$$

where

$$\Upsilon_{22} = \begin{bmatrix} \mathsf{X}^{-1} & \mathsf{X}^{-1} - \mathsf{Y}^{-1} \\ \mathsf{X}^{-1} - \mathsf{Y}^{-1} & \mathsf{X}^{-1} - \mathsf{Y}^{-1} \end{bmatrix}$$

$$\Upsilon_{11} = \begin{bmatrix} \Upsilon_{22} & 0 & 0 & 0 & 0 \\ 0 & \mathrm{dg}[X_1^{-1}] & 0 & 0 & 0\,0 \\ 0 & 0 & \ddots & 0 & 0\,0 \\ 0 & 0 & 0 & \mathrm{dg}[X_N^{-1}] & 0\,0 \\ 0 & 0 & 0 & 0 & I\,0 \\ 0 & 0 & 0 & 0 & 0\,I \end{bmatrix}$$

and

$$\Upsilon_{12} = \begin{bmatrix} \Phi & \begin{matrix} \mathsf{D}_1 \mathrm{dg}[X_1^{-1}] \\ 0 \end{matrix} & \cdots & \begin{matrix} \mathsf{D}_N \mathrm{dg}[X_N^{-1}] \\ 0 \end{matrix} & \begin{matrix} G \\ 0 \end{matrix} & \begin{matrix} 0 \\ -\mathsf{Y}^{-1}FH \end{matrix} \end{bmatrix}$$

$$\Phi = \begin{pmatrix} \mathsf{A} & 0 \\ -\mathsf{Y}^{-1}\mathsf{F}\mathsf{L}\,\mathsf{A} + \mathsf{Y}^{-1}\mathsf{F}\mathsf{L} \end{pmatrix} \Upsilon_{22}.$$

From the Schur complement, this is equivalent to

$$\begin{bmatrix} \mathsf{X}^{-1} & \mathsf{X}^{-1} - \mathsf{Y}^{-1} \\ \mathsf{X}^{-1} - \mathsf{Y}^{-1} & \mathsf{X}^{-1} - \mathsf{Y}^{-1} \end{bmatrix} >$$

$$\begin{bmatrix} \mathsf{A} & 0 \\ -\mathsf{Y}^{-1}\mathsf{F}\mathsf{L}\,\mathsf{A} + \mathsf{Y}^{-1}\mathsf{F}\mathsf{L} \end{bmatrix} \begin{bmatrix} \mathsf{X}^{-1} & \mathsf{X}^{-1} - \mathsf{Y}^{-1} \\ \mathsf{X}^{-1} - \mathsf{Y}^{-1} & \mathsf{X}^{-1} - \mathsf{Y}^{-1} \end{bmatrix} \begin{bmatrix} \mathsf{A} & 0 \\ -\mathsf{Y}^{-1}\mathsf{F}\mathsf{L}\,\mathsf{A} + \mathsf{Y}^{-1}\mathsf{F}\mathsf{L} \end{bmatrix}^*$$

$$+ \sum_{i=1}^{N} \begin{bmatrix} \mathsf{D}_i \\ 0 \end{bmatrix} \mathrm{dg}[X_i^{-1}] \begin{bmatrix} \mathsf{D}_i^* & 0 \end{bmatrix} + \begin{bmatrix} G \\ 0 \end{bmatrix} \begin{bmatrix} G^* & 0 \end{bmatrix} + \begin{bmatrix} 0 \\ -\mathsf{Y}^{-1}FH \end{bmatrix} \begin{bmatrix} 0 & -\mathsf{Y}^{-1}F^*H^* \end{bmatrix}$$

$$(\text{B.45})$$

Therefore the LMI (5.81) can be rewritten as in (B.45). Consider J_{aux} chosen as in (B.44) and the choice of parameters for the LMIs (5.80) and (5.81) as in Proposition 5.18. Notice that the extra terms ϵI and $2\epsilon I$ in (5.84) and (5.83) ensure that $P_\epsilon > 0, Z_\epsilon > 0$. Subtracting (5.84) from (5.83), we get that

$$(Z_\epsilon - P_\epsilon) = \mathsf{A}(Z_\epsilon - P_\epsilon)\mathsf{A}^* + \mathsf{T}(P_\epsilon)(\mathsf{L}P_\epsilon\mathsf{L}^* + \mathsf{H}\mathsf{H}^*)\mathsf{T}(P_\epsilon)^* + \epsilon I \qquad (B.46)$$

since

$$\mathsf{A}P_\epsilon\mathsf{L}^*(\mathsf{L}P_\epsilon\mathsf{L}^* + \mathsf{H}\mathsf{H}^*)^{-1}\mathsf{L}P_\epsilon\mathsf{A}^* = \mathsf{T}(P_\epsilon)(\mathsf{L}P_\epsilon\mathsf{L}^* + \mathsf{H}\mathsf{H}^*)\mathsf{T}(P_\epsilon)^*.$$

From $r_\sigma(\mathsf{A}) < 1$, it follows that $Z_\epsilon - P_\epsilon > 0$ (and thus $\mathsf{X}^{-1} - \mathsf{Y}^{-1} > 0$). Notice also that

$$\begin{bmatrix} \mathsf{A} & 0 \\ -\mathsf{T}(P_\epsilon)\mathsf{L} & \mathsf{A} + \mathsf{T}(P_\epsilon)\mathsf{L} \end{bmatrix} \begin{bmatrix} Z_\epsilon & Z_\epsilon - P_\epsilon \\ Z_\epsilon - P_\epsilon & Z_\epsilon - P_\epsilon \end{bmatrix} \begin{bmatrix} \mathsf{A} & 0 \\ -\mathsf{T}(P_\epsilon)\mathsf{L} & \mathsf{A} + \mathsf{T}(P_\epsilon)\mathsf{L} \end{bmatrix}^* =$$
$$\begin{bmatrix} \mathsf{A}Z_\epsilon\mathsf{A}^* & \mathsf{A}(Z_\epsilon - P_\epsilon)\mathsf{A}^* + \mathsf{A}P_\epsilon\mathsf{L}^*\mathsf{T}(P_\epsilon)^* \\ \mathsf{A}(Z_\epsilon - P_\epsilon)\mathsf{A}^* + \mathsf{T}(P_\epsilon)\mathsf{L}P_\epsilon\mathsf{A}^* & \mathsf{A}(Z_\epsilon - P_\epsilon)\mathsf{A}^* + \mathsf{T}(P_\epsilon)\mathsf{L}P_\epsilon\mathsf{L}^*\mathsf{T}(P_\epsilon)^* \end{bmatrix} \qquad (B.47)$$

and

$$-\mathsf{A}P_\epsilon\mathsf{L}^*\mathsf{T}(P_\epsilon)^* = \mathsf{T}(P_\epsilon)(\mathsf{L}P_\epsilon\mathsf{L}^* + \mathsf{H}\mathsf{H}^*)\mathsf{T}(P_\epsilon)^*. \qquad (B.48)$$

From (5.84), (B.48), (B.47) and (B.46), we get that

$$\begin{bmatrix} Z_\epsilon & Z_\epsilon - P_\epsilon \\ Z_\epsilon - P_\epsilon & Z_\epsilon - P_\epsilon \end{bmatrix}$$
$$= \begin{bmatrix} \mathsf{A} & 0 \\ -\mathsf{T}(P_\epsilon)\mathsf{L} & \mathsf{A} + \mathsf{T}(P_\epsilon)\mathsf{L} \end{bmatrix} \begin{bmatrix} Z_\epsilon & Z_\epsilon - P_\epsilon \\ Z_\epsilon - P_\epsilon & Z_\epsilon - P_\epsilon \end{bmatrix} \begin{bmatrix} \mathsf{A} & 0 \\ -\mathsf{T}(P_\epsilon)\mathsf{L} & \mathsf{A} + \mathsf{T}(P_\epsilon)\mathsf{L} \end{bmatrix}^*$$
$$+ \sum_{i=1}^{N} \begin{bmatrix} \mathsf{D}_i \\ 0 \end{bmatrix} \mathrm{dg}[Z_{\epsilon i}^{-1}] \begin{bmatrix} \mathsf{D}_i^* & 0 \end{bmatrix} + \begin{bmatrix} \mathsf{G} \\ 0 \end{bmatrix} \begin{bmatrix} \mathsf{G}^* & 0 \end{bmatrix}$$
$$+ \begin{bmatrix} 0 \\ \mathsf{T}(P_\epsilon)\mathsf{H} \end{bmatrix} \begin{bmatrix} 0 & \mathsf{H}^*\mathsf{T}(P_\epsilon)^* \end{bmatrix} + \begin{bmatrix} 2\epsilon I & \epsilon I \\ \epsilon I & \epsilon I \end{bmatrix} \qquad (B.49)$$

so that (B.45) is satisfied. Finally notice that

$$A_f = -\mathsf{Y}^{-1}R(I - \mathsf{Y}^{-1}\mathsf{X})^{-1} = \mathsf{A} + \mathsf{T}(P_\epsilon)\mathsf{L}$$
$$B_f = -\mathsf{T}(P_\epsilon)$$
$$J_f = \begin{bmatrix} J & \cdots & J \end{bmatrix}.$$

$\square$

C

Auxiliary Results for the H_2-control Problem

Consider $\mathbb{T} = \{\ldots, -1, 0, 1, \ldots\}$. On the stochastic basis $(\Omega, \{\mathfrak{F}_k\}_{k\in\mathbb{T}}, \mathfrak{F}, \mathcal{P})$ consider the system

$$\mathcal{G} = \begin{cases} x(k+1) = A_{\theta(k)}x(k) + B_{\theta(k)}\varrho(k) \\ z(k) = C_{\theta(k)}x(k) + D_{\theta(k)}\varrho(k) \end{cases} \tag{C.1}$$

with the input sequence $\varrho(k)$ of dimension m and the output sequence $z(k)$ of dimension q. Recursively we can write $x(k)$ as

$$x(k) = \sum_{l=-\infty}^{k-1} \left[A_{\theta(k-1)} \cdots A_{\theta(l+1)}\right] B_{\theta(l)}\varrho(l).$$

Thus, we can write $z(k)$ as

$$z(k) = C_{\theta(k)}\left\{ \sum_{l=-\infty}^{k-1} \left[A_{\theta(k-1)} \cdots A_{\theta(l+1)}\right] B_{\theta(l)}\varrho(l)\right\} + D_{\theta(k)}\varrho(k). \tag{C.2}$$

Let the operator $\mathcal{G}$ on $\mathcal{C}^m$ be defined as $\mathcal{G}(\varrho) = (\ldots, z(-1), z(0), z(1), \ldots), \varrho \in \mathcal{C}^m$, and $\mathcal{G}(\varrho)(k) = z(k)$.

Proposition C.1. *Suppose that System (C.1) is MSS. Then $\mathcal{G} \in \mathbb{B}(\mathcal{C}^m, \mathcal{C}^q)$.*

Proof. The proof of this result parallels the proof of Theorem 3.34. By the triangle inequality and (C.2) we have

$$\|z(k)\|_2 \leq \sum_{l=-\infty}^{k-1} \|C_{\theta(k)}\left[A_{\theta(k-1)} \cdots A_{\theta(l+1)}\right] B_{\theta(l)}\varrho(l)\|_2 + \|D_{\theta(k)}\varrho(k)\|_2$$

$$\leq \|C\|_{\max} \sum_{l=-\infty}^{k-1} \|\left[A_{\theta(k-1)} \cdots A_{\theta(l+1)}\right] B_{\theta(l)}\varrho(l)\|_2 + \|D\|_{\max}\|\varrho(k)\|_2.$$

$$\tag{C.3}$$

As seen in the proof of Theorem 3.34,

$$\left\| A_{\theta(k-1)} \cdots A_{\theta(l+1)} B_{\theta(l)} \varrho(l) \right\|_2^2 \le n \, \|B\|_{\max}^2 \left\| \mathcal{T}^{k-l-1} \right\|_1 \|w(l)\|_2^2 .$$

From MSS and Theorem 3.9 in Chapter 3, there exists $0 < \zeta < 1$ and $\beta \ge 1$ such that $\left\| \mathcal{T}^k \right\|_1 \le \beta \zeta^k$, and therefore, from (C.3),

$$\|z(k)\|_2 \le \sum_{l=-\infty}^{k} \zeta_{k-l} \beta_l ,$$

where $\zeta_\iota \overset{\triangle}{=} (\zeta^{1/2})^{(\iota-1)}$ for $\iota \ge 1$, $\zeta_0 \overset{\triangle}{=} 1$, and

$$\beta_\iota \overset{\triangle}{=} (1 + \|C\|_{\max})(\|D\|_{\max} + (n\beta)^{1/2} \|B\|_{\max}) \|\varrho(\iota)\|_2 .$$

Set $a \overset{\triangle}{=} (\zeta_0, \zeta_1, \ldots)$ and $b \overset{\triangle}{=} (\ldots, \beta_{-1}, \beta_0, \beta_1, \ldots)$. Since $a \in \ell_1$ (that is, $\sum_{\iota=0}^{\infty} |\zeta_\iota| < \infty$) and $b \in \ell_2$ (that is, $\sum_{\iota=-\infty}^{\infty} |\beta_\iota|^2 < \infty$) it follows that the convolution $c \overset{\triangle}{=} a * b = (c_0, c_1, \ldots)$, $c_k \overset{\triangle}{=} \sum_{\iota=-\infty}^{k} \zeta_{k-\iota} \beta_\iota$, lies itself in ℓ_2 with $\|c\|_2 \le \|a\|_1 \|b\|_2$ (cf. [103], p. 529). Hence,

$$\|z\|_2 = \left\{ \sum_{k=-\infty}^{\infty} E(\|z(k)\|^2) \right\}^{1/2} \le \left\{ \sum_{\iota=-\infty}^{k} c_\iota^2 \right\}^{1/2} = \|c\|_2 < \infty.$$

Therefore, for some $d > 0$,

$$\|z\|_2 \le d\|\varrho\|_2.$$

Finally we have that $z_k = (\ldots, z(k-1), z(k)) \in \mathcal{C}_k^q$ for each $k \in \mathbb{T}$, and therefore $z = (\ldots, z(0), \ldots) \in \mathcal{C}^q$. □

In what follows we define

$$\Pi A(k,l) = A_{\theta(k)} \cdots A_{\theta(l+1)}, \quad k \ne l, \quad \Pi A(k,k) = I, \tag{C.4}$$

and

$$\Sigma(\varrho)(s) = \sum_{t=-\infty}^{s} \left[\Pi A(s,t)) \right] B_{\theta(t)} \varrho(t).$$

We will need to use the adjoint operator $\mathcal{G}^*$ of $\mathcal{G}$, which is such that for any $\varrho \in \mathcal{C}^m$ and any $v \in \mathcal{C}^q$ we have that

$$\langle \mathcal{G}(\varrho); v \rangle = \langle \varrho; \mathcal{G}^*(v) \rangle \tag{C.5}$$

(see Definition 5.22.1 in [181]). Moreover, $\mathcal{G}^*$ is a bounded linear operator (see Theorem 5.22.2 in [181]), and thus

$$\mathcal{G}^* \in \mathbb{B}(\mathcal{C}^q, \mathcal{C}^m).$$

From (C.5),

$$\langle \mathcal{G}(\varrho); v \rangle = \sum_{k=-\infty}^{\infty} E(z(k)^* v(k))$$

$$= E\Big[\sum_{k=-\infty}^{\infty} \Big\{ \sum_{l=-\infty}^{\infty} \Big(\varrho(l)^* B_{\theta(l)}^* \big[\Pi A(k-1,l) \big]^* C_{\theta(k)}^* v(k) \, \mathbf{1}_{\{l \le k-1\}} \Big)$$

$$+ \varrho(k)^* D_{\theta(k)}^* v(k) \Big\} \Big]$$

$$= E\Big[\sum_{l=-\infty}^{\infty} \varrho(l)^* E\Big(\sum_{k=l+1}^{\infty} B_{\theta(l)}^* \big[\Pi A(k-1,l) \big]^* C_{\theta(k)}^* v(k) | \mathfrak{F}_l \Big)$$

$$+ \sum_{l=-\infty}^{\infty} \varrho(l)^* D_{\theta(l)} v(l) \Big]$$

$$= \sum_{l=-\infty}^{\infty} E\Big[\varrho(l)^* \Big\{ \sum_{k=l+1}^{\infty} E\Big(B_{\theta(l)}^* \big[\Pi A(k-1,l) \big]^* C_{\theta(k)}^* v(k) | \mathfrak{F}_l \Big)$$

$$+ D_{\theta(l)}^* v(l) \Big\} \Big]$$

$$= \langle \varrho; \mathcal{G}^*(v) \rangle$$

and therefore

$$\mathcal{G}^*(v)(l) = \sum_{k=l+1}^{\infty} E\Big(B_{\theta(l)}^* \big[A_{\theta(k-1)} \cdots A_{\theta(l+1)} \big]^* C_{\theta(k)}^* v(k) | \mathfrak{F}_l \Big) + D_{\theta(l)}^* v(l).$$

$$(C.6)$$

Proposition C.2. *Suppose that System (C.1) is MSS, and that* $X = (X_1, X_2, \ldots, X_N) \in \mathbb{H}^n, X_i = X_i^*$, *satisfies for each* $i \in \mathbb{N}$

$$A_i^* \mathcal{E}_i(X) A_i - X_i + C_i^* C_i = 0 \tag{C.7}$$

$$D_i^* C_i + B_i^* \mathcal{E}_i(X) A_i = 0. \tag{C.8}$$

Then

$$\mathcal{G}^* \mathcal{G}(\varrho)(k) = (D_{\theta(k)}^* D_{\theta(k)} + B_{\theta(k)}^* \mathcal{E}_{\theta(k)}(X) B_{\theta(k)}) \varrho(k). \tag{C.9}$$

Proof. From (C.6) and (C.2), we have

$$\mathcal{G}^* \mathcal{G}(\varrho)(l) = \sum_{k=l+1}^{\infty} E\Big(B_{\theta(l)}^* \big[\Pi A(k-1,l) \big]^* C_{\theta(k)}^* C_{\theta(k)} \Sigma(\varrho)(k-1)$$

$$+ B_{\theta(l)}^* \big[\Pi A(k-1,l) \big]^* C_{\theta(k)}^* D_{\theta(k)} \varrho(k) | \mathfrak{F}_l \Big)$$

$$+ D_{\theta(l)}^* C_{\theta(l)} \Sigma(\varrho)(l-1) + D_{\theta(l)}^* D_{\theta(l)} \varrho(l).$$

From (C.7) we have that

$$\sum_{k=0}^{\infty} E\left(B_{\theta(l)}^{*}\left[\Pi A(k+l,l)\right]^{*} C_{\theta(k+l+1)}^{*} C_{\theta(k+l+1)} \Sigma(\varrho)(k+l)|\mathfrak{F}_{l}\right)$$

$$= \sum_{k=0}^{\infty} E\left(B_{\theta(l)}^{*}\left\{\left[\Pi A(k+l,l)\right]^{*}\left(X_{\theta(k+l+1)}\right.\right.\right.$$

$$\left.\left.\left. - A_{\theta(k+l+1)}^{*} X_{\theta(k+l+2)} A_{\theta(k+l+1)}\right) \Sigma(\varrho)(k+l)\right\}|\mathfrak{F}_{l}\right)$$

$$= \sum_{k=0}^{\infty} E\left(B_{\theta(l)}^{*}\left\{\left[\Pi A(k+l,l)\right]^{*} X_{\theta(k+l+1)} \Sigma(\varrho)(k+l)\right.\right.$$

$$- \left[\Pi A(k+l+1,l)\right]^{*} X_{\theta(k+l+2)} \Sigma(\varrho)(k+l+1)$$

$$+ \left.\left.\left[\Pi A(k+l+1,l)\right]^{*} X_{\theta(k+l+2)} B_{\theta(k+l+1)} \varrho(k+l+1)\right\}|\mathfrak{F}_{l}\right)$$

$$= E\left(B_{\theta(l)}^{*} X_{\theta(l+1)} \Sigma(\varrho)(l)|\mathfrak{F}_{l}\right)$$

$$+ \sum_{k=0}^{\infty} E\left(B_{\theta(l)}^{*}\left[\Pi A(k+l+1,l)\right]^{*} X_{\theta(k+l+2)} B_{\theta(k+l+1)} \varrho(k+l+1)|\mathfrak{F}_{l}\right)$$

On the other hand, from (C.8), we conclude that for $k \geq 0$,

$$E\left(B_{\theta(l)}^{*}\left[\Pi A(k+l+1,l)\right]^{*} C_{\theta(k+l+1)}^{*} D_{\theta(k+l+1)} \varrho(k+l+1)|\mathfrak{F}_{l}\right)$$

$$= -E\left(B_{\theta(l)}^{*}\left[\Pi A(k+l,l)\right]^{*} A_{\theta(k+l+1)}^{*} X_{\theta(k+l+2)} B_{\theta(k+l+1)} \varrho(k+l+1)|\mathfrak{F}_{l}\right)$$

$$= -E\left(B_{\theta(l)}^{*}\left[\Pi A(k+l+1,l)\right]^{*} X_{\theta(k+l+2)} B_{\theta(k+l+1)} \varrho(k+l+1)|\mathfrak{F}_{l}\right)$$

and

$$D_{\theta(l)}^{*} C_{\theta(l)} \Sigma(\varrho)(l-1) = E\left(D_{\theta(l)}^{*} C_{\theta(l)} \Sigma(\varrho)(l-1)|\mathfrak{F}_{l}\right)$$

$$= -E\left(B_{\theta(l)}^{*} X_{\theta(l+1)} A_{\theta(l)} \Sigma(\varrho)(l-1)|\mathfrak{F}_{l}\right).$$

Noticing that

$$E\left(B_{\theta(l)}^{*} X_{\theta(l+1)}\left\{\Sigma(\varrho)(l) - A_{\theta(l)} \Sigma(\varrho)(l-1)\right\}|\mathfrak{F}_{l}\right) = B_{\theta(l)}^{*} \mathcal{E}_{\theta(l)}(X) B_{\theta(l)} \varrho(l)$$

we conclude, after adding up all the previous equations, that

$$\mathcal{G}^{*} \mathcal{G}(\varrho)(l) = \left(D_{\theta(l)}^{*} D_{\theta(l)} + B_{\theta(l)}^{*} \mathcal{E}_{\theta(l)}(X) B_{\theta(l)}\right) \varrho(l)$$

completing the proof of the proposition. □

Suppose that there exists $X = (X_1, \ldots, X_N) \in \mathbb{H}^{n+}$ the mean square stabilizing solution (see Definition 4.4) of the optimal control CARE (4.35), and let $F = (F_1, \ldots, F_N)$ with $F_i = \mathcal{F}_i(X)$ and $\mathcal{F}_i(X)$ as in (4.36). Set also $R_i = (D_i^* D_i + B_i^* \mathcal{E}_i(X) B_i)$, $\tilde{A}_i = A_i + B_i F_i$, and $\tilde{C}_i = C_i + D_i F_i$ for each $i \in \mathbb{N}$, and $\Pi \tilde{A}(k, l)$ as in (C.4) replacing A_i by $\tilde{A}_i$. We can now prove the following result.

Proposition C.3. *Let $\mathcal{G}_c$ and $\mathcal{G}_U$ be as in (6.17) and (6.18) respectively. Then for any $\varrho \in \mathcal{C}^m$,*

$$\mathcal{G}_U^* \mathcal{G}_U = I \tag{C.10}$$

and

$$\mathcal{G}_U^* \mathcal{G}_c(\varrho)(t) = R_{\theta(t)}^{-1/2} B_{\theta(t)}^* \sum_{l=t}^{\infty} E\left[\left[\tilde{A}_{\theta(l)} \dots \tilde{A}_{\theta(t+1)}\right]^* \mathcal{E}_{\theta(l)(X)} G_{\theta(l)} \varrho(l) | \mathfrak{F}_l \right]. \tag{C.11}$$

Proof. From the control CARE (4.35) we have that

$$\tilde{A}_i^* \mathcal{E}_i(X) \tilde{A}_i - X_i + \tilde{C}_i^* \tilde{C}_i = 0 \tag{C.12}$$

and since $R_i F_i = -B_i^* \mathcal{E}_i(X) A_i$, we have that

$$\begin{aligned} R_i^{-1/2}(D_i^* \tilde{C}_i + B_i^* \mathcal{E}_i(X) \tilde{A}_i) &= R_i^{-1/2}(D_i^*(C_i + D_i F_i) + B_i^* \mathcal{E}_i(X)(A_i + B_i F_i)) \\ &= R_i^{-1/2}((D_i^* D_i + B_i^* \mathcal{E}_i(X) B_i) F_i + B_i^* \mathcal{E}_i(X) A_i) \\ &= R_i^{-1/2}(R_i F_i + B_i^* \mathcal{E}_i(X) A_i) = 0. \end{aligned} \tag{C.13}$$

Thus from (C.12), (C.13), and Proposition C.2 we have that

$$\begin{aligned} \mathcal{G}_U^* \mathcal{G}_U(v)(k) &= (R_{\theta(k)}^{-1/2} D_{\theta(k)}^* D_{\theta(k)} R_{\theta(k)}^{-1/2} + R_{\theta(k)}^{-1/2} B_{\theta(k)}^* \mathcal{E}_{\theta(k)}(X) B_{\theta(k)} R_{\theta(k)}^{-1/2}) v(k) \\ &= R_{\theta(k)}^{-1/2}(B_{\theta(k)}^* \mathcal{E}_{\theta(k)}(X) B_{\theta(k)} + D_{\theta(k)}^* D_{\theta(k)}) R_{\theta(k)}^{-1/2} v(k) \\ &= R_{\theta(k)}^{-1/2} R_{\theta(k)} R_{\theta(k)}^{-1/2} v(k) = v(k) \end{aligned}$$

completing the proof of (C.10) of the proposition.

Let us calculate now $\mathcal{G}_c^* \mathcal{G}_U(v)(k)$. We have that

$$\mathcal{G}_c^*(\varrho)(l) = \sum_{k=l+1}^{\infty} E\left(G_{\theta(k)}^* \left[\Pi \tilde{A}(k-1,l)\right]^* \tilde{C}_{\theta(k)}^* \varrho(k) | \mathfrak{F}_l \right)$$

and

$$\mathcal{G}_U(v)(k) = \tilde{C}_{\theta(k)} \left(\sum_{l=-\infty}^{k-1} \left[\Pi \tilde{A}(k-1,l)\right] B_{\theta(l)} R_{\theta(l)}^{-1/2} v(l)\right) + D_{\theta(k)} R_{\theta(k)}^{-1/2} v(k)$$

and thus

$$\mathcal{G}_c^*\mathcal{G}_U(v)(l) = \sum_{k=l+1}^{\infty} E\left\{G_{\theta(l)}^*\left[\Pi\widetilde{A}(k-1,l)\right]^*\widetilde{C}_{\theta(k)}^*\widetilde{C}_{\theta(k)}\right.$$

$$\times\left.\left[\sum_{t=-\infty}^{k-1}\left[\Pi\widetilde{A}(k-1,l)\right]B_{\theta(t)}R_{\theta(t)}^{-1/2}v(t) + D_{\theta(k)}R_{\theta(k)}^{-1/2}v(k)\right]\Big|\mathfrak{F}_l\right\}$$

$$= \sum_{k=0}^{\infty} E\left\{G_{\theta(l)}^*\left[\Pi\widetilde{A}(k+l,l)\right]^*\widetilde{C}_{\theta(k+l+1)}^*\widetilde{C}_{\theta(k+l+1)}\right.$$

$$\times\left.\sum_{t=-\infty}^{k+l}\left[\Pi\widetilde{A}(k+l,l)\right]B_{\theta(t)}R_{\theta(t)}^{-1/2}v(t)\Big|\mathfrak{F}_l\right\}$$

$$+ \sum_{k=0}^{\infty} E\left\{G_{\theta(l)}^*\left[\Pi\widetilde{A}(k+l,l)\right]^*\widetilde{C}_{\theta(k+l+1)}^*D_{\theta(k+l+1)}R_{\theta(k+l+1)}^{-1/2}v(k+l+1)\Big|\mathfrak{F}_l\right\}$$

that is,

$$\mathcal{G}_c^*\mathcal{G}_U(v)(l) = \sum_{k=0}^{\infty} E\left\{G_{\theta(l)}^*\left[\Pi\widetilde{A}(k+l,l)\right]^*\left[X_{\theta(k+l+1)}\right.\right.$$

$$\left.- \widetilde{A}_{\theta(k+l+1)}^*X_{\theta(k+l+2)}\widetilde{A}_{\theta(k+l+1)}\right]$$

$$\times\left.\sum_{t=-\infty}^{k+l}\left[\Pi\widetilde{A}(k+l,l)\right]B_{\theta(t)}R_{\theta(t)}^{-1/2}v(t)\Big|\mathfrak{F}_l\right\}$$

$$- \sum_{k=0}^{\infty} E\left\{G_{\theta(l)}^*\left[\Pi\widetilde{A}(k+l,l)\right]^*\widetilde{A}_{\theta(k+l+1)}^*\right.$$

$$\times\left.X_{\theta(k+l+2)}B_{\theta(k+l+1)}R_{\theta(k+l+1)}^{-1/2}v(k+l+1)\Big|\mathfrak{F}_l\right\}$$

$$= \sum_{k=0}^{\infty} E\left\{G_{\theta(l)}^*\left[\Pi\widetilde{A}(k+l,l)\right]^*X_{\theta(k+l+1)}\right.$$

$$\times\left.\sum_{t=-\infty}^{k+l}\left[\Pi\widetilde{A}(k+l,l)\right]B_{\theta(t)}R_{\theta(t)}^{-1/2}v(t)\Big|\mathfrak{F}_l\right\}$$

$$- \sum_{k=0}^{\infty} E\left\{G_{\theta(l)}^*\left[\Pi\widetilde{A}(k+l,l)\right]^*\widetilde{A}_{\theta(k+l+1)}^*X_{\theta(k+l+2)}\widetilde{A}_{\theta(k+l+1)}\right.$$

$$\times\left.\sum_{t=-\infty}^{k+l}\left[\Pi\widetilde{A}(k+l,l)\right]B_{\theta(t)}R_{\theta(t)}^{-1/2}v(t)\Big|\mathfrak{F}_l\right\}$$

$$- \sum_{k=0}^{\infty} E\left\{G_{\theta(l)}^*\left[\Pi\widetilde{A}(k+l,l)\right]^*X_{\theta(k+l+2)}B_{\theta(k+l+1)}R_{\theta(k+l+1)}^{-1/2}v(k+l+1)\Big|\mathfrak{F}_l\right\}$$

and thus,

$$\mathcal{G}_c^* \mathcal{G}_U(v)(l) = \sum_{k=0}^{\infty} E\left\{ G_{\theta(l)}^* \left[\Pi \widetilde{A}(k+l,l) \right]^* X_{\theta(k+l+1)} \right.$$

$$\times \sum_{t=-\infty}^{k+l} \left[\Pi \widetilde{A}(k+l,t) \right] B_{\theta(t)} R_{\theta(t)}^{-1/2} v(t) | \mathfrak{F}_l \Big\}$$

$$- \sum_{k=0}^{\infty} E\left\{ G_{\theta(l)}^* \left[\Pi \widetilde{A}(k+l+1,l) \right]^* X_{\theta(k+l+2)} \right.$$

$$\times \sum_{t=-\infty}^{k+l+1} \left[\Pi \widetilde{A}(k+l+1,t) \right] B_{\theta(t)} R_{\theta(t)}^{-1/2} v(t) | \mathfrak{F}_l \Big\}$$

$$+ \sum_{k=0}^{\infty} E\left\{ G_{\theta(l)}^* \left[\Pi \widetilde{A}(k+l+1,l) \right]^* X_{\theta(k+l+2)} B_{\theta(k+l+1)} R_{\theta(k+l+1)}^{-1/2} v(k+l+1) | \mathfrak{F}_l \right\}$$

$$- \sum_{k=0}^{\infty} E\left\{ G_{\theta(l)}^* \left[\Pi \widetilde{A}(k+l+1,l) \right]^* X_{\theta(k+l+2)} B_{\theta(k+l+1)} R_{\theta(k+l+1)}^{-1/2} v(k+l+1) | \mathfrak{F}_l \right\}$$

$$= \sum_{k=0}^{\infty} E\left\{ G_{\theta(l)}^* \left[\Pi \widetilde{A}(k+l,l) \right]^* X_{\theta(k+l+1)} \sum_{t=-\infty}^{k+l} \left[\Pi \widetilde{A}(k+l,t) \right] B_{\theta(t)} R_{\theta(t)}^{-1/2} v(t) | \mathfrak{F}_l \right\}$$

$$- \sum_{k=1}^{\infty} E\left\{ G_{\theta(l)}^* \left[\Pi \widetilde{A}(k+l,l) \right]^* X_{\theta(k+l+1)} \sum_{t=-\infty}^{k+l} \left[\Pi \widetilde{A}(k+l,t) \right] B_{\theta(t)} R_{\theta(t)}^{-1/2} v(t) | \mathfrak{F}_l \right\}$$

$$= E\left\{ G_{\theta(l)}^* X_{\theta(l+1)} \sum_{t=-\infty}^{l} \left[\Pi \widetilde{A}(l,t) \right] B_{\theta(t)} R_{\theta(t)}^{-1/2} v(t) | \mathfrak{F}_l \right\}$$

$$= G_{\theta(l)}^* \mathcal{E}_{\theta(l)}(X) \sum_{t=-\infty}^{l} \left[\Pi \widetilde{A}(l,t) \right] B_{\theta(t)} R_{\theta(t)}^{-1/2} v(t).$$

Therefore,

$$\langle \mathcal{G}_c^* \mathcal{G}_U(v); \varrho \rangle = \sum_{l=-\infty}^{\infty} E\left\{ (\mathcal{G}_c^* \mathcal{G}_U(v)(l))^* \varrho(l) \right\}$$

$$= E\left[\sum_{l=-\infty}^{\infty} \left\{ E\left[G_{\theta(l)}^* X_{\theta(l+1)} \sum_{t=-\infty}^{l} \left[\Pi \widetilde{A}(l,t) \right] B_{\theta(t)} R_{\theta(t)}^{-1/2} v(t)/\mathfrak{F}_l \right] \right\}^* \varrho(l) \right]$$

$$= E\left[\sum_{l=-\infty}^{\infty} \sum_{t=-\infty}^{\infty} E\left[v^*(t) R_{\theta(t)}^{-1/2} B_{\theta(t)}^* \left[\Pi \widetilde{A}(l,t) \right]^* X_{\theta(l+1)} G_{\theta(l)} \varrho(l) \mathrm{i}_{\{t \leq l\}} | \mathfrak{F}_l \right] \right]$$

$$= E\left[\sum_{t=-\infty}^{\infty} v^*(t) \sum_{l=t}^{\infty} E\left[R_{\theta(t)}^{-1/2} B_{\theta(t)}^* \left[\Pi \widetilde{A}(l,t) \right]^* X_{\theta(l+1)} G_{\theta(l)} \varrho(l) | \mathfrak{F}_l \right] \right]$$

$$= \langle v; \mathcal{G}_U^* \mathcal{G}_c(\varrho) \rangle$$

and

$$\mathcal{G}_U^*\mathcal{G}_c(\varrho)(t) = R_{\theta(t)}^{-1/2}B_{\theta(t)}^* \sum_{l=t}^{\infty} E\left[\left[\Pi \widetilde{A}(l,t)\right]^* \mathcal{E}_{\theta(l)}(X)G_{\theta(l)}\varrho(l)|\mathfrak{F}_l\right]$$

completing the proof of (C.11) of the proposition.   □

References

1. H. Abou-Kandil, G. Freiling, and G. Jank. Solution and asymptotic behaviour of coupled Riccati equations in jump linear systems. *IEEE Transactions on Automatic Control*, 39:1631–1636, 1994.
2. H. Abou-Kandil, G. Freiling, and G. Jank. On the solution of discrete-time Markovian jump linear quadratic control problems. *Automatica*, 31:765–768, 1995.
3. G. A. Ackerson and K. S. Fu. On the state estimation in switching environments. *IEEE Transactions on Automatic Control*, 15:10–17, 1970.
4. R. Akella and P. R. Kumar. Optimal control of production rate in a failure prone manufacturing system. *IEEE Transactions on Automatic Control*, 31:116–126, 1986.
5. S. Allam, F. Dufour, and P. Bertrand. Discrete-time estimation of a Markov chain with marked point process observations. applications to Markovian jump filtering. *IEEE Transactions on Automatic Control*, 46:903–908, 2001.
6. B. D. O. Anderson and J. B. Moore. *Optimal Control: Linear Quadratic Methods*. Prentice Hall International, 1989.
7. J. G. Van Antwerp and R. D. Braatz. A tutorial on linear and bilinear matrix inequalities. *Journal of Process Control*, 10:363–385, 2000.
8. H. Arai and S. Tachi. Position control of a manipulator with passive joints using dynamic coupling. *IEEE Transactions on Robotics and Automation*, 7:528–534, 1991.
9. K. J. Åstrom and B. Wittenmark. *Adaptive Control*. Addison Wesley, 1989.
10. M. Athans. Command and control (c2) theory: A challenge to control science. *IEEE Transactions on Automatic Control*, 32:286–293, 1987.
11. R. El Azouzi, M. Abbad, and E. Altman. Perturbation of multivariable linear quadratic systems with jump parameters. *IEEE Transactions on Automatic Control*, 46:1666–1671, 2001.
12. T. Başar and P. Bernhard. H_∞-*Optimal Control and Related Minimax Problems*. Birkhäuser, 1990.
13. J. Baczynski, M. D. Fragoso, and E. P. Lopes. On a discrete time linear jump stochastic dynamic game. *Int. J. of Syst. Science*, 32:979–988, 2001.
14. A. V. Balakrishnan. *Kalman Filtering Theory*. Optimization Software, Publications Division, 1987.

15. Y. Bar-Shalom. Tracking methods in a multitarget environment. *IEEE Transactions on Automatic Control*, 23:618–626, 1978.

16. Y. Bar-Shalom and K. Birmiwal. Variable dimension for maneuvering target tracking. *IEEE Transactions on Automatic Control*, 18:621–628, 1982.

17. Y. Bar-Shalom and X. R. Li. *Estimation and Tracking. Principles, Techniques and Software*. Artech House, 1993.

18. G.K. Basak, A. Bisi, and M.K. Ghosh. Stability of a random diffusion with linear drift. *Journal of Mathematical Analysis and Applications*, 202:604–622, 1996.

19. A. Ben-Tal and A. Nemirovski. *Lectures on Modern Convex Optimization: Analysis, Algorithms, and Engineering Applications*. SIAM Press, 2001.

20. K. Benjelloun and E. K. Boukas. Mean square stochastic stability of linear time-delay systems with Markovian jump parameters. *IEEE Transactions on Automatic Control*, 43:1456–1460, 1998.

21. A. Bensoussan. *Stochastic Control of Partially Observable Systems*. Cambridge University Press, 1992.

22. F. Bernard, F. Dufour, and P. Bertrand. On the JLQ problem with uncertainty. *IEEE Transactions on Automatic Control*, 42:869–872, 1997.

23. T. Bielecki and P. R. Kumar. Necessary and sufficient conditions for a zero inventory policy to be optimal in an unreliable manufacturing system. In *Proceedings of the 25th IEEE Conference on Decision and Control*, pages 248–250, Athens, 1986.

24. P. Billingsley. *Probability and Measure*. John Wiley, 1986.

25. R. R. Bitmead and B. O. Anderson. Lyapunov techniques for the exponential stability of linear difference equations with random coefficients. *IEEE Transactions on Automatic Control*, 25:782–787, 1980.

26. S. Bittanti, A. J. Laub, and J. C. Willems. *The Riccati Equation*. Springer-Verlag, 1991.

27. W. P. Blair and D. D. Sworder. Continuous-time regulation of a class of econometric models. *IEEE Transactions on Systems, Man and Cybernetics*, SMC5:341–346, 1975.

28. W. P. Blair and D. D. Sworder. Feedback control of a class of linear discrete systems with jump parameters and quadratic cost criteria. *International Journal of Control*, 21:833–844, 1975.

29. G. Blankenship. Stability of linear differential equations with random coefficients. *IEEE Transactions on Automatic Control*, 22:834–838, 1977.

30. H. A. P. Blom. An efficient filter for abruptly changing systems. In *Proceedings of the 23rd IEEE Conference on Decision and Control*, pages 656–658, Las Vegas, NV, 1984.

31. H. A. P. Blom. Continuous-discrete filtering for the systems with Markovian switching coefficients and simultaneous jump. In *Proceedings of the 21st Asilomar Conference on Signals, Systems and Computers*, pages 244–248, Pacific Grove, 1987.

32. H. A. P. Blom and Y. Bar-Shalom. The interacting multiple model algorithm for systems with Markovian jumping parameters. *IEEE Transactions on Automatic Control*, 33:780–783, 1988.

33. S. Bohacek and E. Jonckheere. Nonlinear tracking over compact sets with linear dynamically varying h^∞ control. *SIAM Journal on Control and Optimization*, 40:1042–1071, 2001.

34. S. Bohacek and E. A. Jonckheere. Relationships between linear dynamically varying systems and jump linear systems. *Mathematics of Control, Signals and Systems*, 16:207–224, 2003.

35. E. K. Boukas and K. Benjelloun. On stabilization of uncertain linear systems with jump parameters. *International Journal of Control*, 72:842–850, 2001.

36. E. K. Boukas and A. Benzaouia. Stability of discrete-time linear systems with Markovian jumping parameters and constrained control. *IEEE Transactions on Automatic Control*, 47:516–521, 2002.

37. E. K. Boukas and Z. K. Liu. Jump linear quadratic regulator with controlled jump rates. *IEEE Transactions on Automatic Control*, 46:301–305, 2001.

38. E. K. Boukas and Z. K. Liu. Robust H_∞-control of discrete-time Markovian jump linear systems with mode-dependent time-delays. *IEEE Transactions on Automatic Control*, 46:1918–1924, 2001.

39. E. K. Boukas, Z. K. Liu, and G. X. Liu. Delay-dependent robust stability and H_∞-control of jump linear systems with time-delay. *International Journal of Control*, 74:329–340, 2001.

40. E. K. Boukas and P. Shi. H_∞-control for discrete-time linear systems with Markovian jump parameters. In *Proceedings of the 36th IEEE Conference on Decision and Control*, pages 4143–4148, 1997.

41. E. K. Boukas and H. Yang. Stability of discrete-time linear systems with Markovian jumping parameters. *Mathematics of Control, Signals and Systems*, 8:390–402, 1995.

42. S. Boyd, L. El Ghaoui, E. Feron, and V. Balakrishnan. *Linear Matrix Inequalities in System and Control Theory*. SIAM Press, 1994.

43. J. W. Brewer. Kronecker products and matrix calculus in system theory. *IEEE Transactions on Circuits and Systems*, 25:772–781, 1978.

44. P. E. Caines. *Linear Stochastic Systems*. Wiley, London, 1988.

45. P. E. Caines and H. F. Chen. Optimal adaptive LQG control for systems with finite state process parameters. *IEEE Transactions on Automatic Control*, 30:185–189, 1985.

46. P. E. Caines and J. Zhang. On the adaptive control of jump parameter systems via nonlinear filtering. *SIAM Journal on Control and Optimzation*, 33:1758–1777, 1995.

47. D. O. Cajueiro. Stochastic optimal control of jumping Markov parameter processes with applications to finance. *PhD Thesis*, Instituto Tecnológico de Aeronáutica - ITA, Brazil, 2002.

48. F. M. Callier and C. A. Desoer. *Linear System Theory*. Springer-Verlag, 1991.

49. S. L. Campbell and C. D. Meyer, Jr. *Generalized Inverses of Linear Transformations*. Dover Press, 1991.

50. Y. Y. Cao and J. Lam. Robust H_∞ control of uncertain Markovian jump systems with time-delay. *IEEE Transactions on Automatic Control*, 45:77–83, 2000.

51. C. G. Chang and M. Athans. State estimation for discrete systems with switching parameters. *IEEE Transactions on Aerospace and Electronic Systems*, 14:418–424, 1978.

52. H. J. Chizeck and Y. Ji. Optimal quadratic control of jump linear systems with Gaussian noise in discrete-time. In *Proceedings of the 27th IEEE Conference on Decision and Control*, pages 1989–1992, 1988.

53. H. J. Chizeck, A. S. Willsky, and D. Castañon. Discrete-time Markovian jump linear quadratic optimal control. *International Journal of Control*, 43:213–231, 1986.

54. E. F. Costa and J. B. R. do Val. On the detectability and observability of discrete-time Markov jump linear systems. *Systems and Control Letters*, 44:135–145, 2001.

55. E. F. Costa and J. B. R. do Val. On the detectability and observability of continuous-time Markov jump linear systems. *SIAM Journal on Control and Optimization*, 41:1295–1314, 2002.

56. E. F. Costa and J. B. R. do Val. Weak detectability and the linear-quadratic control problem of discrete-time Markov jump linear systems. *International Journal of Control*, 75:1282–1292, 2002.

57. E. F. Costa, J. B. R. do Val, and M. D. Fragoso. Weak detectability of LQ problem of discrete-time infinite Markov jump linear systems. *Stochastic Analysis and Applications*, 23, 2005.

58. O. L. V. Costa. Linear minimum mean square error estimation for discrete-time Markovian jump linear systems. *IEEE Transactions on Automatic Control*, 39:1685–1689, 1994.

59. O. L. V. Costa. Discrete-time coupled Riccati equations for systems with Markov switching parameters. *Journal of Mathematical Analysis and Applications*, 194:197–216, 1995.

60. O. L. V. Costa. Mean square stabilizing solutions for discrete-time coupled algebraic Riccati equations. *IEEE Transactions on Automatic Control*, 41:593–598, 1996.

61. O. L. V. Costa and E. K. Boukas. Necessary and sufficient conditions for robust stability and stabilizability of continuous-time linear systems with Markovian jumps. *Journal of Optimization Theory and Applications*, 99:359–379, 1998.

62. O. L. V. Costa, J. B. do Val, and J. C. Geromel. Continuous-time state-feedback H_2-control of Markovian jump linear systems via convex analysis. *Automatica*, 35:259–268, 1999.

63. O. L. V. Costa and J. B. R. do Val. Full information H_∞-control for discrete-time infinite Markov jump parameter systems. *Journal of Mathematical Analysis and Applications*, 202:578–603, 1996.

64. O. L. V. Costa and J. B. R. do Val. Jump LQ-optimal control for discrete-time Markovian systems with l_2-stochastic inputs. *Stochastic Analysis and Applications*, 16:843–858, 1998.

65. O. L. V. Costa, J. B. R. do Val, and J. C. Geromel. A convex programming approach to H_2-control of discrete-time Markovian jump linear systems. *International Journal of Control*, 66:557–579, 1997.

66. O. L. V. Costa and M. D. Fragoso. Stability results for discrete-time linear systems with Markovian jumping parameters. *Journal of Mathematical Analysis and Applications*, 179:154–178, 1993.

67. O. L. V. Costa and M. D. Fragoso. Discrete-time LQ-optimal control problems for infinite Markov jump parameter systems. *IEEE Transactions on Automatic Control*, 40:2076–2088, 1995.

68. O. L. V. Costa and M. D. Fragoso. Comments on stochastic stability of jump linear systems. *IEEE Transactions on Automatic Control*, 49:1414–1416, 2004.

69. O. L. V. Costa and S. Guerra. Robust linear filtering for discrete-time hybrid Markov linear systems. *International Journal of Control*, 75:712–727, 2002.

70. O. L. V. Costa and S. Guerra. Stationary filter for linear minimum mean square error estimator of discrete-time Markovian jump systems. *IEEE Transactions on Automatic Control*, 47:1351–1356, 2002.

71. O. L. V. Costa and R. P. Marques. Mixed H_2/H_∞-control of discrete-time Markovian jump linear systems. *IEEE Transactions on Automatic Control*, 43:95–100, 1998.

72. O. L. V. Costa and R. P. Marques. Robust H_2-control for discrete-time Markovian jump linear systems. *International Journal of Control*, 73:11–21, 2000.

73. O. L. V. Costa and E. F. Tuesta. Finite horizon quadratic optimal control and a separation principle for Markovian jump linear systems. *IEEE Transactions on Automatic Control*, 48:1836–1842, 2003.

74. O. L. V. Costa and E. F. Tuesta. H_2-control and the separation principle for discrete-time Markovian jump linear systems. *Mathematics of Control, Signals and Systems*, 16:320–350, 2004.

75. O.L.V. Costa, E.O. Assumpção, E.K. Boukas, and R.P.Marques. Constrained quadratic state feedback control of discrete-time Markovian jump linear systems. *Automatica*, 35:617–626, 1999.

76. O.L.V. Costa and J.C.C. Aya. Monte-Carlo TD(λ)-methods for the optimal control of discrete-time Markovian jump linear systems. *Automatica*, 38:217–225, 2002.

77. A. Czornik. *On Control Problems for Jump Linear Systems*. Wydawnictwo Politechniki Slaskiej, 2003.

78. A. Czornik and A. Swierniak. On the discrete JLQ and JLQG problems. *Nonlinear Analysis*, 47:423–434, 2001.

79. M. H. A. Davis. *Linear Estimation and Stochastic Control*. Chapman and Hall, 1977.

80. M. H. A. Davis and R. B. Vinter. *Stochastic Modelling and Control*. Chapman and Hall, 1985.

81. M.H.A. Davis and S.I. Marcus. An introduction to nonlinear filtering. In M. Hazewinkel and J.C. Willems, editors, *Stochastic Systems: The Mathematics of Filtering and Identification and Aplications*, volume 2, pages 53–75. D. Reidel Publishing Company, 1981.

82. D. P. de Farias, J. C. Geromel, and J. B. R. do Val. A note on the robust control of Markov jump linear uncertain systems. *Optimal Control Applications & Methods*, 23:105–112, 2002.

83. D. P. de Farias, J. C. Geromel, J. B. R. do Val, and O. L. V. Costa. Output feedback control of Markov jump linear systems in continuous-time. *IEEE Transactions on Automatic Control*, 45:944–949, 2000.

84. C. E. de Souza and M. D. Fragoso. On the existence of a maximal solution for generalized algebraic Riccati equation arising in stochastic control. *Systems and Control Letters*, 14:233–239, 1990.

85. C. E. de Souza and M. D. Fragoso. H_∞-control for linear systems with Markovian jumping parameters. *Control Theory and Advanced Technology*, 9:457–466, 1993.

86. C. E. de Souza and M. D. Fragoso. H_∞ filtering for Markovian jump linear systems. *International Journal of Systems Science*, 33:909–915, 2002.

87. C. E. de Souza and M. D. Fragoso. Robust H_∞ filtering for uncertain Markovian jump linear systems. *International Journal of Robust and Nonlinear Control*, 12:435–446, 2002.

88. C. E. de Souza and M. D. Fragoso. H_∞ filtering for discrete-time systems with Markovian jump parameters. *International Journal of Robust and Nonlinear Control*, 14:1299–1316, 2003.

89. J. B. R. do Val and T. Başar. Receding horizon control of jump linear systems and a macroeconomic policy problem. *Journal of Economic Dynamics and Control*, 23:1099–1131, 1999.

90. J. B. R. do Val and E. F. Costa. Numerical solution for linear-quadratic control problems of Markov jump linear systems and weak detectability concept. *Journal of Optimization Theory and Applications*, 114:69–96, 2002.

91. J. B. R. do Val, J. C. Geromel, and O. L. V. Costa. Uncoupled Riccati iterations for the linear quadratic control problem of discrete-time Markov jump linear systems. *IEEE Transactions on Automatic Control*, 43:1727–1733, 1998.

92. J. B. R. do Val, J. C. Geromel, and A. P. C. Gonçalves. The H_2-control for jump linear systems: Cluster observations of the Markov state. *Automatica*, 38:343–349, 2002.

93. J. B. R. do Val, C. Nespoli, and Y. R. Z. Caceres. Stochastic stability for Markovian jump linear systems associated with a finite number of jump times. *Jounal of Mathematical Analysis and Applications*, 285:551–563, 2003.

94. A. Doucet and C. Andrieu. Iterative algorithms for state estimation of jump Markov linear systems. *IEEE Transactions on Automatic Control*, 49:1216–1227, 2000.

95. A. Doucet, A. Logothetis, and V. Krishnamurthy. Stochastic sampling algorithms for state estimation of jump Markov linear systems. *IEEE Transactions on Automatic Control*, 45:188–202, 2000.

96. J. C. Doyle, K. Glover, P. P. Khargonekar, and B. A. Francis. State-space solutions to standard H_2 and H_∞-control problems. *IEEE Transactions on Automatic Control*, 34:831–847, 1989.

97. V. Dragan, P. Shi, and E. K. Boukas. Control of singularly perturbed systems with Markovian jump parameters: an H_∞ approach. *Automatica*, 35:1369–11378, 1999.

98. F. Dufour and P. Bertrand. Stabilizing control law for hybrid models. *IEEE Transactions on Automatic Control*, 39:2354–2357, 1994.

99. F. Dufour and P. Bertrand. An image based filter for discrete-time Markovian jump linear systems. *Automatica*, 32:241–247, 1996.

100. F. Dufour and R.J. Elliot. Filtering with discrete state observations. *Applied Mathematics and Optimization*, 40:259–272, 1999.

101. F. Dufour and R. J. Elliott. Adaptive control of linear systems with Markov perturbations. *IEEE Transactions on Automatic Control*, 43:351–372, 1997.

102. F. Dufour and D. Kannan. Discrete time nonlinear filtering with marked point process observations. *Stochatic Analysis and Applications*, 17:99–115, 1999.

103. N. Dunford and J. T. Schwartz. *Linear Operators - Part I: General Theory*. Wiley, 1958.

104. R. Elliot, F. Dufour, and F. Sworder. Exact hybrid filters in discrete-time. *IEEE Transactions on Automatic Control*, 41:1807–1810, 1996.

105. R. J. Elliot, L. Aggoun, and J. B. Moore. *Hidden Markov Models: Estimation and Control*. Springer-Verlag, NY, 1995.

106. M.H.C Everdij and H.A.P. Blom. Embedding adaptive JLQG into LQ martingale control with a completely observable stochastic control matrix. *IEEE Transactions on Automatic Control*, 41:424–430, 1996.

107. Y. Fang. A new general sufficient condition for almost sure stability of jump linear systems. *IEEE Transactions on Automatic Control*, 42:378–382, 1997.

108. Y. Fang, K. A. Loparo, and X. Feng. Almost sure and δ-moment stability of jump linear systems. *International Journal of Control*, 59:1281–1307, 1994.

109. X. Feng, K. A. Loparo, Y. Ji, and H. J. Chizeck. Stochastic stability properties of jump linear systems. *IEEE Transactions on Automatic Control*, 37:38–53, 1992.

110. W. H. Fleming and W. N. MacEneaney. Risk sensitive optimal control and differential games. In K. S. Lawrence, editor, *Stochastic Theory and Adaptive Control*, volume 184 of *Lecture Notes in Control and Information Science*, pages 185–197. Springer-Verlag, 1992.

111. W.H. Fleming. *Future Directions in Control Theory - A Mathematical Perspective*. SIAM Philadelphia, 1988.

112. W.H. Fleming and R.W. Rishel. *Deterministic and Stochastic Optimal Control*. Springer-Verlag, 1975.

113. J. J. Florentin. Optimal control of continuous-time Markov stochastic systems. *Journal of Electronic Control*, 10:473–488, 1961.

114. M. D. Fragoso. On a partially observable LQG problem for systems with Markovian jumping parameters. *Systems and Control Letters*, 10:349–356, 1988.

115. M. D. Fragoso. On a discrete-time jump LQG problem. *International Journal of Systems Science*, 20:2539–2545, 1989.

116. M. D. Fragoso. A small random perturbation analysis of a partially observable LQG problem with jumping parameter. *IMA Journal of Mathematical Control and Information*, 7:293–305, 1990.

117. M. D. Fragoso and J. Baczynski. Optimal control for continuous time LQ problems with infinite Markov jump parameters. *SIAM Journal on Control and Optimization*, 40:270–297, 2001.

118. M. D. Fragoso and J. Baczynski. Lyapunov coupled equations for continuous-time infinite Markov jump linear systems. *Journal of Mathematical Analysis and Applications*, 274:319–335, 2002.

119. M. D. Fragoso and J. Baczynski. Stochatic versus mean square stability in continuous-time linear infinite Markov jump parameters. *Stochastic Analysis and Applications*, 20:347–356, 2002.

120. M. D. Fragoso and O. L. V. Costa. Unified approach for mean square stability of continous-time linear systems with Markovian jumping parameters and additive disturbances. In *Proceedings of the 39th Conference on Decision and Control*, pages 2361–2366, Sydney, Australia, 2000.

121. M. D. Fragoso and O. L. V. Costa. Mean square stabilizability of continuous-time linear systems with partial information on the Markovian jumping parameters. *Stochastic Analysis and Applications*, 22:99–111, 2004.

122. M. D. Fragoso, O. L. V. Costa, and C. E. de Souza. A new approach to linearly perturbed Riccati equations arising in stochastic control. *Journal of Applied Mathematics and Optimization*, 37:99–126, 1998.

123. M. D. Fragoso, J. B. R. do Val, and D. L. Pinto Junior. Jump linear H_∞-control: the discrete-time case. *Control Theory and Advanced Technology*, 10:1459–1474, 1995.

124. M. D. Fragoso, E. C. S. Nascimento, and J. Baczynski. H_∞ control for continuous-time linear systems systems with infinite Markov jump parame-

ters via semigroup. In *Proceedings of the 39th Conference on Decision and Control*, pages 1160–1165, Sydney, Australia, 2001.

125. M.D. Fragoso and E.M. Hemerly. Optimal control for a class of noisy linear systems with markovian jumping parameters and quadratic cost. *International Journal of Systems Science*, 22:2553–3561, 1991.

126. B. A. Francis. *A Course in H_∞-Control Theory*, volume 88 of *Lecture Notes in Control and Information Sciences*. Springer-Verlag, 1987.

127. Z. Gajic and R. Losada. Monotonicity of algebraic Lyapunov iterations for optimal control of jump parameter linear systems. *Systems and Control Letters*, 41:175–181, 2000.

128. J. Gao, B. Huang, and Z. Wang. LMI-based robust H_∞ control of uncertain linear jump systems with time-delay. *Automatica*, 37:1141–1146, 2001.

129. J. C. Geromel. Optimal linear filtering under parameter uncertainty. *IEEE Transactions on Signal Processing*, 47:168–175, 1999.

130. J. C. Geromel, P. L. D. Peres, and S. R. Souza. A convex approach to the mixed H_2/H_∞-control problem for discrete-time uncertain systems. *SIAM Journal on Control and Optimization*, 33:1816–1833, 1995.

131. L. El Ghaoui and S. I. Niculescu, editors. *Advances in Linear Matrix Inequality Methods in Control*. SIAM Press, 1999.

132. M.K. Ghosh, A. Arapostathis, and S. I. Marcus. A note on an LQG regulator with Markovian switching and pathwise average cost. *IEEE Transactions on Automatic Control*, 40:1919–1921, 1995.

133. K. Glover and J. C. Doyle. State-space formulae for stabilizing controllers that satisfy an H_∞-norm bound and relations to risk sensitivity. *Systems and Control Letters*, 11:167–172, 1988.

134. W. S. Gray and O. González. Modelling electromagnetic disturbances in closed-loop computer controlled flight systems. In *Proceedings of the 1998 American Control Conference*, pages 359–364, Philadelphia, PA, 1998.

135. W. S. Gray, O. R. González, and M. Doğan. Stability analysis of digital linear flight controllers subject to electromagnetic disturbances. *IEEE Transactions on Aerospace and Electronic Systems*, 36:1204–1218, 2000.

136. M. Green and D. J. N. Limebeer. *Linear Robust Control*. Prentice Hall, 1995.

137. B. E. Griffiths and K. A. Loparo. Optimal control of jump linear quadratic Gaussian systems. *International Journal of Control*, 42:791–819, 1985.

138. M. T. Hadidi and C. S. Schwartz. Linear recursive state estimators under uncertain observations. *IEEE Transactions on Automatic Control*, 24:944–948, 1979.

139. R. Z. Hasminskii. *Stochastic Stability of Differential Equations*. Sijthoff and Noordhoff, 1980.

140. J. W. Helton and M. R. James. *Extending H_∞ Control to Nonlinear Systems*. Frontiers in Applied Mathematics, SIAM, 1999.

141. A. Isidori and A. Astolfi. Disturbance attenuation and H_∞-control via measurement feedback in nonlinear systems. *IEEE Transactions on Automatic Control*, 37:1283–1293, 1992.

142. M. R. James. Asymptotic analysis of nonlinear stochastic risk-sensitive control and differential games. *Mathematical Control Signals and Systems*, 5:401–417, 1991.

143. Y. Ji and H. J. Chizeck. Controllability, observability and discrete-time Markovian jump linear quadratic control. *International Journal of Control*, 48:481–498, 1988.

144. Y. Ji and H. J. Chizeck. Optimal quadratic control of jump linear systems with separately controlled transition probabilities. *International Journal of Control*, 49:481–491, 1989.

145. Y. Ji and H. J. Chizeck. Controllability, observability and continuous-time Markovian jump linear quadratic control. *IEEE Transactions on Automatic Control*, 35:777–788, 1990.

146. Y. Ji and H. J. Chizeck. Jump linear quadratic Gaussian control: steady state solution and testable conditions. *Control Theory and Advanced Technology*, 6:289–319, 1990.

147. Y. Ji, H. J. Chizeck, X. Feng, and K. A. Loparo. Stability and control of discrete-time jump linear systems. *Control Theory and Advanced Technology*, 7:247–270, 1991.

148. M. Kac. *Discrete Thoughts : Essays on Mathematics, Science and Philosophy.* Birkhäuser, 1992.

149. G. Kallianpur. *Stochastic Filtering Theory.* Springer-Verlag, 1980.

150. T. Kazangey and D. D. Sworder. Effective federal policies for regulating residential housing. In *Proceedings of the Summer Computer Simulation Conference*, pages 1120–1128, 1971.

151. B. V. Keulen, M. Peters, and R. Curtain. H_∞-control with state feedback: The infinite dimensional case. *Journal of Mathematical Systems, Estimation and Control*, 3:1–39, 1993.

152. M. Khambaghi, R. Malhamé, and M. Perrier. White water and broke recirculation policies in paper mills via Markovian jump linear quadratic control. In *Proceedings of the American Control Conference*, pages 738 –743, Philadelphia, 1998.

153. P. P. Khargonekar and M. A. Rotea. Mixed H_2/H_∞-control: a convex optimization approach. *IEEE Transactions on Automatic Control*, 36:824–837, 1991.

154. F. Kozin. A survey of stability of stochastic systems. *Automatica*, 5:95–112, 1969.

155. N. N. Krasovskii and E. A. Lidskii. Analytical design of controllers in systems with random attributes I, II, III. *Automatic Remote Control*, 22:1021–1025, 1141–1146, 1289–1294, 1961.

156. C. S. Kubrusly. Mean square stability for discrete bounded linear systems in Hilbert spaces. *SIAM Journal on Control and Optimization*, 23:19–29, 1985.

157. C. S. Kubrusly. On discrete stochastic bilinear systems stability. *Journal of Mathematical Analysis and Applications*, 113:36–58, 1986.

158. C. S. Kubrusly and Costa O. L. V. Mean square stability conditions for discrete stochastic bilinear systems. *IEEE Transactions on Automatic Control*, 36:1082–1087, 1985.

159. H. Kushner. *Stochastic Stability and Control.* Academic Press, New York, 1967.

160. A. Leizarowitz. Estimates and exact expressions for Lyapunov exponents of stochastic linear differential equations. *Stochastics*, 24:335–356, 1988.

161. Z. G. Li, Y. C. Soh, and C. Y. Wen. Sufficient conditions for almost sure stability of jump linear systems. *IEEE Transactions on Automatic Control*, 45:1325–1329, 2000.

162. D. J. N. Limebeer, B. D. O. Anderson, P. P. Khargonekar, and M. Green. A game theoretic approach to H_∞ control for time varying systems. *SIAM Journal on Control and Optimization*, 30:262–283, 1992.

163. K. A. Loparo. Stochastic stability of coupled linear systems: A survey of methods and results. *Stochastic Analysis and Applications*, 2:193–228, 1984.

164. K. A. Loparo, M. R. Buchner, and K. Vasudeva. Leak detection in an experimental heat exchanger process: a multiple model approach. *IEEE Transactions on Automatic Control*, 36:167–177, 1991.

165. D. G. Luenberger. *Introduction to Dynamic Systems: Theory, Models & Applications*. John Wiley & Sons, 1979.

166. M. S. Mahmoud and P. Shi. Robust Kalman filtering for continuous time-lag systems with Markovian jump parameters. *IEEE Transactions on Circuits and Systems*, 50:98–105, 2003.

167. M. S. Mahmoud and P. Shi. Robust stability, stabilization and H_∞ control of time-delay systems with Markovian jump parameters. *International Journal of Robust and Nonlinear Control*, 13:755–784, 2003.

168. R. Malhame and C. Y. Chong. Electric load model synthesis by diffusion approximation in a high order hybrid state stochastic systems. *IEEE Transactions on Automatic Control*, 30:854–860, 1985.

169. M. Mariton. Almost sure and moments stability of jump linear systems. *Systems and Control Letters*, 30:1145–1147, 1985.

170. M. Mariton. On controllability of linear systems with stochastic jump parameters. *IEEE Transactions on Automatic Control*, 31:680–683, 1986.

171. M. Mariton. *Jump Linear Systems in Automatic Control*. Marcel Decker, 1990.

172. M. Mariton and P. Bertrand. Output feedback for a class of linear systems with jump parameters. *IEEE Transactions on Automatic Control*, 30:898–900, 1985.

173. M. Mariton and P. Bertrand. Robust jump linear quadratic control: A mode stabilizing solution. *IEEE Transactions on Automatic Control*, 30:1145–1147, 1985.

174. M. Mariton and P. Bertrand. Reliable flight control systems: Components placement and feedback synthesis. In *Proceedings of the 10th IFAC World Congress*, pages 150–154, Munich, 1987.

175. R. Morales-Menéndez, N. de Freitas, and D. Poole. Estimation and control of industrial processes with particle filters. In *Proceedings of the 2003 American Control Conference*, pages 579–584, Denver, 2003.

176. T. Morozan. Optimal stationary control for dynamic systems with Markov perturbations. *Stochastic Analysis and Applications*, 1:299–325, 1983.

177. T. Morozan. Stabilization of some stochastic discrete-time control systems. *Stochastic Analysis and Applications*, 1:89–116, 1983.

178. T. Morozan. Stability and control for linear systems with jump Markov perturbations. *Stochastic Analysis and Applications*, 13:91–110, 1995.

179. A. S. Morse. *Control Using Logic-Based Switching*. Springer-Verlag, London, 1997.

180. N. E. Nahi. Optimal recursive estimation with uncertain observation. *IEEE Transactions on Information Theory*, 15:457–462, 1969.

181. A. W. Naylor and G. R. Sell. *Linear Operator Theory in Engineering and Science*. Springer-Verlag, second edition, 1982.

182. Y. Nesterov and A. Nemirovskii. *Interior Point Polynomial Algorithms in Convex Programming*. SIAM Press, 1994.

183. K. Ogata. *Discrete-Time Control Systems*. Prentice Hall, 2nd. edition, 1995.

184. J. Oostveen and H. Zwart. Solving the infinite-dimensional discrete-time algebraic Riccati equation using the extended symplectic pencil. Internal Report W-9511, University of Groningen, Department of Mathematics, 1995.

185. G. Pan and Y. Bar-Shalom. Stabilization of jump linear Gaussian systems without mode observations. *International Journal of Control*, 64:631–661, 1996.

186. Z. Pan and T. Başar. Zero-sum differential games with random structures and applications in H_∞-control of jump linear systems. In *6th International Symposium on Dynamic Games and Applications*, pages 466–480, Quebec, 1994.

187. Z. Pan and T. Başar. H_∞-control of large scale jump linear systems via averaging and aggregation. *International Journal of Control*, 72:866–881, 1999.

188. B. Park and W. H. Kwon. Robust one-step receding horizon control of discrete-time Markovian jump uncertain systems. *Automatica*, 38:1229–1235, 2002.

189. S. Patilkulkarni, W. S. Gray H. Herencia-Zapana, and O. R. González. On the stability of jump-linear systems driven by finite-state machines with Markovian inputs. In *Proceedings of the 2004 American Control Conference*, 2004.

190. R. Patton, P. Frank, and R. Clark, editors. *Fault Diagnosis in Dynamic Systems*. Prentice Hall, 1989.

191. M. A. Rami and L. El Ghaoui. Robust stabilization of jump linear systems using linear matrix inequalities. In *IFAC Symposium on Robust Control Design*, pages 148–151, Rio de Janeiro, 1994.

192. M. A. Rami and L. El Ghaoui. LMI optimization for nonstandard Riccati equations arising in stochastic control. *IEEE Transactions on Automatic Control*, 41:1666–1671, 1996.

193. A. C. Ran and R. Vreugdenhil. Existence and comparison theorems for the algebraic Riccati equations for continuous and discrete-time systems. *Linear Algebra and its Applications*, 99:63–83, 1988.

194. S. Ross. *Applied Probability Models with Optimization Applications*. Holden-Day, 1970.

195. A. Saberi, P. Sannuti, and B. M. Chen. H_2-*Optimal Control*. Prentice-Hall, 1995.

196. P. A. Samuelson. Interactions between the multiplier analysis and the principle of accelleration. *Review of Economic Statistics*, 21:75–85, 1939.

197. P. Shi and E. K. Boukas. H_∞ control for Markovian jumping linear systems with parametric uncertainty. *Journal of Optimization Theory and Application*, 95:75–99, 1997.

198. P. Shi, E. K. Boukas, and R. K. Agarwal. Control of Markovian jump discrete-time systems with norm bounded uncertainty and unknown delay. *IEEE Transactions on Automatic Control*, 44:2139–2144, 1999.

199. P. Shi, E. K. Boukas, and R. K. Agarwal. Robust control for Markovian jumping discrete-time systems. *International Jounal of Systems Science*, 30:787–797, 1999.

200. A. A. G. Siqueira and M. H. Terra. Nonlinear and Markovian H_∞ controls of underactuated manipulators. *IEEE Transactions on Control System Technology*, to appear.

201. E. D. Sontag. *Mathematical Control Theory*. Springer-Verlag, 1990.

202. A. Stoica and I. Yaesh. Jump Markovian-based control of wing deployment for an uncrewed air vehicle. *Journal of Guidance*, 25:407–411, 2002.

203. A. A. Stoorvogel. *The H_∞-Control Problem: A State Space Approach*. Prentice-Hall, New York, 1992.

204. A. A. Stoorvogel and A. J. T. M. Weeren. The discrete-time Riccati equation related to the H_∞-control problem. *IEEE Transactions on Automatic Control*, 39:686–91, 1994.

205. D. D. Sworder. On the stochastic maximum principle. *Journal of Mathematical Analysis and Applications*, 24:627–640, 1968.

206. D. D. Sworder. Feedback control of a class of linear systems with jump parameters. *IEEE Transactions on Automatic Control*, 14:9–14, 1969.

207. D. D. Sworder and J. E. Boyd. *Estimation Problems in Hybrid Systems*. Cambridge University Press, London, 1999.

208. D. D. Sworder and R. O. Rogers. An LQG solution to a control problem with solar thermal receiver. *IEEE Transactions on Automatic Control*, 28:971–978, 1983.

209. G. Tadmor. Worst case design in the time domain: The maximum principle and the standard H_∞-problem. *Mathematics of Control, Signals and Systems*, 3:301–324, 1990.

210. J. K. Tugnait. Adaptive estimation and identification for discrete systems with Markov jump parameters. *IEEE Transactions on Automatic Control*, 27:1054–1064, 1982.

211. J. K. Tugnait. Detection and estimation for abruptly changing systems. *Automatica*, 18:607–615, 1982.

212. A. J. van der Schaft. Nonlinear state-space H_∞-control theory. In H. L. Trentelman and J. C. Willems, editors, *Essays on Control: Perspectives in the Theory and its Applications*. Birkhäuser, 1993.

213. M. Vidyasagar. *Nonlinear Systems Analysis*. Prentice-Hall, 2nd. edition, 1993.

214. R. Vinter. *Optimal Control*. Birkhauser, 2000.

215. Z. Wang, H. Qiao, and K. J. Burnham. On stabilization of bilinear uncertain time-delay stochastic systems with Markovian jumping parameters. *IEEE Transactions on Automatic Control*, 47:640–646, 2002.

216. J. Weidmann. *Linear Operators in Hilbert Spaces*. Springer-Verlag, 1980.

217. P. Whittle. *Risk-sensitive Optimal Control*. John Wiley & Sons, 1990.

218. A.S. Willsky. A survey of design methods for failure detection in dynamic systems. *Automatica*, 12:601–611, 1976.

219. W. H. Wonham. Random differential equations in control theory. In A. T. Bharucha-Reid, editor, *Probabilistic Methods in Applied Mathematics*, volume 2, pages 131–212. Academic Press, 1970.

220. W. M. Wonham. On a matrix Riccati equation of stochastic control. *SIAM Journal of Control*, 6:681–697, 1968.

221. W.H. Wonham. On the separation theorem of stochastic control. *SIAM Journal on Control and Optimization*, 6:312–326, 1968.

222. V. A. Yakubovich. A frequency theorem for the case in which the state and control spaces are Hilbert spaces with an application to some problems in the synthesis of optimal controls I. *Siberian Mathematical Journal*, 15:457–476, 1974.

223. C. Yang, P. Bertrand, and M. Mariton. Adaptive control in the presence of Markovian parameter jumps. *International Journal of Control*, 52:473–484, 1990.

224. J. Yong and X. Y. Zhou. *Stochastic Controls: Hamiltonian Systems and HJB Equations*. Springer-Verlag, 1999.

225. G. Zames. Feedback and optimal sensitivity: Mode reference transformations, multiplicative seminorms and approximate inverses. *IEEE Transactions on Automatic Control*, 26:301–320, 1981.

226. Q. Zhang. Optimal filtering of discrete-time hybrid systems. *Journal of Optimization Theory and Applications*, 100:123–144, 1999.

227. Q. Zhang. Hybrid filtering for linear systems with non-Gaussian disturbances. *IEEE Transactions on Automatic Control*, 45:50–61, 2000.

228. K. Zhou and J. C. Doyle. *Essentials of Robust Control*. Prentice Hall, 1998.

229. K. Zhou, J. C. Doyle, and K. Glover. *Robust and Optimal Control*. Prentice Hall, 1996.

Notation and Conventions

As a general rule, lowercase Greek and Roman letters are used for vector, scalar variables and functions, while uppercase Greek and Roman letters are used for matrix variables and functions as well as sequences of matrices. Sets and spaces are denoted by blackboard uppercase characters (such as $\mathbb{R}$, $\mathbb{C}$) and operators by calligraphic characters (such as $\mathcal{L}$, $\mathcal{T}$).

Sometimes it is not possible or convenient to adhere completely to this rule, but the exceptions should be clearly perceived based on their specific context.

The following lists present the main symbols and general notation used throughout the book, followed by a brief explanation and the number of the page of their definition or first appearance.

Symbol	Description	Page
$\emptyset$	Empty set.	222
$\|\cdot\|$	Any norm.	16
$\|\cdot\|_1$	1-norm.	16
$\|\cdot\|_1$	Induced 1-norm.	16
$\|\cdot\|_{1t}$	$1t$-norm.	219
$\|\cdot\|_2$	Euclidean norm.	16
$\|\cdot\|_2$	Induced Euclidean norm.	16
$\|\cdot\|_{max}$	max-norm.	16
$\|\mathcal{G}_F\|_2^2$	H_2-cost for system $\mathcal{G}_F$. Also denotes the H_2-norm.	80, 83
$\|\bar{\mathcal{G}}\|_2^2$	Optimal H_2-cost for system $\mathcal{G}_F$	80
$\oplus$	Direct sum of subspaces.	21
$\otimes$	Kronecker product.	17
$\star$	Conjugate transpose.	15
$'$	Transpose.	15
$\mathbf{1}$	Indicator function.	31
Γ	N-sequence of system matrices.	31

Symbol	Description	Page
Δ	N-sequence of uncertainties.	175
ΔA	N-sequence of uncertainties.	175
ΔB	N-sequence of uncertainties.	175
δ	Upper bound for H_∞-control.	146
Θ_0	The set of all $\mathfrak{F}_0$-measurable variables in $\mathbb{N}$.	21
$\theta(k)$ $\theta_0 = \theta(0)$	Markov state at time k.	2
$\hat{\theta}(k)$	Estimate for $\theta(k)$.	4, 59
$\tilde{\theta}_\nu(k)$	A Markov state at time k.	63
$\lambda_i(.)$	i-th eigenvalue.	16
$\mu(k) = E(x(k))$ $\mu_0 = \mu(0)$	Expected value of $x(k)$.	31
$\nu(k)$	Estimation error.	134
$\pi_i(k),\ \pi_i$	Probabilities.	48
υ	N-sequence of distributions.	9
Φ	A set.	58
$\phi(.)$	Lyapunov function.	22
$\varphi(.)$	Linear operator.	17
$\hat{\varphi}(.)$	Linear operator.	17
Ψ	A set related with H_2-control.	86
$\hat{\Psi}$	A set related with H_2-control.	86
Ψ	A set related with robust H_2-control.	175
Ω	Sample space.	20
$\tilde{\Omega}_k$	A set.	20
A	N-sequence of system matrices.	8
$\hat{A}$	N-sequence of filter matrices.	103
B	N-sequence of system matrices.	8
$\hat{B}$	N-sequence of filter matrices.	103
C	N-sequence of system matrices.	8
$\hat{C}$	N-sequence of filter matrices.	103
C^m	A Hilbert space.	21
D	N-sequence of system matrices.	8
$\mathrm{diag}[.]$	Block diagonal matrix.	34
$\mathrm{dg}[.]$	Block diagonal matrix.	119
$E(.)$	Expected value.	21
F	N-sequence.	57
G	N-sequence of system matrices.	8
H	N-sequence of system matrices.	8
I, I_n	Order n identity matrix.	16
i	Index (usually operation mode).	2
j	Index (usually operation mode).	4
k	Time.	2

Symbol	Description	Page
M	Riccati gain (filter).	108
N	Number of operation modes.	9
P	Transition probability matrix.	4
p_{ij}	Probability of transition from mode i to mode j.	4
$Q(k)$	N-sequence of second moment matrices at time k.	31
$Q_i(k)$	Second moment of $x(k)$ at mode i.	31
$q(k)$	Vector of first moments at time k.	31
$q_i(k)$	First moment of $x(k)$ at mode i.	31
Re	Real part.	94
$r_\sigma(.)$	Spectral radius.	15
$S(k)$	A second moment matrix.	121
S	A stationary second moment matrix.	122
S_c	Controllability Grammian.	24, 83
S_o	Observability Grammian.	25, 83
$s = \{s_{ij}\}$	Positive numbers (a generalization of p_{ij}).	203
$\mathrm{tr}(.)$	Trace operator.	16
$U(k, l)$	N-sequence.	49
$U_j(k, l)$	A second moment matrix of x at times k, l and mode j.	49
$u(k)$	Control variable at time k.	8
$\bar{u}(k)$	An optimal control law.	80
$v(k)$	Filtering error at time k.	103
X	Solution of the control CARE.	78
$\widehat{X}$	Stabilizing solution of the control CARE.	226
$X_F(.)$	A linear operator.	146
$X_F^0(.)$	A special case of $X_F(.)$.	146
X^l	An N-sequence.	213
X^+	Maximal solution of the control CARE.	213
$x(k)$	System state at time k.	4
$x_0 = x(0)$		
$\widehat{x}(k)$	Estimated state at time k.	103
$\widehat{x}_0 = \widehat{x}(0)$		
$x_e(k)$	Augmented state.	120
$\widehat{x}_e(k)$	Estimated state at time k.	104
$\widetilde{x}_e(k)$	Estimation error at time k.	105
$w(k)$	Additive disturbance at time k.	8
Y	Solution to the filtering CARE.	108
$y(k)$	Measured variable at time k.	8
Z_F	A linear operator.	146
$Z_F^0(.)$	A special case of $Z_F(.)$.	146
$z(k)$	System output at time k.	8

Symbol	Description	Page
$z(k)$	A vector.	114
z_s	An output sequence.	83
$\mathbb{B}(\mathbb{X},\mathbb{Y})$	Banach space of $\mathbb{X}$ into $\mathbb{Y}$.	15
$\mathbb{B}(\mathbb{X}) = \mathbb{B}(\mathbb{X},\mathbb{X})$		
$\mathbb{C}^n$	n-dimensional complex Euclidean space.	15
$\mathbb{F}$	Set of stabilizing controllers.	79
$\mathbb{H}^{n,m}$	Linear space.	16
$\mathbb{H}^n = \mathbb{H}^{n,n}$		
$\mathbb{H}^{n*}$	Linear space.	17
$\mathbb{H}^{n+}$	Cone of positive semidefinite N matrices.	17
$\mathbb{L}$	A set.	208
$\mathbb{M}$	A set.	208
$\mathbb{N}$	Set of operation modes.	9
$\mathbb{N}^\nu$	ν-fold product space of $\mathbb{N}$.	63
$\mathbb{N}_k$	Copies of $\mathbb{N}$.	20
$\mathbb{P}$	Polytope of transition probability matrices.	175
$\mathbb{Q}(k)$	Second moment of $x(k)$.	31
$\mathbb{Q}_0 = \mathbb{Q}(0)$		
$\mathbb{R}^n$	n-dimensional real Euclidean space.	15
$\mathbb{X}$	Banach space.	15
$\mathbb{Y}$	Banach space.	15
$\mathbb{T}$	Discrete-time set.	20
$\mathbb{T}_k = \{i \in \mathbb{T}; i \le k\}$		
$\mathbb{U}(k,l)$	A second moment matrix of x at times k, l.	49
$\mathbb{W}$	A set.	204
$\mathcal{A}_1, \ldots, \mathcal{A}_4$	Matrices associated to the second moment of x.	34
$\mathcal{B}$	Matrix associated to the first moment of x.	34
$\mathcal{C}$	A matrix.	34
$\mathcal{C}_k^m$	A linear space.	21
$\mathcal{E}(.)$	A linear operator.	33
$\mathcal{F}(X)$	Riccati gain (control).	78
$\mathcal{F}(.)$	Riccati gain.	205
$\mathcal{F}(.,.)$	Riccati gain.	205
$\mathcal{G}$	A system.	2
$\mathcal{G}_c$	A decomposed system.	139
$\mathcal{G}_{cl}$	A closed loop system.	110
$\mathcal{G}_F$	A closed loop system stabilized by a controller F.	80
$\mathcal{G}_K$	A Markov jump controller.	133
$\mathcal{G}_U$	A decomposed system.	139
$\mathcal{G}_v$	A system.	140

Symbol	Description	Page
$\mathcal{I}(.)$	Identity operator.	15
$\mathcal{J}(.)$	A linear operator.	33
$\mathcal{K}(.)$	A mapping.	87
$\mathcal{L}(.)$	A linear operator.	33
$\mathcal{L}^+(.)$	A linear operator.	213
$\mathcal{N}$	A matrix.	34
$\mathcal{O}(.)$	An operator.	83
$\mathcal{P}(.)$	Probability.	4, 20
$\mathcal{T}(.)$	A linear operator.	33
$\mathcal{U}$	Set of admissible controllers.	79
$\mathcal{U}_c$	Set of admissible controllers.	73
$\mathcal{V}(.)$	A linear operator.	33
$\widetilde{\mathcal{W}}(.)$	A linear operator.	150
$\mathcal{X}(.)$	A mapping.	205
$\mathcal{X}(.,.)$	A mapping.	205
$\mathcal{Y}(.)$	A mapping.	87
$\mathcal{Y}(.)$	A mapping.	205
$\mathcal{Y}(.,.)$	A mapping.	205
$\mathcal{Z}(.)$	A mapping.	217
$\mathfrak{D}(.)$	A mapping (H_∞ Riccati equation).	157
$\mathfrak{F}$	σ-field.	20
$\mathfrak{F}_k$	σ-field.	20
$\mathfrak{G}_k$	σ-field.	73
$\tilde{\mathfrak{F}}_\ell$	σ-field.	20
$\mathfrak{J}$	Cost.	26
$\mathfrak{J}_T$	Cost.	26
$\mathfrak{J}(\theta_0, x_0, u)$	Expected cost.	74
$\bar{\mathfrak{J}}(\theta_0, x_0)$	Minimal expected cost.	74
$\mathfrak{J}_{av}(\theta_0, x_0, u)$	Long run average cost.	79
$\tilde{\mathfrak{J}}_{av}(\theta_0, x_0)$	Optimal long run average cost.	79
$\mathfrak{M}(.)$	Affine operator (H_∞ Riccati equation).	157
$\mathfrak{N}$	σ-field.	20
$\mathfrak{R}(.)$	A mapping (H_∞ Riccati equation).	157
$\mathfrak{U}(.)$	A mapping (H_∞ Riccati equation).	157
$\mathfrak{V}(., k)$	A linear operator.	116

Abbreviation	Description	Page
APV	Analytical Point of View.	12
ARE	Algebraic Riccati Equation.	27
ASC	Almost Sure Convergence.	63
AWSS	Asymptotically Wide Sense Stationary.	49
CARE	Coupled Algebraic Riccati Equations.	78
DC	Direct Current.	189

Abbreviation	Description	Page
HMM	Hidden Markov Model.	12
i.i.d.	Independent and Identically Distributed.	13
IMM	Interacting Multiple Models.	102
JLQ	Jump Linear Quadratic.	13
LMI	Linear Matrix Inequality.	27
LMMSE	Linear Minimal Mean Square Error.	115
LQG	Linear Quadratic Gaussian.	131
LQR	Linear Quadratic Regulator.	26
LTI	Linear Time-Invariant.	21
MJLS	Markov Jump Linear System(s).	1
MSS	Mean Square Stability.	36
MM	Multiple Models.	12
MWe	Electrical power in MW.	167
rms	Root Mean Square.	198
SS	Stochastic Stability.	37
w.p.1	With Probability One.	65

Index